Nancy Lau

1971

numerical algorithms: origins and applications

numerical algorithms: origins and applications

bruce w. arden
professor of computer and communication sciences,
and associate director, computing center, university of michigan

kenneth n. astill
professor of mechanical engineering, tufts university

addison-wesley publishing company
reading, massachusetts · menlo park, california · london · don mills, ontario

This book is published
under the editorship of
Michael A. Harrison

*To our wives Patty and Pat
and our children,
for their patience and understanding*

preface

This book has its origins in a previous text written by one of the authors, entitled *An Introduction to Digital Computing*. The earlier book was published by Addison-Wesley in 1963 and, as the name implies, it was intended to be a general introduction to digital computing. In the intervening years there has been an enormous increase in the subject matter of computing with the result that, increasingly, the four general topics of computers, programming, numerical algorithms, and nonnumerical algorithms are introduced separately. As always, such specialization has both benefits and detriments. On the positive side, it permits the production of books in which it can reasonably be assumed that the reader has already been introduced to machines and programming. On the negative side, the specialization may result in deemphasizing the need for a detailed consideration of algorithms, by setting it apart from the highly motivating material associated with programming.

This book, while it assumes a minimal computing background, attempts to retain a connection between the development and formulation of numerical algorithms and their programmed implementation. Also, for additional motivation many of the methods are introduced by the statement of a physical problem. In effect, the book is intermediate between a "FORTRAN Computing Course" and a formal book on numerical analysis. As such, it should be helpful for upper-class undergraduates and first-year graduate students in science, engineering, and computing. The authors strongly believe that anyone, regardless of discipline, who wishes to be regarded as educated in computing should recognize the essentiality of the necessary-number and function approximation (indeed, finitude in general) in the development and use of algorithms for numerical computation. Hopefully, this book will be useful in the acquisition of such understanding, as well as in providing factual knowledge about a carefully selected variety of useful numerical algorithms. It would be gratifying if, in addition, some readers achieve sufficient motivation to study in even more analytic detail the relationships of the concept of infinitude to numerical algorithms.

The prerequisites for a course using this book would include integral calculus, some differential equations, and the ability to program for a digital computer. While the emphasis is on numerical methods, it should be recognized that normally the ultimate objective in formulating a numerical solution is to carry it out on a digital computer. Consequently emphasis is given to program development.

In the typical pursuit of this course, one would expect that a student would program and execute four to eight problems on the computer in the course of a semester. With these objectives in mind, the text contains a liberal sprinkling of examples requiring the development of programs to solve problems with the methods being treated. Each program is written in FORTRAN IV language, and each is normally accompanied by a flow chart. In the actual formulation, the program was developed directly from the flow chart. Flow charts are a convenient mechanism for translating the mathematical problems into programs, and they represent a useful way for scientists and engineers to communicate their problems to professional programmers. Symbols used in the flow charts are defined in Appendix A. The manner in which the information flows is evident from the inspection of one or two flow charts, and students quickly develop skill in interpreting and developing flow charts if they do not already have this facility.

All of the FORTRAN programs included are complete in the sense that they have been compiled and run with sample data. Where an independent test program is required, it is included with the program. The input and output lines shown with the programs are those actually used and produced by the programs. The various FORTRAN languages do differ in small details. These programs were run on an IBM System/360, Model 67, at The University of Michigan under the MTS operating system. To the best of the authors' knowledge the programs are "standard" in form, with the single exception that a free-format input was often used. In such cases input values could be arbitrarily spaced, separated by commas, as long as each number occupied a field the same size as (or smaller than) the field description in the associated format statement.

The authors would like to express thanks to Mrs. Arthur Wallace and Mrs. Thomas Steel for their help in typing the manuscript, and to Mr. Joseph Paster for his help in preparing some of the programs.

October, 1969   B. W. A.
                                                              K. N. A.

In the realm of acknowledgments it is probably unusual for an author to comment on the efforts of a co-author. Considering the current pace of academic life, a joint effort of this type might easily be frustrated by the difficulties of distance and demanding schedules. That such frustrations did not arise is due largely to my colleague in this work who completely undertook the pressing and often tedious task of final editing and proofreading. Without his willingness to overcome delays by additional personal effort, this book would not be a reality at this time.

Ann Arbor, Michigan   B. W. A.

contents

chapter one

computational error

The solution of problems through numerical computation often entails many repeated arithmetic operations. Small errors arising from input numbers or approximations in computer operations can propagate in the process to magnitudes which are unacceptable in the result. In this section we shall examine some of the sources of errors and how they accumulate.

Real numbers, with their infinite string of digits, cannot be represented in the finite number of storage locations in a digital computer; they must be approximated by rational numbers, for example,

$$\pi \approx \frac{31415926}{10000000} = 3.1415926.$$

This inability to represent real numbers is not the only source of error in a computational problem. The four categories listed below are in the realm of conscious error as opposed to mistakes. Since these errors are known to exist, the assumption is that they cannot be eliminated but that, hopefully, their magnitudes may be estimated.

1. The equations and expressions which are used to describe physical processes are, in general, approximations or idealizations which at the outset introduce a disparity between the physical problem and its computational analog. Such errors could be called *formulation* errors.

2. In actual fact, a digital computer is limited to performing the arithmetic operations on a limited set of rational numbers. Or, stated differently, only rational functions can be evaluated, where *rational function* is defined to mean any function that can be evaluated by operating on numbers using only addition, subtraction, multiplication, and division. The function

$$g(x) = \frac{4x^2 + 3}{3x^3 - 2x^2 + x - 8}$$

is an example. Functions which are defined by limiting processes, such as

$$\ln x = \int_1^x \frac{dt}{t},$$

must somehow be represented by a rational function, and thus an error is introduced. This error is called a *truncation* error since the rational function is often obtained by truncating (i.e., terminating) an infinite series after a specified number of terms.

3. The already mentioned fact that an infinite number of digits cannot be used to represent a number gives rise to what is called *round-off* errors. Remembering that a number is a polynomial, one sees that this type of error arises from the truncation of terms in this polynomial. However, it should be kept in mind that the term "truncation error" refers to the error arising from functional approximation, not number approximation.

4. Physical quantities can be measured only to a limited accuracy. When such values are used in computations, their indefinite value, or error, is frequently more restrictive than the limited number of digits available to express these measures. Such errors could be called *measurement* errors.

There is no way to estimate formulation error other than to check by observation how well the mathematical expression predicts the physical case. When the formulation error is admittedly large, then, occasionally, a great deal of effort is expended to reduce the other types of error. It would seem that the following corruption of an old adage would apply in such cases: "A thing not worth doing at all is not worth doing well."

Estimation of truncation errors is one of the main tasks of numerical analysis. Examples of such estimation are given in later sections.

Round-off and measurement errors have a similar effect even though the causes differ. The precision with which a number can be expressed is limited. The magnitude of the error caused by this imprecision is relatively simply determined in detail for individual operations and is the principal concern of this chapter. However, the determination of the error in the result of a complicated calculation (even though the error in the original quantities is known) is a vexing problem which cannot always be solved. As the last statement implies, error is propagated; that is, if one or both operands in an operation are approximate, then the result of the operation is approximate or in error. If this result is then used as an operand, the error is propagated to the value of another expression, etc. In addition to the error propagated at each step of a calculation, an error may also be *generated* because the operations themselves (not the operands) are only approximations. The most obvious example of an inexact operation is division, where, unless the divisor divides evenly into the dividend, the quotient will have an infinite number of digits. Restricting the number of quotient digits makes the operation an approximation. One also uses approximate multiplication, i.e., one retains not all the product digits but only a specified number of the most significant digits (generally, as many digits as the word size permits). In floating-point arithmetic, even addition and subtraction are often approximated. When numbers with different scale factors (or possibly even with the same scale factor) are added, not all the sum digits can be accommodated in the basic number size of the machine, and hence the least significant sum digits are dropped. In the following discussion such *generated* errors are not considered; the discussion is limited to the propagation of errors due to inaccurate operands.

Significant numbers

The most common way of indicating the degree of precision of a number is to write only those digits which are known accurately or, in other words, are *significant*. The rightmost digit that is written can be in error by at most half a unit because a greater error would mean that rounding would produce a different final digit. Writing the significant number 1.2932 implies that the "true" value represented by this number is less than or equal to 1.293249999 ... and greater than or equal to 1.29315. If these extreme values are rounded or "half-adjusted" to four decimal places by adding .00005 and then dropping the digits from the fifth decimal place on, the result is 1.2932. Alternatively, this range can be indicated by writing 1.2932 ± 0.00005, which in turn can be written 1.2932 (.5), where the number enclosed in parentheses is understood to be in units of the rightmost place. Another alternative expression which allows the error amount to be written as an integer is 1.29320 (5). The shortcomings of significant number notation become apparent if one now supposes that the example number is not known quite so accurately and the range of indefiniteness is, say, 7 units in the first place dropped, i.e. 1.29320 (7). But this representation is no longer a significant number, and the next larger significant number which includes this range (1.29313 to 1.29327) is 1.2930 (5) = 1.29300 (50), whose range is 1.29250 to 1.29350. This notation forces one to designate more variation than is known to exist in the given approximation. An alternative approach is to deal directly with the range of the numbers, as was done above, rather than use the significant number notation. Before the discussion turns to *range numbers*, it should be noted that when an integer with trailing zeros, say 92600, is written, it is not clear how many of the digits are significant. To clear up this ambiguity one must either explicitly state the significance or adopt some writing convention, such as 9260×10^1, which indicates, in this example, that the rightmost zero is not significant.

Range numbers

In range-number form, a number N is replaced by a pair, the largest and smallest possible values in its range. These high and low values are bracketed and displayed one above the other as shown below.

$$\begin{bmatrix} N_H \\ N_L \end{bmatrix}$$

The significant number 1.2932 becomes

$$\begin{bmatrix} 1.29325 \\ 1.29315 \end{bmatrix}.$$

If a number is known exactly, the extremes of the range are the same:

$$6\tfrac{3}{8} = \begin{bmatrix} 6.375 \\ 6.375 \end{bmatrix}.$$

The arithmetic operations expressed in terms of range numbers and some numerical examples are shown below.

Addition

$$x + y = \begin{bmatrix} x_H \\ x_L \end{bmatrix} + \begin{bmatrix} y_H \\ y_L \end{bmatrix} = \begin{bmatrix} x_H + y_H \\ x_L + y_L \end{bmatrix},$$

$$1.29(6) + 7.81(4) = \begin{bmatrix} 1.35 \\ 1.23 \end{bmatrix} + \begin{bmatrix} 7.85 \\ 7.77 \end{bmatrix} = \begin{bmatrix} 9.20 \\ 9.00 \end{bmatrix}.$$

Subtraction

If the signs are considered to be a part of the numbers, the problem becomes one of addition.

$$1.29(6) - 7.81(4) = \begin{bmatrix} 1.35 \\ 1.23 \end{bmatrix} + \begin{bmatrix} -7.77 \\ -7.85 \end{bmatrix} = \begin{bmatrix} -6.42 \\ -6.62 \end{bmatrix}$$

Multiplication

$$x \cdot y = \begin{bmatrix} x_H \\ x_L \end{bmatrix} \times \begin{bmatrix} y_H \\ y_L \end{bmatrix} = \begin{bmatrix} x_H \cdot y_H \\ x_L \cdot y_L \end{bmatrix},$$

$$3.46(7) \times 2.120(5) = \begin{bmatrix} 3.53 \\ 3.39 \end{bmatrix} \times \begin{bmatrix} 2.125 \\ 2.115 \end{bmatrix} = \begin{bmatrix} 7.50125 \\ 7.16985 \end{bmatrix}.$$

In this instance negative signs, if any, should precede the brackets since the objective is to produce the products that are largest and smallest in absolute value. As an example consider

$$\begin{bmatrix} -3.39 \\ -3.53 \end{bmatrix} \quad \text{to be} \quad -\begin{bmatrix} 3.53 \\ 3.39 \end{bmatrix}.$$

Division

$$x \div y = \begin{bmatrix} x_H \\ x_L \end{bmatrix} \div \begin{bmatrix} y_H \\ y_L \end{bmatrix} = \begin{bmatrix} x_H \div y_L \\ x_L \div y_H \end{bmatrix}.$$

Here again signs should be placed outside the brackets so that the values within are highest and lowest in absolute value.

$$4.246(8) \div 0.120(5) = \begin{bmatrix} 4.254 \\ 4.238 \end{bmatrix} \div \begin{bmatrix} 0.125 \\ 0.115 \end{bmatrix} = \begin{bmatrix} 36.992 \\ 33.904 \end{bmatrix}.$$

The quotients should be truncated so that no possible quotients are excluded from the range. Thus the quotient $4.254 \div 0.115 = 36.9913\ldots$ was adjusted to 36.992 to be included in the range $36.9913\ldots$ (Note that 36.991 would have excluded this value.)

If the extremes of the range are of different sign, zero is included in the range. This inclusion is not permissible if the range number is a divisor, and will require adjustments in the rules above for the other cases. However, all the rules given can be subsumed under one general rule prescribing the procedure of obtaining the range number of the result of an operation: Of the four possible combinations of the range number pairs which are operands, select the two giving the largest range to designate the resulting range number.

Range numbers are useful to demonstrate the propagation of errors. The following evaluation employing significant numbers illustrates this point.

$$y = 0.12 \times 236.4 - (63.8 \times 2.01) \div 25$$

$$= \begin{bmatrix} 0.125 \\ 0.115 \end{bmatrix} \times \begin{bmatrix} 236.45 \\ 236.35 \end{bmatrix} - \begin{bmatrix} 63.85 \\ 63.75 \end{bmatrix} \times \begin{bmatrix} 2.015 \\ 2.005 \end{bmatrix} \div \begin{bmatrix} 25.5 \\ 24.5 \end{bmatrix}$$

$$= \begin{bmatrix} 29.55625 \\ 27.18025 \end{bmatrix} - \begin{bmatrix} 128.65775 \\ 127.81875 \end{bmatrix} \div \begin{bmatrix} 25.5 \\ 24.5 \end{bmatrix}$$

$$= \begin{bmatrix} 29.55625 \\ 27.18025 \end{bmatrix} - \begin{bmatrix} 5.251336 \\ 5.012500 \end{bmatrix}$$

$$= \begin{bmatrix} 24.543750 \\ 21.928914 \end{bmatrix}.$$

If this result is adjusted to two decimal places, it becomes

$$\begin{bmatrix} 24.55 \\ 21.92 \end{bmatrix}.$$

As a significant number this range is represented by 2×10^1, i.e., by only one significant figure. Except for limited hand calculation, the utility of range numbers is restricted to such demonstrations. For the determination of error bounds, it is desirable to express the amount of the error explicitly. Moreover, the error from the approximation of infinite decimal numbers does not appear in range numbers, i.e.,

$$\tfrac{1}{3} \approx 0.3333 = \begin{bmatrix} 0.3333 \\ 0.3333 \end{bmatrix}.$$

Since the error is known, there is no range; the *approximation-error* form corrects these deficiencies.

Approximation-error numbers

As the name implies, numbers in approximation-error form consist of two parts, the approximation and the error. For instance, $\tfrac{1}{3} = 0.3333 + \tfrac{1}{3} \times 10^{-5}$, and, in general, $x = \bar{x} + \epsilon$. More often than not the actual error is not known; only the range is known. Hence, the 1.2932, the significant number previously used as an example, is in this form:

$$1.2932 + \epsilon, \qquad -0.00005 \leq \epsilon < 0.00005.$$

Thus a complete statement is $x = \bar{x} + \epsilon$, where $-\eta \leq \epsilon < \eta$. The basic arithmetic operations are expressed in this form as follows.

Addition

$$x + y = (\bar{x} + \epsilon_1) + (\bar{y} + \epsilon_2) = (\bar{x} + \bar{y}) + (\epsilon_1 + \epsilon_2).$$

The errors are additive. If n numbers are added,

$$x_1 + x_2 + x_3 + \cdots + x_n,$$

the error of the sum will be

$$\epsilon_1 + \epsilon_2 + \epsilon_3 + \cdots + \epsilon_n.$$

If, in addition, the individual errors have a common range, that is, $-\eta \leq \epsilon_i < \eta$ for $i = 1, 2, \ldots, n$, then

$$\epsilon_1 + \epsilon_2 + \cdots + \epsilon_n = \sum_{i=1}^{n} \epsilon_i < n\eta.$$

As a simple example, suppose that the significant numbers 11, 12, 13, . . . , 20 are added. The error for each number is less than 0.5, and the total error for the sum of the ten numbers is less than ten times that upper bound:

$$\epsilon_{\text{total}} < 10 \cdot 0.5 = 5.$$

Subtraction

$$x - y = (\bar{x} + \epsilon_1) - (\bar{y} + \epsilon_2) = (\bar{x} - \bar{y}) + (\epsilon_1 - \epsilon_2).$$

Since the values of ϵ_1 and ϵ_2 may be positive or negative, the determination of the maximum error is the same as in addition.

Multiplication

$$x \cdot y = (\bar{x} + \epsilon_1)(\bar{y} + \epsilon_2) = (\bar{x} \cdot \bar{y}) + (\epsilon_2 \bar{x} + \epsilon_1 \bar{y} + \epsilon_1 \epsilon_2)$$

Since the errors are usually small compared to the approximation numbers, the $\epsilon_1\epsilon_2$-term is very often neglected. As before, to determine the maximum error that can be made in a multiplication, the positive upper bounds for the errors are used, together with the absolute values of the approximating numbers. Assuming that $-\eta_1 \leq \epsilon_1 < \eta_1$ and $-\eta_2 \leq \epsilon_2 < \eta_2$, one has

$$\epsilon_2 \bar{x} + \epsilon_1 \bar{y} < \eta_2 |\bar{x}| + \eta_1 |\bar{y}|.$$

Division

This operation is a little more complicated.

$$\frac{x}{y} = \frac{\bar{x} + \epsilon_1}{\bar{y} + \epsilon_2} = \frac{\bar{x}(1 + \epsilon_1/\bar{x})}{\bar{y}(1 + \epsilon_2/\bar{y})} = \frac{\bar{x}}{\bar{y}}\left(1 + \frac{\epsilon_1}{\bar{x}}\right)\left(1 + \frac{\epsilon_2}{\bar{y}}\right)^{-1}$$

The rightmost term can be expanded by means of the binomial theorem to yield the common series

$$\frac{1}{1+z} = 1 - z + z^2 - z^3 + \cdots$$

Then

$$\frac{x}{y} = \frac{\bar{x}}{\bar{y}}\left(1 + \frac{\epsilon_1}{\bar{x}}\right)\left(1 - \frac{\epsilon_2}{\bar{y}} + \frac{\epsilon_2^2}{\bar{y}^2} - \cdots\right).$$

If all second- and higher-order terms are neglected, i.e., those involving products of the ϵ's, then

$$\frac{x}{y} \cong \frac{\bar{x}}{\bar{y}}\left(1 + \frac{\epsilon_1}{\bar{x}} - \frac{\epsilon_2}{\bar{y}}\right) = \left(\frac{\bar{x}}{\bar{y}}\right) + \left(\frac{\epsilon_1 \bar{y} - \epsilon_2 \bar{x}}{\bar{y}^2}\right).$$

The maximum error can be obtained by replacing the errors with their upper bounds and using the absolute values of the approximations:

$$\frac{\epsilon_1 \bar{y} - \epsilon_2 \bar{x}}{\bar{y}^2} < \frac{\eta_1 |\bar{y}| + \eta_2 |\bar{x}|}{\bar{y}^2},$$

where

$$-\eta_1 \leq \epsilon_1 < \eta_1 \quad \text{and} \quad -\eta_2 \leq \epsilon_2 < \eta_2.$$

To illustrate this kind of analysis, the error in evaluating a second-degree polynomial, $a_2 x^2 + a_1 x + a_0$, will be computed. This error estimation will be carried out for the evaluation done in nested form, that is, $(a_2x + a_1)x + a_0$. The approximation-error forms used are:

$$a_2 = \bar{a}_2 + \epsilon_2$$
$$a_1 = \bar{a}_1 + \epsilon_1$$
$$a_0 = \bar{a}_0 + \epsilon_0$$
$$x = \bar{x} + \epsilon_x$$

$$
\begin{aligned}
(a_2x + a_1)x + a_0 &= \left((\bar{a}_2 + \epsilon_2)(\bar{x} + \epsilon_x) + \bar{a}_1 + \epsilon_1\right)(\bar{x} + \epsilon_x) + \bar{a}_0 + \epsilon_0 \\
&\approx (\bar{a}_2\bar{x} + \epsilon_2\bar{x} + \epsilon_x\bar{a}_2 + \bar{a}_1 + \epsilon_1)(\bar{x} + \epsilon_x) + \bar{a}_0 + \epsilon_0 \\
&\approx \bar{a}_2\bar{x}^2 + \bar{a}_1\bar{x} + \bar{a}_0 + [\epsilon_2\bar{x}^2 + 2\epsilon_x\bar{a}_2\bar{x} + \epsilon_1\bar{x} + \bar{a}_1\epsilon_x + \epsilon_0].
\end{aligned}
$$

The second-order terms (those involving products of errors) are neglected. Applying this result to a specific case,

$$6.80x^2 + 3.25x + 8.24,$$

one can compute the maximum error. Assume that x is in the range $0.00 < x \le 10.00$ and that all the numbers are significant numbers. Then

$$\text{Error} = \epsilon_2\bar{x}^2 + 2\epsilon_x\bar{a}_2\bar{x} + \epsilon_1\bar{x} + \bar{a}_1\epsilon_x + \epsilon_0.$$

Since all the maximum errors are the same (0.005) and all the numbers are positive, the maximum error of the expression can be obtained by substituting 0.005 for $\epsilon_2, \epsilon_1, \epsilon_0, \epsilon_x$ and the maximum value 10.00 for $\bar{x}$.

$$
\begin{aligned}
\text{Max error} &= 0.005(10^2 + 2(6.80)10 + 10 + 3.25 + 1) \\
&= 0.005(100 + 136 + 14.25) \\
&= 1.25125.
\end{aligned}
$$

An error-limited problem

A final numerical example is included here to illustrate that computations can be limited by the lack of precision of the input numbers and, moreover, that carrying extra digits in the various steps of calculation does not eliminate the difficulty. Consider the simultaneous equations

$$2.00x + 3.00y = 5.00,$$
$$0.65x + 1.07y = 1.68.$$

Assuming that the coefficients and the terms on the right-hand sides are significant numbers, one computes the solutions twice, once to determine the maximum value of x and y and once to determine the minimum. The solutions are obtained by determinants (Cramer's rule):

$$x_{max} = \frac{\begin{vmatrix} 5.005 & 2.995 \\ 1.675 & 1.075 \end{vmatrix}}{\begin{vmatrix} 1.995 & 3.005 \\ 0.655 & 1.065 \end{vmatrix}} = 2.325, \qquad y_{max} = \frac{\begin{vmatrix} 2.005 & 4.995 \\ 0.645 & 1.685 \end{vmatrix}}{\begin{vmatrix} 1.995 & 3.005 \\ 0.655 & 1.065 \end{vmatrix}} = 1.001.$$

$$x_{min} = \frac{\begin{vmatrix} 4.995 & 3.005 \\ 1.685 & 1.065 \end{vmatrix}}{\begin{vmatrix} 2.005 & 2.995 \\ 0.645 & 1.075 \end{vmatrix}} = 1.146, \qquad y_{min} = \frac{\begin{vmatrix} 1.995 & 5.005 \\ 0.655 & 1.675 \end{vmatrix}}{\begin{vmatrix} 2.005 & 2.995 \\ 0.645 & 1.075 \end{vmatrix}} = 0.283.$$

The imprecision of the original coefficients leads to this wide range of results. The calculation is sensitive to the changes in values permitted by the range of the numbers, and no better solution is possible unless the original numbers are known to higher accuracy. This problem corresponds geometrically to the intersection of two lines which are almost parallel. Any slight change in coefficients (and therefore slopes) causes a marked change in the location of the point of intersection (Fig. 1.1). In such cases, one equation is "almost" the same as the other, except for a multiplicative constant. Systems of equations of this type are called *near-singular*.

Fig. 1.1

Exercises

1. In range-number calculations, divisors of the form

$$\begin{bmatrix} 0.1 \\ -0.1 \end{bmatrix}$$

are not permitted. Explain.

2. Programs (even machines) have been proposed which carry out all arithmetic operations using some number-with-error form. Evaluate the quadratic expression

$$3.25x^2 - 7.32x + 1.50$$

using range-number arithmetic. Assume that the coefficients are significant numbers and that the value of $x = 1.5$ is exact.

3. Evaluate the expression of Problem 2 by means of arithmetic operations using approximation-error numbers.

4. It is tempting to say that the average of the upper and lower bounds could be used to represent the range number. Show that this approach leads to a contradiction, i.e., the "average" result does not correspond to the range-number result. Consider, for example, the square of the range number

$$\begin{bmatrix} x_H \\ x_L \end{bmatrix}$$

5. When round-off error is the factor limiting accuracy in a computing process, carrying more digits in the numbers will alleviate the problem. In a fixed-word length machine, one must, in such instances, represent one number by more than one word. The rules for arithmetic operations using *double-precision* numbers can be obtained in a manner similar to the approximation-error operations. Thus, if the fraction a has two ten-digit parts, $a = a_1 + a_2 \cdot 10^{-10}$, then both a_1 and a_2 are fractions. Adding a similar double-precision number b gives

$$\begin{aligned} a + b &= (a_1 + a_2 \cdot 10^{-10}) + (b_1 + b_2 \cdot 10^{-10}) \\ &= (a_1 + b_1) + (a_2 + b_2) \cdot 10^{-10}. \end{aligned}$$

What would the two parts of the product of the double-precision number a and b be?

chapter two

Taylor's series and divided differences

The subject of numerical analysis has been facetiously subtitled, "Numerical Analysis or 1001 Applications of Taylor's Series." Although this is a gross over-simplification, it is true that the Taylor series representation of functions is a very common point of departure when estimates of the magnitude of truncation errors are sought. The terms of a series are generally rational functions, and the sum of the neglected terms, or error, provides a measure of the "goodness" of the rational approximation represented by the retained terms. To develop and illustrate the use of such error terms for (1) the case when the function and its derivatives are known and (2) the case when only values of the function are known, three background theorems from calculus are needed.

1. The differential mean-value theorem *If a function $f(x)$ is continuous in $a \leq x \leq b$ and differentiable in $a < x < b$, then there exists at least one point, ζ, in the interval at which*

$$f'(\zeta) = \frac{f(b) - f(a)}{b - a},$$

where $f'(\zeta)$ denotes the derivative evaluated at the intermediate point ζ.

In less formal terms, the theorem states that at some point in the interval the tangent to the graph of $f(x)$ has the same slope as the straight line passing through the points $(a, f(a))$ and $(b, f(b))$. See Fig. 2.1.

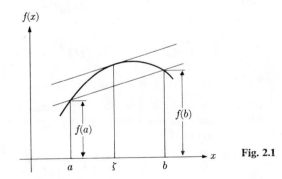

Fig. 2.1

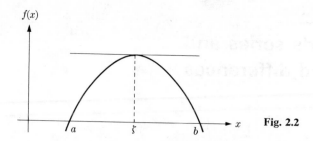

Fig. 2.2

A variant of this theorem is Rolle's theorem which states, in addition, that

$$\text{if} \quad f(a) = f(b) = 0, \quad \text{then} \quad f'(\zeta) = 0$$

(see Fig. 2.2). A repeated application of Rolle's theorem permits one to say that if there are n abscissas for which the function is zero,

$$f(a_1) = f(a_2) = f(a_3) = \cdots = f(a_n) = 0,$$

then there are $n - 1$ intermediate values at which the derivative vanishes:

$$f'(\zeta_1) = f'(\zeta_2) = \cdots = f'(\zeta_{n-1}) = 0$$

(see Fig. 2.3).

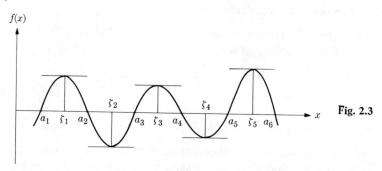

Fig. 2.3

2. The integral mean-value theorem *If the functions $f(x)$ and $p(x)$ are continuous in $a \leq x \leq b$ and $p(x)$ is nonnegative in the interval, then*

$$\int_a^b f(x)p(x)\,dx = f(\zeta)\int_a^b p(x)\,dx,$$

where $a \leq \zeta \leq b$.

The graphical version of this theorem for the special case $p(x) = 1$ is shown in Fig. 2.4, where M is the maximum value of $f(x)$ in $a \leq x \leq b$, and m is the minimum value in the interval. Certainly

$$m\int_a^b dx = m(b - a) \leq \int_a^b f(x)\,dx \leq M\int_a^b dx = M(b - a),$$

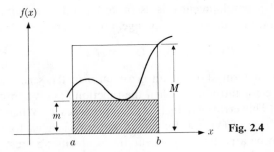

Fig. 2.4

and since $f(x)$ takes all the values between m and M,

$$\int_a^b f(x)\, dx = f(\zeta)(b - a)$$

where $a \le \zeta \le b$. In the same way, so long as $\int_a^b p(x)\, dx$ is known to be positive or zero,

$$m \int_a^b p(x)\, dx \le \int_a^b f(x)p(x)\, dx \le M \int_a^b p(x)\, dx$$

or

$$\int_a^b f(x)p(x)\, dx = f(\zeta) \int_a^b p(x)\, dx, \qquad a \le \zeta \le b$$

3. A mean-value theorem *If the function $f(x)$ is continuous in $a \le x \le b$, and c and d are nonnegative constants, then*

$$cf(a) + df(b) = (c + d)f(\zeta),$$

where $a \le \zeta \le b$. That this result is true can be seen by rewriting it:

$$cf(a) + df(b) = (c + d)\left[\frac{c}{c + d}f(a) + \frac{d}{c + d}f(b)\right].$$

The sum of the two fractions is unity so that

$$f(a) \le \frac{c}{c + d}f(a) + \frac{d}{c + d}f(b) \le f(b)$$

or

$$f(b) \le \frac{c}{c + d}f(a) + \frac{d}{c + d}f(b) \le f(a).$$

Since the function is continuous, there is some intermediate abscissa at which the function is equal to this sum.

The result of the first of these three theorems can be rewritten to show its application to approximating functions:

$$f(b) = f(a) + (b - a)f'(\zeta), \qquad a < \zeta \le b.$$

In the further development it is useful to label the endpoints of the interval (z, x) instead of (a, b) and, with this change, one obtains

$$f(x) = f(z) + (x - z)f'(\zeta), \qquad z \le \zeta \le x.$$

The rightmost term above is an error term in the sense that its value gives the amount of error introduced when the function $f(x)$ is approximated by the single value $f(z)$. This error is reduced as z approaches x. When $f(x)$ is constant, there is no error since $f'(x) = 0$ for any x, but for more elaborate functions this does not seem to be a particularly useful approach since z must usually be quite close to x to keep the error small. A more usable approximation can be constructed by using Theorem 1 to introduce a second derivative:

$$f(x) = f(z) + (x - z)[f'(z) + f'(\zeta) - f'(z)]$$

$$= f(z) + (x - z)f'(z) + (x - z)(\zeta - z)\left[\frac{f'(\zeta) - f'(z)}{\zeta - z}\right]$$

$$= f(z) + (x - z)f'(z) + (x - z)(\zeta - z)f''(\zeta_1),$$

where $z \le \zeta_1 \le \zeta \le x$. If the interval is pictured as a line segment, these four abscissas are spaced as shown in Fig. 2.5. The intermediate points ζ_1 and ζ are not constants but will vary as x or z is varied. If x is regarded as fixed and z is varied, there is no effect on $f(x)$. In symbols,

$$\frac{d}{dz}f(x) = \frac{d}{dz}[f(z) + (x - z)f'(z) + (x - z)(\zeta - z)f''(\zeta_1)] = 0$$

or

$$\frac{d}{dz}f(x) = f'(z) - f'(z) + (x - z)f''(z) + \frac{d}{dz}[(x - z)(\zeta - z)f''(\zeta_1)] = 0.$$

Fig. 2.5

From this last result it is seen that

$$\frac{d}{dz}[(\zeta - z)(x - z)f''(\zeta_1)] = -(x - z)f''(z).$$

Now, if both sides are integrated, with the lower limit of integration, z, treated as a variable, the correspondence between these two forms becomes clearer:

$$\int_z^x \left\{\frac{d}{dt}((\zeta - t)(x - t)f''(\zeta_1))\right\} dt = \int_z^x - (x - t)f''(t)\, dt.$$

Theorem 2 can be applied to the right-hand side of this expression thus:

$$-(\zeta - z)(x - z)f''(\zeta_1) = -f''(\zeta_2)\int_z^x (x - t)\,dt$$

$$= \left[f''(\zeta_2)\frac{(x - t)^2}{2}\right]_z^x,$$

$$(\zeta - z)(x - z)f''(\zeta_1) = \frac{(x - z)^2}{2}f''(\zeta_2), \qquad z \le \zeta_2 \le x.$$

Using this result in the improved approximation yields

$$f(x) = f(z) + (x - z)f'(z) + \frac{(x - z)^2}{2}f''(\zeta_2).$$

In this case, if $f(x)$ were a straight line, for example, $f(x) = ax + b$, the rightmost error term would be zero, since the second derivative would be zero, and the first two terms would not be approximate. It is possible to add higher derivatives to this approximation, which is an elaboration of the differential mean-value theorem, but it is simpler to start with the general form.

Taylor's series

In the preceding section an approximation for $f(x)$ was obtained which contained terms involving $(x - z)$ and $(x - z)^2$. Extending this approach to include higher powers of $x - z$, let us suppose that an arbitrary function $f(x)$ can be represented by the $n + 1$ terms of a polynomial,

$$f(x) = a_0 + a_1(x - z) + a_2(x - z)^2 + \cdots + a_n(x - z)^n + R_{n+1},$$

where the sum of the terms dropped from this presumably infinite series is represented by a remainder term R_{n+1}. To determine the values of the coefficients $a_0, a_1, \ldots, a_n$, the series is first differentiated n times:

$$f'(x) = 1 \cdot a_1 + 2 \cdot a_2(x - z) + \cdots + na_n(x - z)^{n-1} + R'_{n+1}$$

$$f''(x) = 2 \cdot a_2 + 3 \cdot 2a_3(x - z) + \cdots + n(n - 1)a_n(x - z)^{n-2} + R''_{n+1}$$

$$f'''(x) = 3 \cdot 2 \cdot a_3 + 4 \cdot 3 \cdot 2 \cdot a_4(x - z) + \cdots$$

$$+ n(n - 1)(n - 2)a_n(x - z)^{n-3} + R'''_{n+1}$$

$$\vdots$$

$$f^{(n)}(x) = n!a_n + R^{(n)}_{n+1}$$

By taking advantage of the fact that evaluating the expressions above at $x = z$ causes all terms on the right except the first to vanish, the coefficients can be simply determined. Even the R^k_{n+1}-terms vanish, since terms of this kind represent an indefinite sum of terms of the same form, i.e., each having as a factor a power of $(x - z)$. Making the substitution $x = z$, one then has

$$f(z) = a_0, \qquad\qquad a_0 = f(z)$$

$$f'(z) = a_1, \qquad\qquad a_1 = f'(z)$$

$$f''(z) = 2 \cdot a_2, \qquad\qquad a_2 = \frac{f''(z)}{2}$$

$$f'''(z) = 3 \cdot 2 \cdot a_3, \qquad a_3 = \frac{f'''(z)}{3 \cdot 2}$$

$$\vdots \qquad\qquad\qquad \vdots$$

$$f^{(n)}(z) = n! a_n, \qquad\qquad a_n = \frac{f^{(n)}(z)}{n!}$$

Making these substitutions for the coefficients in the original form leads to the Taylor series,

$$f(x) = f(z) + f'(z)(x - z) + \frac{f''(z)}{2!}(x - z)^2 + \frac{f'''(z)}{3!}(x - z)^3 + \cdots$$

$$+ \frac{f^{(n)}(z)}{n!}(x - z)^n + R_{n+1}$$

$$= \sum_{i=0}^{n} \frac{f^{(i)}(z)}{i!}(x - z)^i + R_{n+1}.$$

If one chooses

$$x = z + h,$$

then another common representation of Taylor's series is obtained:

$$f(z + h) = f(z) + f'(z)h + \frac{f''(z)}{2!}h^2 + \cdots + \frac{f^{(n)}(z)}{n!}h^n + R_{n+1}$$

$$= \sum_{i=0}^{n} \frac{f^{(i)}(z)}{i!}h^i + R_{n+1}.$$

When $z = 0$, the resultant series is called Maclaurin's series:

$$f(h) = f(0) + f'(0)h + \frac{f''(0)h^2}{2!} + \cdots + \frac{f^{(n)}(0)}{n!}h_n + R_{n+1}$$

$$= \sum_{i=0}^{n} \frac{f^{(i)}(0)}{i!}h^i + R_{n+1}.$$

The first representation may be regarded as a method of determining $f(x)$ from information known about the function at a point z [that is, $f(z), f'(z), f''(z), \ldots$].

The second form is, of course, the same; only the point of view has shifted. The series provides a way of determining the functional value at some point $(z + h)$ near the central point z, again using the known information at z. Both expressions are "expansions about z," but the second one gives z the position of central importance and the question, always present when series are considered, of how far from the known values at z, values of the function may be evaluated, can be expressed in terms of h. The range of h for which a series converges, $-r < h < r$, must be determined for each series and, although it is not always conclusive, the Cauchy ratio test is useful in the determination of the "circle of convergence." To make this test, the ratio of the absolute values of the $(n + 1)$ and the nth terms is formed. If the limit of this ratio as $n \to \infty$ is less than unity, the series converges; if it is greater, it diverges. The situation when the ratio is unity is left undetermined.

The remainder term R_{n+1} is an infinite series, but to estimate the truncation error introduced by dropping the remainder, a method is needed which permits one to estimate the magnitude of R_{n+1} without adding terms of the series it represents. Proceeding as before, one can produce such an estimate by differentiating the first form of Taylor's series:

$$f(x) = f(z) + f'(z)(x - z) + \frac{f''(z)}{2!}(x - z)^2 + \cdots + \frac{f^{(n)}(z)}{n!}(x - z)^n + R_{n+1}.$$

Differentiating with respect to z yields

$$0 = \underbrace{f'(z) - f'(z)}_{= 0} + \underbrace{(x - z)f''(z) - \frac{2f''(z)(x - z)}{2!}}_{= 0}$$

$$\underbrace{+ (x - z)^2 \frac{f'''(z)}{2!} + \cdots + n\frac{f^{(n)}(z)}{n!}(x - z)^{n-1}}_{= 0}$$

$$+ (x - z)^n \frac{f^{(n+1)}(z)}{n!} + R'_{n+1},$$

and hence

$$R'_{n+1} = -\frac{(x - z)^n f^{(n+1)}(z)}{n!}.$$

To obtain an expression for R_{n+1}, both sides of the equation above are integrated. As in the earlier error-term determination, integrating the derivative with respect to z of R_{n+1} (with the lower limit of integration taken to be z) gives

$$\int_z^x R'_{n+1}\, dt = -R_{n+1}.$$

This is a function of z, and the constant obtained from the evaluation at the upper limit x is zero since R_{n+1} has a factor of $(x - z)^n$. Accordingly,

$$-R_{n+1} = -\int_z^x \frac{(x - t)^n}{n!} f^{(n+1)}(t) \, dt = \left[f^{(n+1)}(\zeta) \frac{(x - t)^{n+1}}{(n + 1)!} \right]_z^x$$

$$= -f^{(n+1)}(\zeta) \frac{(x - z)^{n+1}}{(n + 1)!}$$

where $z \le \zeta \le x$. If the substitution $x = z + h$ is made, then

$$R_{n+1} = f^{(n+1)}(\zeta) \frac{h^{n+1}}{(n + 1)!}$$

where $z \le \zeta \le z + h$. The integration above is justified by the integral mean-value theorem. For the case of $x < z$ some additional argument is required since $p(x) = (x - z)^n$ is not necessarily nonnegative in the interval.

It is apparent that if $n + 1$ terms of a Taylor series are retained (this count includes the constant term), the truncation error is proportional to the $(n + 1)$-derivative and the $(n + 1)$-power of the interval h. If the function $f(x)$ were an nth-degree polynomial, there would be no error in a Taylor series representation of it, since the $(n + 1)$-derivative of an nth-degree polynomial is zero. For example, the Taylor series representation of a second-degree polynomial terminates with the third term:

$$f(x) = a_0 + a_1 x + a_2 x^2,$$

$$f(x + h) = (a_0 + a_1 x + a_2 x^2) + (a_1 + 2a_2 x)h + \frac{(2a_2)h^2}{2}$$

$$= a_0 + a_1(x + h) + a_2(x + h)^2.$$

As an example, consider the approximation of $f(x) = \cos(x)$ by the first three terms of Taylor's series. The necessary derivatives are

$$f'(x) = -\sin(x), \qquad f''(x) = -\cos(x), \qquad \text{and} \qquad f'''(x) = \sin(x).$$

Therefore,

$$\cos(x + h) = \cos(x) - h\sin(x) - \frac{h^2}{2}\cos(x) + \frac{h^3}{6}\sin(\zeta)$$

where $x \le \zeta \le x + h$.

To illustrate numerically the application of this formula, compute $\cos(\pi/4) = \cos(\pi/6 + \pi/12)$:

$$x = \frac{\pi}{6} = 0.5236 \text{ rad};$$

$$h = \frac{\pi}{12} = 0.2618 \text{ rad};$$

$$\cos\left(\frac{\pi}{4}\right) \approx \cos(0.5236) - 0.2618\sin(0.5236) - (0.5)(0.2618)^2\cos(0.5236)$$

$$\approx 0.8660 - (0.2618)(0.5) - (0.5)(0.2618)^2(0.8660)$$

$$\approx 0.7054;$$

$$|R_3| \leq \left|\frac{(0.2618)^3}{6}\max\sin(\zeta)\right|, \qquad 0.5236 \leq \zeta \leq 0.5236 + 0.2618,$$

where

$$\max\sin(\zeta) = \sin(0.7854) = 0.7071;$$

$$|R_3| \leq \frac{(0.2618)^3}{6}\,0.7071 = 0.0021.$$

To four decimal places the correct result is $\cos(\pi/4) = 0.7071$ so that the actual error is $(0.7071 - 0.7054) = 0.0017$. The error bound, 0.0021, is larger than the actual error, as one would expect.

Divided-difference polynomials

The Taylor expansion utilized known values of the approximated function and its derivatives at a specific point. Very often such information is not available for the function of interest, but there exists only a set of tabulated values such as would be obtained from an experiment or from mathematical tables. The question then is: given a set of tabulated values,

x_0	$f(x_0)$
x_1	$f(x_1)$
x_2	$f(x_2)$
$\vdots$	$\vdots$
x_n	$f(x_n)$

can some expression analogous to the Taylor series be developed? From the definition of a derivative,

$$f'(x) = \lim_{h \to 0}\frac{f(x+h) - f(x)}{h} \quad \text{and} \quad f'(x_0) = \lim_{x_1 \to x_0}\frac{f(x_1) - f(x_0)}{x_1 - x_0},$$

it appears that the quotient formed from the *finite differences* $f(x_1) - f(x_0)$ and $x_1 - x_0$ would provide an approximation to the first derivative which could replace the first derivative in Taylor's series. Similarly, the second derivative can be approximated by such *finite divided differences:*

$$\lim_{\substack{x_1 \to x_0 \\ x_2 \to x_0}}\frac{[f(x_2) - f(x_1)]/(x_2 - x_1) - [f(x_1) - f(x_0)]/(x_1 - x_0)}{x_2 - x_0} = \frac{f''(x_0)}{2}.$$

However, the substitution of the divided differences is not sufficient in itself, because one has no longer a single interval $(x - z)$ but many:

$$(x - x_0), \qquad (x - x_1), \qquad (x - x_2), \qquad \ldots$$

A more revealing approach to the divided-difference polynomial is to proceed in a fashion analogous to the Taylor series development.

It is probably appropriate at this point to make a short digression on the subject of getting something for nothing. To produce an estimate of the truncation error it was necessary in the case of Taylor's series to use some information that was not used in the truncated series, i.e., knowledge about the $(n + 1)$-derivative. Without this information, no statements about the accuracy of the approximation could be made. Similarly, in the finite-difference case, additional information, not used in producing the rational approximation, will be required to say something about error. If one is given only two points, $(x_0, f(x_0))$ and $(x_1, f(x_1))$, and a straight line is made to pass through these points, nothing can be said about how closely this line approximates the "true" function $f(x)$. However, if a third point, $(x_2, f(x_2))$, is also known, then the proximity of this "unused" point to the line is some measure of the degree of approximation of the line (Fig. 2.6).

The differential mean-value theorem can be used to replace derivatives by divided differences. As an illustration, let us replace the first derivative in the three-term Taylor series. (The symbol x_0 is substituted for z to achieve consistency with the tabular notation.)

$$f(x) = f(x_0) + (x - x_0)f'(x_0) + \frac{(x - x_0)^2}{2}f''(\zeta), \qquad x_0 \leq \zeta \leq x.$$

For an arbitrary $x_1 < x_0$,

$$\frac{f(x_1) - f(x_0)}{x_1 - x_0} = f'(\zeta_1), \qquad x_1 \leq \zeta_1 \leq x_0$$

and therefore

$$f(x) = f(x_0) + (x - x_0)\left[\frac{f(x_1) - f(x_0)}{x_1 - x_0} - (f'(\zeta_1) - f'(x_0))\right]$$
$$+ \frac{(x - x_0)^2}{2}f''(\zeta).$$

Applying Theorem 2 again, one obtains

$$\frac{f'(\zeta_1) - f'(x_0)}{\zeta_1 - x_0} = f''(\zeta_2), \qquad \zeta_1 \leq \zeta_2 \leq x_0.$$

Making this substitution and writing $-(\zeta_1 - x_0)$ as $(x_0 - \zeta_1)$ yields

$$f(x) = f(x_0) + (x - x_0)\frac{f(x_1) - f(x_0)}{x_1 - x_0}$$
$$+ (x - x_0)(x_0 - \zeta_1)f''(\zeta_2) + \frac{(x - x_0)^2}{2}f''(\zeta).$$

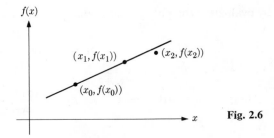

Fig. 2.6

Since the multipliers of the derivatives in the two right-hand terms are nonnegative, Theorem 3 can be used to simplify the above result, and one obtains

$$f(x) = f(x_0) + (x - x_0)\frac{f(x_1) - f(x_0)}{x_1 - x_0}$$

$$+ \left[(x - x_0)(x_0 - \zeta_1) + \frac{(x - x_0)^2}{2} \right] f''(\zeta_3),$$

where $\zeta_2 \le \zeta_3 \le \zeta$.

The derivative has now been replaced at the cost of considerable complication of the error term, but some further simplification can be made if it is required that this term must be zero when $x = x_1$. At $x = x_1$,

$$\left[(x_1 - x_0)(x_0 - \zeta_1) + \frac{(x_1 - x_0)^2}{2} \right] f''(\zeta_3) = 0$$

from which

$$(x_0 - \zeta_1) + \frac{x_1 - x_0}{2} = 0 \quad \text{or} \quad \zeta_1 = \frac{x_0 + x_1}{2},$$

and

$$f(x) = f(x_0) + (x - x_0)\frac{f(x_1) - f(x_0)}{x_1 - x_0} + (x - x_0)(x - x_1)\frac{f''(\zeta_3)}{2}.$$

The intermediate point ζ_3 is in the interval containing x, x_0, and x_1.

Now that this form for the sum of products is established, the next step is to postulate that the function can be represented by n terms of this form. Assume that an arbitrary function, $f(x)$, for which there are $n + 1$ values, $f(x_0)$, $f(x_1)$, ..., $f(x_n)$, can be represented by the polynomial

$$\begin{aligned} f(x) = a_0 &+ a_1(x - x_0) + a_2(x - x_0)(x - x_1) \\ &+ a_3(x - x_0)(x - x_1)(x - x_2) + \cdots \\ &+ a_n(x - x_0)(x - x_1) \cdots (x - x_{n-1}) \\ &+ (x - x_0)(x - x_1) \cdots (x - x_n)R_{n+1}. \end{aligned}$$

The complete error term is composed of two factors: the product, $\prod_{i=0}^{n}(x - x_i)$, and R_{n+1}, whose specific form is yet to be determined. The coefficients may be

determined by evaluating the polynomial at the given points.
At $x = x_0$:

$$f(x_0) = a_0, \qquad a_0 = f(x_0).$$

At $x = x_1$:

$$f(x_1) = a_0 + a_1(x_1 - x_0), \qquad a_1 = \frac{f(x_1) - f(x_0)}{x_1 - x_0}.$$

At $x = x_2$:

$$f(x_2) = a_0 + a_1(x_2 - x_0) + a_2(x_2 - x_0)(x_2 - x_1),$$

$$a_2 = \frac{[f(x_2) - f(x_1)]/(x_2 - x_1) - [f(x_1) - f(x_0)]/(x_1 - x_0)}{x_2 - x_0}.$$

The values of the a's become more complicated at each evaluation, but each is systematically formed. The general form is

$$a_k = \frac{(a_{k-1} \text{ with subscripts increased by } 1) - a_{k-1}}{x_k - x_0}.$$

The coefficient a_k depends upon all the x's and functional values with subscripts k and less, that is, $a_k = f(x_k, x_{k-1}, \ldots, x_0)$. This notation is a convenient means of designating the coefficients since the f denotes the function involved and the x's enclosed in parentheses label the abscissas and functional values upon which the coefficient depends. To distinguish it from ordinary function notation, square brackets will be used instead of parentheses. Thus

$$a_0 = f[x_0]$$
$$a_1 = f[x_1, x_0]$$
$$a_2 = f[x_2, x_1, x_0]$$
$$\vdots$$
$$a_n = f[x_n, x_{n-1}, \ldots, x_0]$$

With this notation, it is a simple matter to express the recursive relation between successive coefficients:

$$a_0 = f[x_0]$$

$$a_1 = \frac{f[x_1] - f[x_0]}{x_1 - x_0} = f[x_1, x_0]$$

$$a_2 = \frac{f[x_2, x_1] - f[x_1, x_0]}{x_2 - x_0} = f[x_2, x_1, x_0]$$

$$a_3 = \frac{f[x_3, x_2, x_1] - f[x_2, x_1, x_0]}{x_3 - x_0}$$

$$\vdots$$

$$a_n = \frac{f[x_n, \ldots, x_1] - f[x_{n-1}, \ldots, x_0]}{x_n - x_0} = f[x_n, x_{n-1}, \ldots, x_0]$$

The coefficients are called *divided differences*, and $a_k = f[x_k, x_{k-1}, \ldots, x_0]$ is the kth divided difference. Note that the x's in the denominators are those which are not common to the two numerator terms. With this notation, the original polynomial is written

$$f(x) = f[x_0] + f[x_1, x_0](x - x_0) + f[x_2, x_1, x_0](x - x_0)(x - x_1) + \cdots$$

$$+ f[x_n, x_{n-1}, \ldots, x_0](x - x_0)(x - x_1) \cdots (x - x_{n-1})$$

$$+ \prod_{i=0}^{n} (x - x_i)R_{n+1}.$$

Differentiation is not available as a tool for determining a value for R_{n+1}, as it was in the Taylor series case. But the expression for R_{n+1} can be recursively simplified into a single term:

$$\prod_{i=0}^{n} (x - x_i)R_{n+1} = \underbrace{f(x) - f(x_0)} - f[x_1, x_0](x - x_0)$$

$$- f[x_2, x_1, x_0](x - x_0)(x - x_1) - \cdots$$

$$= \underbrace{(x - x_0)f[x, x_0] - f[x_1, x_0](x - x_0)} - f[x_2, x_1, x_0](x - x_0)(x - x_1) - \cdots$$

$$= \underbrace{(x - x_0)(x - x_1)f[x, x_1, x_0]} - f[x_2, x_1, x_0](x - x_0)(x - x_1) - \cdots$$

$$= (x - x_0)(x - x_1)(x - x_2)f[x, x_2, x_1, x_0] - \cdots$$

Continuing in this way, one obtains

$$\prod_{i=0}^{n} (x - x_i)R_{n+1} = (x - x_0)(x - x_1) \cdots (x - x_n)f[x, x_n, x_{n-1}, \ldots, x_0]$$

and

$$R_{n+1} = f[x, x_n, \ldots, x_0].$$

Here the variable x appears as one of the arguments of a divided difference, that is, x is not one of the given abscissas. Since x is just "some point" in the interval containing $x_0, \ldots, x_n$, this formula is not an effective means of estimating R_{n+1}. It is worth noting at this time, however, that if x is replaced by x_{n+1}, the right-hand side of the expression above would be the first term dropped from the divided-difference polynomial for $f(x)$. As a practical procedure, then, evaluating this first neglected term (provided the necessary additional $(n + 1)$-point is available) yields an *estimate* of the magnitude of the error. However, a little analysis produces an expression for the error which is closer to the Taylor series form.

If the nth-degree divided-difference polynomial is designated $p_n(x)$, then

$$f(x) = p_n(x) + R_{n+1} \prod_{i=0}^{n} (x - x_i).$$

At the given points, both $f(x) - p_n(x)$ and $\prod_{i=0}^{n} (x - x_i)$ are zero. The first expression is zero at these points because $p_n(x)$ was constructed to equal $f(x_0)$ at $x = x_0$, $f(x_1)$ at $x = x_1$, etc., and the second expression has a zero factor at each of the given points. However, if the equality above is to hold at some other value of x, there must be some value for R_{n+1} which makes the equality possible. Consider

$$F(x) = f(x) - p_n(x) - R_{n+1} \prod_{i=0}^{n} (x - x_i) .$$

In this case, $F(x)$ is zero at the $n + 1$ points $x_0, x_1, \ldots, x_n$, and for a given value of x, it is also zero at at least one other abscissa. Applying Rolle's theorem, one finds that the derivative $F'(x)$ is zero at $n + 1$ points included in the interval containing the given points. By the same reasoning, $F''(x)$ has n zeros, $F'''(x)$ has $n - 1$ zeros, etc.:

$$F'(x) = f'(x) - p_n'(x) - R_{n+1} \frac{d}{dx} \left(\prod_{i=0}^{n} (x - x_i) \right),$$

$$F''(x) = f''(x) - p_n''(x) - R_{n+1} \frac{d^2}{dx^2} \left(\prod_{i=0}^{n} (x - x_i) \right).$$

Since $p_n(x)$ is an nth-degree polynomial, its $(n + 1)$-derivative is zero. The product $\prod_{i=0}^{n} (x - x_i)$ is an $(n + 1)$-degree polynomial,

$$x^{n+1} + b_{n-1}x^n + \cdots + b_0,$$

and its $(n + 1)$-derivative is a constant,

$$(n + 1)(n)(n - 1) \cdots (1) = (n + 1)!.$$

From these observations, it follows that

$$F^{(n+1)}(x) = f^{(n+1)}(x) - R_{n+1}(n + 1)!.$$

Since $F^{(1)}$ has $n + 1$ zeros, $F^{(2)}$ has n zeros, $F^{(3)}$ has $n - 1$ zeros, etc., $F^{(n+1)}$ has 1 zero. Assume that this zero is at the point $x = \zeta$, that is, $F^{(n+1)}(\zeta) = 0$. Then

$$0 = f^{(n+1)}(\zeta) - R_{n+1}(n + 1)!$$

and

$$R_{n+1} = \frac{f^{(n+1)}(\zeta)}{(n + 1)!},$$

where ζ is in the interval containing $x_0, x_1, \ldots, x_n$. Using this result, one obtains the complete error term

$$E_{n+1}(x) = \prod_{i=0}^{n} (x - x_i) \frac{f^{(n+1)}(\zeta)}{(n + 1)!},$$

which is the Taylor series result except for the replacement of h^{n+1} by the $n + 1$ factors which are the intervals from x to the given points. For computational purposes, this form of the remainder has even less utility than in the Taylor series since, in general, no information on the magnitude of derivatives is available.

However, this form is useful in the analytic expression of truncation errors. For example, it shows very clearly that if $f(x)$ is a polynomial of degree n or less, the divided-difference polynomial $p_n(x)$ is exact, that is, the error is zero since $f^{(n+1)}(x)$ is zero.

The systematic rules for computing divided differences suggest a tabular arrangement.

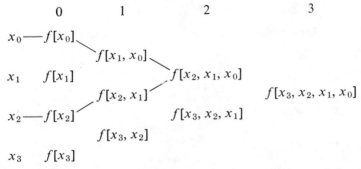

The first two columns in a divided-difference table are given, and the remaining entries are formed according to a very simple rule: To compute an entry in the kth column, subtract the adjacent entries in the $(k - 1)$-column and divide that difference by the difference of the abscissas intercepted by proceeding diagonally upward and diagonally downward to the left. For example,

$$f[x_2, x_1, x_0] = \frac{f[x_2, x_1] - f[x_1, x_0]}{x_2 - x_0}.$$

The coefficients entering into the divided-difference polynomial which has been derived are those on the downward diagonal of the table:

$$f(x) \approx f[x_0] + f[x_1, x_0](x - x_0) + f[x_2, x_1, x_0](x - x_0)(x - x_1)$$
$$\approx f[x_0] + (x - x_0)\{f[x_1, x_0] + (x - x_1)f[x_2, x_1, x_0]\}$$

Note that no restrictions have been placed on the order or even the spacing between points, so that it is possible (by rearranging the three points used) to use the same formula with different divided differences and increments. The result would be the same, however, since only one second-degree polynomial can pass through the three points. If there is no rearrangement, then any zigzag "path" through the table which terminates on the same divided difference can be used to determine the coefficients of the polynomial.

For this table, there are four possible paths. If, for example, the path shown above is chosen, then the approximating polynomial for $f(x)$ is written:

$$f(x) \approx f[x_1] + f[x_2, x_1](x - x_1) + f[x_2, x_1, x_0](x - x_1)(x - x_2)$$

To illustrate some of these properties, a simple example is included at this point. Suppose that

$$f(x) = x^3 - 12x + 2$$

Some values of $f(x)$ are listed below, and the divided-difference table is appended to the right of these columns.

II	I	x	$f[x]$	1	2	3	4
x_3	x_0	-1	13				
				-11			
x_2	x_1	0	2		1		
				-8		1	
x_0	x_2	2	-14		6		0
				16		1	
x_1	x_3	4	18		11		
				49			
x_4	x_4	5	67				

The fact that the fourth divided difference is zero verifies that all the given points are evaluations of a third-degree polynomial. Writing

$$f(x) \approx f[x_0] + f[x_1, x_0](x - x_0) + f[x_2, x_1, x_0](x - x_0)(x - x_1)$$
$$+ f[x_3, x_2, x_1, x_0](x - x_0)(x - x_1)(x - x_2),$$

and considering the points to be labeled as in column I, one obtains the numerical form:

$$f(x) = 13 + (-11)(x + 1) + (1)(x + 1)(x) + (1)(x + 1)(x)(x - 2)$$
$$= 13 - 11x - 11 + x^2 + x + x^3 - x^2 - 2x$$
$$= x^3 - 12x + 2.$$

Relabeling the points as in column II leads to the illustrated zigzag path, and

$$f(x) = -14 + (16)(x - 2) + (6)(x - 2)(x - 4) + 1(x - 2)(x - 4)(x)$$
$$= -14 + 16x - 32 + 6x^2 - 36x + 48 + x^3 - 6x^2 + 8x$$
$$= x^3 - 12x + 2.$$

The representation of a function which is not a polynomial is illustrated by the following table listing values for $f(x) = \cos [(\pi/1.8)x]$. To keep the divided differences from being very small, and hence difficult to represent in tabular form, the argument x is expressed in hundreds of degrees.

	x	$\cos\left[(\pi/1.8)x\right]$	1	2	3
x_0	0.00	1.0000000			
			-0.44658201		
x_1	0.30	0.8660254		-1.2891711	
			-1.2200847		0.60540858
x_2	0.60	0.5000000		-0.74430335	
			-1.6666667		
x_3	0.90	0.0000000			

$$\cos\left(\frac{\pi}{1.8}x\right) \approx 1.0000000 - 0.44658201x - 1.2891711(x)(x - 0.30)$$

$$E_3(x) \approx (x)(x - 0.30)(x - 0.60)0.60540858$$

As a specific numerical example, let $x = 0.45$. Then

$$\cos\left(\frac{0.45}{1.8}\pi\right) \approx 0.712019$$

$$E_3(0.45) \approx 0.006130$$

In this case, since the function is known, a bound for the derivative part of the error can be computed. The relevant derivatives are:

$$f(x) = \cos\left(\frac{\pi}{1.8}x\right),$$

$$f'(x) = -\frac{\pi}{1.8}\sin\left(\frac{\pi}{1.8}x\right),$$

$$f''(x) = -\left(\frac{\pi}{1.8}\right)^2\cos\left(\frac{\pi}{1.8}x\right),$$

$$f'''(x) = \left(\frac{\pi}{1.8}\right)^3\sin\left(\frac{\pi}{1.8}x\right).$$

The derivative part of the error term, $R_3 = f'''(\zeta)/3!$, where $0.0 \le \zeta \le 0.9$, will have its greatest value when $\zeta = 0.9$. Accordingly,

$$R_3 \le \left(\frac{\pi}{1.8}\right)^3 \frac{\sin(0.9\pi/1.8)}{6} = \left(\frac{\pi}{1.8}\right)^3 \frac{1}{6} = 0.886090.$$

When this factor is multiplied by $0.45(0.45 - 0.30)(0.45 - 0.60) = -0.010125$, an error bound for the case $x = 0.45$ is determined:

$$|E_3(0.45)| \le 0.010125(0.866090) = 0.00897166$$

To four decimal places, $\cos[(\pi/1.8)0.45] = 0.7071$, and hence the error is $(0.7120 - 0.7071) = 0.0049$. The value of $E_3(0.45) = 0.0061$ provides only an estimate of the magnitude of error and is not a bound. The derivative form, however, did provide a bound, $|E_3(0.45)| \le 0.0090$.

Exercises

1. Approximate $f(x) = x^2 + 4x + 2$ by a straight line in the interval $(1, 2)$. For this approximation, choose the Taylor series that makes the maximum error in the interval as small as possible.

2. In a certain computation it is desired to approximate $f(x) = 2^x$ in the interval $(0, 1)$. How many terms of the Taylor series approximation are needed to ensure that the maximum error is less than 0.001 ? It is useful here to remember that

$$\left(\frac{d}{dx}\right)a^x = \log_e a \cdot a^x,$$

and that

$$\log_e 2 \approx 0.7.$$

3. The truncated Taylor-series approximation $\sin x = x$ is often used for small angles. As a function of x, what is the maximum error in this approximation? When some terms of a series are zero, it is best, in terms of the error estimate, to truncate the series after a zero term. Why?

4. In some small computers there is no built-in division operation. However, one method of effectively dividing is to multiply the dividend by the reciprocal of the divisor and to write the reciprocal as a series involving only multiplication and addition; that is,

$$\frac{y}{x} = y\left(\frac{1}{x}\right).$$

Obtain such a reciprocal expression by finding the Taylor-series expansion of

$$\frac{1}{x} = \frac{1}{1 - (1 - x)}, \qquad 0 < x < 2.$$

In these cases, the accuracy of the quotient is determined by analysis of the error term.

5. What conclusion may be drawn when the kth divided differences are constant? Can any conclusive statement be made about the "true" function from which the points were taken? Explain your answer in terms of the divided-difference error term.

6. In calibrating a thermometer, the following points were obtained:

Reading	Temperature
0	2
10	14
40	38

In a data-processing program, readings are to be converted to temperature. Hence a function must be generated which represents temperature as a function of reading. Find a function which, when evaluated at readings of 0, 10, and 40, gives the tabulated temperatures.

7. Construct a second-degree divided-difference polynomial, using the last three points from the following set.

x	$f(x)$
−1	4
0	2
2	4
3	20

Evaluate the polynomial at $x = 1$. What is the approximate error?

8. a) Given the variables r, road resistance, in pounds per ton weight of an automobile, and v, velocity, mph, construct a difference table for the following data:

r	4.79	5.85	7.19	8.80	12.85	18.00	24.25	31.60
v	5	10	15	20	30	40	50	60

b) Fit an interpolating polynomial through the data, using the highest order practical to interpolate for the resistance at 17 miles per hour.

chapter three

interpolation

It is probably difficult to identify interpolation, in the context in which one is first introduced to the word, as a special case of rational approximation. The table-scanning problem of finding an intermediate value between two tabulated values is simply one of approximating the tabulated function by a straight line near the point of interest and then evaluating the straight line by means of the given argument. This process lends itself readily to symbolic description. Suppose that one is given the table,

$$
\begin{array}{cc}
x_0 & f(x_0) \\
x_1 & f(x_1) \\
x_2 & f(x_2) \\
x_3 & f(x_3)
\end{array}
$$

and it is desired to find the approximate functional value which corresponds to an argument x, where $x_1 < x < x_2$. The usual formula is

$$f(x) \approx f(x_1) + \frac{x - x_1}{x_2 - x_1}\,(f(x_2) - f(x_1)).$$

With a little algebraic rearrangement this formula can be written $f(x) \approx Ax + B$, where A and B are constants; hence the approximation is a straight line (or first-degree polynomial). For our purposes, however, it is more interesting to write

$$f(x) \approx f[x_1] + f[x_2, x_1](x - x_1)$$

or to eliminate the "approximately equal":

$$p(x) = f[x_1] + f[x_2, x_1](x - x_1)$$

The familiar linear interpolation, then, is simply representing the function by the first two terms of the divided-difference polynomial $p(x)$. This identification suggests that more generalized interpolation procedures can be produced by simply including more terms of the divided-difference polynomial. If, in addition, the development of the divided-difference polynomial is recalled, a complete definition of "interpolation" can be formulated. The divided-difference polynomial was constructed from tabular values in such a way that it equaled the given functional values at the given arguments; that is,

$$f(x_0) = p(x_0), \quad f(x_1) = p(x_1), \quad f(x_2) = p(x_2), \quad \text{etc.}$$

With this additional qualification, interpolation is defined as follows:

Given $n + 1$ noncoincident points (abscissas and ordinates) and an argument x, then in the interval containing the given abscissas, nth-degree **interpolation** *is accomplished by constructing the nth-degree polynomial which passes through the given points and by evaluating it for the given argument.*

If x is outside the interval containing the abscissas, the process is *extrapolation*. The divided-difference polynomial provides a way of constructing the nth-degree polynomial, but whatever scheme is employed, there is only one nth-degree polynomial which will pass through the given $n + 1$ points. This last statement is true because the assumption that there is more than one such polynomial leads to a contradiction. Suppose that $p(x)$ and $q(x)$ are nth-degree polynomials which are equal for $x = x_0, x_1, \ldots, x_n$; that is,

$$p(x_0) = q(x_0),$$
$$p(x_1) = q(x_1),$$
$$\vdots \qquad \vdots$$
$$p(x_n) = q(x_n).$$

Then the difference, $p(x) - q(x)$, which is at most an nth-degree polynomial, is zero at $n + 1$ points; this is a contradiction, since an nth-degree polynomial can have only n zeros (or roots).

The subject of interpolation seems to have been quickly dispatched since for a given set of points only one approximating polynomial can be produced and the divided-difference interpolation polynomial provides an effective method of producing it. There are some alternative procedures which, of course, produce the same polynomial, but nonetheless deserve attention; either they are obvious, they have computational advantages, or they are useful specializations.

Simultaneous-equation approach

Under the obvious category is the straightforward approach described below. As before, the given data are the tabular values.

x_0	$f(x_0)$
x_1	$f(x_1)$
x_2	$f(x_2)$
$\vdots$	$\vdots$
x_n	$f(x_n)$

What is sought are the coefficients of an nth-degree polynomial, $a_0, a_1, \ldots, a_n$, which passes through these given points; that is,

$$a_0 + a_1x_0 + a_2x_0^2 + \cdots + a_nx_0^n = f(x_0)$$
$$a_0 + a_1x_1 + a_2x_1^2 + \cdots + a_nx_1^n = f(x_1)$$
$$a_0 + a_1x_2 + a_2x_2^2 + \cdots + a_nx_2^n = f(x_2)$$
$$\vdots \qquad\qquad \vdots$$
$$a_0 + a_1x_n + a_2x_n^2 + \cdots + a_nx_n^n = f(x_n).$$

There are $n + 1$ linear equations in this set, and there are $n + 1$ unknowns (the a's, not the x's). Hence, solving this set of equations determines the interpolation polynomial. To illustrate the equivalence of this approach to the one using the divided-difference polynomial, consider the two-point case:

$$a_0 + a_1x_0 = f(x_0)$$
$$a_0 + a_1x_1 = f(x_1)$$

Subtracting the first equation from the second determines a_1:

$$a_1(x_1 - x_0) = f(x_1) - f(x_0),$$

$$a_1 = \frac{f(x_1) - f(x_0)}{x_1 - x_0} = f[x_1, x_0],$$

and substituting this result in the first equation determines a_0:

$$a_0 = f(x_0) - a_1x_0 = f(x_0) - f[x_1, x_0]x_0,$$

so that

$$f(x) = f[x_1, x_0]x + (-f[x_1, x_0]x_0 + f(x_0)).$$

This result is the form $Ax + B$ alluded to earlier, but it is easily recognized as the first two terms of the divided-difference interpolation polynomial.

At this point, the question can be raised, why, since this method is so straightforward, is it not used as the basis for interpolation rather than the notationally more complex divided-difference approach? There are three answers: (1) Solving simultaneous linear equations is computationally (if not conceptually) more complex than forming difference tables and interpolating polynomials from these tables. (2) The difference tables are useful in the estimation of error. (3) The divided differences can be readily specialized to the commonly occurring equal-interval tables.

Lagrangian interpolation

For computational reasons it would be advantageous to have an interpolation method that did not require the computation of a divided-difference table or the solution of simultaneous equations. Such a method would be particularly useful when only one or two interpolations are computed in a given range of tabular values. One way of deriving such a method is to take the divided-difference polynomial and, by algebraic manipulation, write it so that the given ordinates $f(x_0), f(x_1), \ldots, f(x_n)$ do not appear in differences but explicitly. The first step is to rewrite the divided differences in this form:

$$f[x_0] = f(x_0),$$

$$f[x_1, x_0] = \frac{f(x_1) - f(x_0)}{x_1 - x_0} = \frac{f(x_1)}{x_1 - x_0} + \frac{f(x_0)}{x_0 - x_1},$$

$$f[x_2, x_1, x_0]$$

$$= \frac{[f(x_2)/(x_2 - x_1) + f(x_1)/(x_1 - x_2)] - [f(x_1)/(x_1 - x_0) + f(x_0)/(x_0 - x_1)]}{x_2 - x_0}$$

$$= \frac{f(x_2)}{(x_2 - x_1)(x_2 - x_0)} + \frac{f(x_1)}{(x_1 - x_2)(x_1 - x_0)} + \frac{f(x_0)}{(x_0 - x_2)(x_0 - x_1)}.$$

The algebra becomes involved beyond this point but a pattern is apparent:

$$f[x_n, x_{n-1}, \ldots, x_0] = \frac{f(x_n)}{(x_n - x_{n-1}) \cdots (x_n - x_0)}$$

$$+ \frac{f(x_{n-1})}{(x_{n-1} - x_n)(x_{n-1} - x_{n-2}) \cdots (x_{n-1} - x_0)} + \cdots$$

$$+ \frac{f(x_0)}{(x_0 - x_n) \cdots (x_0 - x_1)} = \sum_{i=0}^{n} \frac{f(x_i)(x_i - x_i)}{\prod_{j=0}^{n} (x_i - x_j)}.$$

The (seemingly) zero factor in the numerator $(x_i - x_i)$ is included to cancel the factor $(x_i - x_j)$, where $i = j$ in the denominator. A more satisfying form of expressing this deletion of a term in the denominator is

$$f[x_n, x_{n-1}, \ldots, x_0] = \sum_{i=0}^{n} \frac{f(x_i)}{\prod_{j=0, j \neq i}^{n} (x_i - x_j)}.$$

It is apparent from writing the divided differences in this form that they are symmetric. For instance, reversing the values of x_1 and x_0 in the first divided difference does not change its value since the terms are additive and x_1 and x_0 appear symmetrically in every term. In fact, permutations of the x's in divided differences of any order do not result in a change in value. This result can be stated differently: In computing an nth-order divided difference, one can introduce the $n + 1$ values in any order without affecting the result.

Returning to the original task, one now uses these symmetric forms to replace the divided difference in, say, a second-degree polynomial:

$$p(x) = f[x_0] + (x - x_0)f[x_1, x_0] + (x - x_0)(x - x_1)f[x_2, x_1, x_0]$$

$$= f(x_0) + (x - x_0)\left\{\frac{f(x_1)}{x_1 - x_0} + \frac{f(x_0)}{x_0 - x_1}\right\} + (x - x_0)(x - x_1)$$

$$\times \left\{\frac{f(x_2)}{(x_2 - x_0)(x_2 - x_1)} + \frac{f(x_1)}{(x_1 - x_2)(x_1 - x_0)} + \frac{f(x_0)}{(x_0 - x_2)(x_0 - x_1)}\right\}$$

which, after some algebraic labor, reduces to the *Lagrangian* form:

$$p(x) = \frac{(x - x_1)(x - x_2)}{(x_0 - x_1)(x_0 - x_2)}f(x_0) + \frac{(x - x_0)(x - x_2)}{(x_1 - x_0)(x_1 - x_2)}f(x_1)$$

$$+ \frac{(x - x_0)(x - x_1)}{(x_2 - x_0)(x_2 - x_1)}f(x_2)$$

$$= \sum_{i=0}^{2} \prod_{\substack{j=0 \\ j \neq i}}^{2} \left[\frac{x - x_j}{x_i - x_j}\right]f(x_i).$$

Observe that no difference table is required to evaluate this second-degree polynomial. To be sure, the coefficients of the given ordinates are fairly complex, but their systematic form means that their computation can be simply described by a computer program. The generalization to an nth-degree polynomial is obtained by inserting n, instead of 2, for the terminal values of the indices i and j.

Relating the Lagrangian form to the divided-difference polynomial, as was done here, has the advantage that the error term need not be examined again since it is the same as it was before. However, an alternative development similar to the divided-difference case is perhaps easier to understand. Assume that an nth-degree polynomial has the form

$$p(x) = a_0(x - x_1)(x - x_2) \cdots (x - x_n)$$
$$+ a_1(x - x_0)(x - x_2) \cdots (x - x_n) + \cdots$$
$$+ a_n(x - x_0)(x - x_1) \cdots (x - x_{n-1}).$$

This form, i.e., the sum of factored polynomials, guarantees that all terms but one will be zero when $x = x_0, x_1, x_2, \ldots, x_n$. Thus a_0 may be determined when $x = x_0$:

$$f(x_0) = p(x_0) = a_0(x_0 - x_1)(x_0 - x_2) \cdots (x_0 - x_n),$$

$$a_0 = \frac{f(x_0)}{(x_0 - x_1)(x_0 - x_2) \cdots (x_0 - x_n)}.$$

In general, a_k can be determined by letting $x = x_k$, where $k = 0, 1, 2, \ldots, n$:

$$a_k = \frac{f(x_k)}{(x_k - x_0) \cdots (x_k - x_{k-1})(x_k - x_{k+1}) \cdots (x_k - x_n)}.$$

If these coefficients are substituted into the original polynomial, it is identifiable as the Lagrangian form just described. The computation of this interpolation polynomial provides an interesting programming example since it involves a nested iteration having a conditional computation in its scope. To illustrate Lagrangian interpolation numerically before proceeding to such an example, let us suppose that second-degree interpolation is to be used to find an approximate value for $f(3.5)$ from the following table.

	x	$f(x)$
x_0	2	10
x_1	3	30
x_2	4	68
x_3	5	130

We write

$$f(3.5) \approx \frac{(x - x_1)(x - x_2)}{(x_0 - x_1)(x_0 - x_2)} f(x_0) + \frac{(x - x_0)(x - x_2)}{(x_1 - x_0)(x_1 - x_2)} f(x_1)$$

$$+ \frac{(x - x_0)(x - x_1)}{(x_2 - x_0)(x_2 - x_1)} f(x_2)$$

$$\approx \frac{(3.5 - 3)(3.5 - 4)}{(2 - 3)(2 - 4)} 10 + \frac{(3.5 - 2)(3.5 - 4)}{(3 - 2)(3 - 4)} 30$$

$$+ \frac{(3.5 - 2)(3.5 - 3)}{(4 - 2)(4 - 3)} 68$$

$$\approx 46.75.$$

Using linear interpolation, we obtain

$$f(3.5) \approx \frac{(3.5 - 4)}{(3 - 4)} 30 + \frac{(3.5 - 3)}{(4 - 3)} 68 = 49.$$

The tabulated function happens to be $f(x) = x^3 + x$ for which $f(3.5) = 46.375$.

Example Write a computer program which reads a value for $n \leq 10$ and then n abscissas with the corresponding functional values. After these tabular values have been read, the program should read an argument and, using $(n - 1)$-degree Lagrangian interpolation, compute the corresponding functional value. After this approximate value and the argument are printed, another argument is to be read and the interpolation repeated until the input list of arguments has been exhausted.

If the points are numbered from 1 to n, the formula is

$$p(x) = \sum_{i=1}^{n} \left(\prod_{\substack{j=1 \\ j \neq i}}^{n} \frac{x - x_j}{x_i - x_j} \right) f(x_i).$$

Note that both a repeated product and a repeated sum are to be calculated. The sum must initially be set to zero, but the partial product variable, instead of being initially set to 1, can in this case, be set to $f(x_i)$. The flow chart and program are presented in Fig. 3.1. The result for each argument is tabulated in the output along with N, the number of data points.

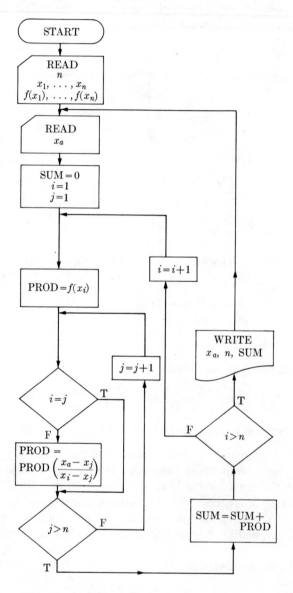

Fig. 3.1 Lagrangian interpolation. (a) Flow chart.

Equal-interval interpolation

Neither the divided difference nor the Lagrangian polynomials imposed any restrictions on the spacing of the given abscissas. In fact, even the order of the points was unimportant. However, it is very often convenient to tabulate a function for equal intervals of the argument, since this technique considerably simplifies both notation and computation.

Forward Differences

The problem is to rewrite the divided-difference polynomial, which, together with the divided difference remainder, is called *Newton's fundamental formula*, in a

```
C       MASTER LIP
C       LAGRANGIAN INTERPOLATION
        DIMENSION X(10),Y(10)
        WRITE(6,2)
2       FORMAT(2X,1HN,5X,8HARGUMENT,10X,17HVALUE OF FUNCTION)
        READ(5,3)N,(X(I),I=1,N)
        READ(5,4)(Y(I),I=1,N)
10      READ(5,5)ARG
3       FORMAT(I9,10F9.0)
4       FORMAT(10F9.0)
5       FORMAT(F9.0)
        SUM=0.
        DO 6 I=1,N
        PROD=Y(I)
        DO 7 J=1,N
        IF(I-J)8,7,8
8       PROD=PROD*(ARG-X(J))/(X(I)-X(J))
7       CONTINUE
6       SUM=SUM+PROD
        WRITE(6,9)N,ARG,SUM
9       FORMAT(I2,F11.3,F26.8)
        GO TO 10
        END

            INPUT

7,-.45,-.15,.15,.45,.75,.90,1.35,
1.0,.86625,.5,0.,-.5,-.707107,-1.0,
-.3,
0.,
.15,
.45,
1.05,
1.20,

            OUTPUT
```

N	ARGUMENT	VALUE OF FUNCTION
7	−0.300	0.96608859
7	0.0	0.70722592
7	0.150	0.49999946
7	0.450	0.0
7	1.050	−0.86587214
7	1.200	−0.96552402

Fig. 3.1 (b) Program and print-out.

form that takes advantage of the fact that

$$x_1 - x_0 = x_2 - x_1 = x_3 - x_2 = \cdots = x_n - x_{n-1} = h.$$

All the denominators of the divided differences are differences of tabulated x's and hence can be represented as multiples of h, i.e.,

$$f[x_1, x_0] = \frac{f(x_1) - f(x_0)}{x_1 - x_0} = \frac{f(x_1) - f(x_0)}{h},$$

$$f[x_2, x_1, x_0] = \frac{[f(x_2) - f(x_1)]/h - [f(x_1) - f(x_0)]/h}{x_2 - x_0}$$

$$= \frac{[f(x_2) - f(x_1)] - [f(x_1) - f(x_0)]}{2h^2}.$$

In fact, the powers of h can be separated and the differences formed from the ordinates only. To express these differences (there is no longer any division involved), one can use the forward-difference notation which is defined in tabular form, called a forward-difference table, as follows:

x_0 $f(x_0)$

$\qquad\qquad \Delta f(x_0) = f(x_1) - f(x_0)$

x_1 $f(x_1)$ $\qquad\qquad\qquad \Delta^2 f(x_0) = \Delta f(x_1) - \Delta f(x_0)$

$\qquad\qquad \Delta f(x_1) = f(x_2) - f(x_1) \qquad\qquad \Delta^3 f(x_0) = \Delta^2 f(x_1) - \Delta^2 f(x_0)$

x_2 $f(x_2)$ $\qquad\qquad\qquad \Delta^2 f(x_1) = \Delta f(x_2) - \Delta f(x_1)$

$\qquad\qquad \Delta f(x_2) = f(x_3) - f(x_2)$

x_3 $f(x_3)$

In general, $\Delta^n f(x_k) = \Delta^{n-1} f(x_{k+1}) - \Delta^{n-1} f(x_k)$. The relationship of this notation to the divided difference follows from our previous remarks.

$$f[x_1, x_0] = \frac{\Delta f(x_0)}{h},$$

$$f[x_2, x_1, x_0] = \frac{\Delta^2 f(x_0)}{2h^2},$$

$$f[x_n, \ldots, x_0] = \frac{\Delta^n f(x_0)}{n! h^n}.$$

Making these substitutions yields Newton's fundamental formula in forward-difference notation:

$$f(x) = f(x_0) + (x - x_0)\frac{\Delta f(x_0)}{h} + (x - x_0)(x - x_1)\frac{\Delta^2 f(x_0)}{2! h^2}$$

$$+ (x - x_0)(x - x_1)(x - x_2)\frac{\Delta^3 f(x_0)}{3! h_3} + \cdots$$

The error term is still

$$(x - x_0) \cdots (x - x_n) \frac{f^{(n+1)}(\xi)}{(n + 1)!}.$$

There is a simple change in variable which produces a further simplification because it cancels out the powers of the increment, h, in the denominator. Due to the equal spacing, all the given abscissas can be written with reference to x_0,

$$x_0 = x_0$$
$$x_1 = x_0 + h$$
$$x_2 = x_0 + 2h$$
$$x_3 = x_0 + 3h$$
$$\vdots$$

and any x can be represented as $x = x_0 + sh$, where s is now a new variable and may be thought of as being expressed in "units of h." Since $x - x_0 = sh$, $x - x_1 = (s - 1)h$, $x - x_2 = (s - 2)h$, etc., this change of variable simplifies the forward-difference form to

$$f(x_0 + sh) = f(x_0) + s \, \Delta f(x_0) + s(s - 1) \frac{\Delta^2 f(x_0)}{2!}$$

$$+ s(s - 1)(s - 2) \frac{\Delta^3 f(x_0)}{3!} + \cdots$$

As a quick illustration, the Lagrangian example is used again.

x	f	Δ	Δ^2	Δ^3
2	10			
		20		
3	30		18	
		38		6
4	68		24	
		62		
5	130			

The required differences are obtained from the forward-difference table above. For $x = 3.5$ and $h = 1$, the value of s is determined as follows:

$$s = \frac{x - x_0}{h} = \frac{3.5 - 2}{1} = 1.5$$

$$f(3.5) \approx 10 + 1.5(20) + 1.5(0.5)\tfrac{18}{2} = 46.75,$$

where the differences employed lie along the diagonal line in the difference table. The computation was much less involved in this instance; however, equal intervals were, of course, required and some computation was necessary to compute the differences.

As with the divided-differences polynomial, the first neglected term provides an estimate of the magnitude of error. In this example,

$$s(s-1)(s-2)\frac{\Delta^3 f(x_0)}{3!} = 1.5(0.5)(-0.5)\frac{6}{3!} = -0.375,$$

which happens to be the exact error since the function was tabulated from the third-degree polynomial $x^3 + x$.

Operator Notation

The primary difficulty with interpolation is notational. The basic idea of finding a polynomial that will pass through a given set of points is straightforward, but the notation required to express these polynomials tends to become cumbersome and does, at times, obscure certain useful relationships. Part of this difficulty can be overcome in the equal-interval case by defining linear operators. The relationships can then be deduced by the algebraic manipulation of the operators. For example, a forward-difference operator, Δ, which can operate on any function, say $f(x)$, can be defined as follows:

$$\Delta f(x) = f(x+h) - f(x)$$

This corresponds to the use of the symbol Δ in the preceding section, where

$$\Delta f(x_0) = f(x_1) - f(x_0)$$

since $x_1 - x_0 = h$. That this operator is linear can be seen by applying the operator to a function which is the sum of two other functions:

$$\begin{aligned}
\Delta[af(x) + bg(x)] &= [af(x+h) + bg(x+h)] - [af(x) + bg(x)] \\
&= a[f(x+h) - f(x)] + b[g(x+h) - g(x)] \\
&= a\,\Delta f(x) + b\,\Delta g(x).
\end{aligned}$$

The application of the operator to a sum of functions is the same as the sum of operators applied to each function individually. This property is the basis for much of the algebraic manipulation of operators.

In interpolation, the emphasis is on tabulated values rather than on functions in symbolic form. The interest is, of necessity, focused on $\Delta f(x_0)$, which is the value of a difference at a specific point, x_0, rather than on $\Delta f(x)$, where x is still not bound to a specific value. It is important to note that the operator Δ is certainly applicable in the latter sense. For instance, if

$$f(x) = ax^2 + bx + c,$$

then

$$\begin{aligned}
\Delta f(x) = \Delta(ax^2 + bx + c) &= a(x+h)^2 + b(x+h) + c - ax^2 - bx - c \\
&= 2ahx + ah^2 + bh.
\end{aligned}$$

Table 3.1

r / k	0	1	2	3	4
0	1				
1	1	−1			
2	1	−2	1		
3	1	−3	3	−1	
4	1	−4	6	−4	1

This expression begins to look like a derivative. Dividing by h and letting $h \to 0$ do produce the derivative:

$$\frac{\Delta f(x)}{h} = 2ax + ah + b,$$

$$\lim_{h \to 0} \frac{\Delta f(x)}{h} = 2ax + b = f'(x).$$

The calculus of finite differences is a subject in its own right, quite independent of the specific problem of interpolation.

As an illustration of the utility of operators, note from the preceding section that

$$\begin{aligned} \Delta^2 f(x_0) &= \Delta f(x_1) - \Delta f(x_0) \\ &= [f(x_2) - f(x_1)] - [f(x_1) - f(x_0)] \\ &= f(x_2) - 2f(x_1) + f(x_0). \end{aligned}$$

The appearance of the coefficients, 1, −2, 1, suggests that they may have been obtained by raising a binomial expression to the second power, that is, $(a - b)^2 = 1 \cdot a^2 - 2ab + 1 \cdot b^2$; or, stated in another way, the coefficients might have come from the row labeled 2 of Pascal's triangle (Table 3.1). This table contains the binomial coefficients

$$C_{k,r} = (-1)^r \left(\frac{k!}{r!(k - r)!} \right).$$

If the suspicion stated above is correct, then

$$\Delta^3 f(x_0) = 1 \cdot f(x_3) - 3f(x_2) + 3f(x_1) - 1 \cdot f(x_0).$$

That this result is true can be easily verified by using operators. First it is convenient to introduce another operator, E, called a shift operator because it simply shifts the argument x by the incremental amount h:

$$Ef(x) = f(x + h).$$

The difference operator may be applied repeatedly:

$$\Delta^2 f(x) = \Delta \, \Delta f(x) = \Delta(f(x + h) - f(x)) = f(x + 2h) - 2f(x + h) + f(x).$$

The repeated application of the shift operator results in a further shift of the argument:

$$E^2 f(x) = EEf(x)$$
$$= Ef(x + h)$$
$$= f(x + 2h).$$

The original goal was to determine the coefficients of the ordinates in a high-order difference, say $\Delta^k f(x_0)$:

$$\Delta = E - 1$$

since

$$(E - 1)f(x) = f(x + h) - f(x)$$

and therefore $\Delta^k = (E - 1)^k$, which essentially demonstrates the desired result since $E - 1$ is a binomial. To give a specific instance,

$$\Delta^3 f(x_0) = (E - 1)^3 f(x_0) = (E^3 - 3E^2 + 3E - 1)f(x_0)$$
$$= f(x_0 + 3h) - 3f(x_0 + 2h) + 3f(x_0 + h) - f(x_0)$$
$$= f(x_3) - 3f(x_2) + 3f(x_1) - f(x_0).$$

This result is much more easily proved by means of operators than by manipulation of the differences themselves.

As a last illustration of the concise expression which operators permit, a Taylor's series expansion is presented in the very compact form made possible by the use of the well-known expansion of e^x:

$$e^x = 1 + x + \frac{x^2}{2!} + \frac{x^3}{3!} + \frac{x^4}{4!} + \cdots$$

Now, if x is formally replaced by $h(d/dx)$ and the entire expansion is treated as an operator, one has

$$f(x + h) = \{e^{h(d/dx)}\} f(x)$$
$$= \left\{ 1 + h\frac{d}{dx} + \frac{h^2}{2!}\frac{d^2}{dx^2} + \frac{h^3}{3!}\frac{d^3}{dx^3} + \frac{h^4}{4!}\frac{d^4}{dx^4} + \cdots \right\} f(x)$$
$$= f(x) + hf'(x) + \frac{h^2}{2!}f''(x) + \frac{h^3}{3!}f'''(x) + \cdots$$

Just the reduction in the amount of writing would be useful, but here again certain manipulations with the operator itself are permissible.

Backward-Differences

The forward-difference notation is constructed with reference to the first of a given set of points. As before, paths through the difference table other than the upper diagonal can be used to form an interpolating polynomial, but it is convenient to

Table 3.2

---

x_{k-3} $f(x_{k-3})$

$\qquad \nabla f(x_{k-2}) = f(x_{k-2}) - f(x_{k-3})$

x_{k-2} $f(x_{k-2})$ $\qquad\qquad\qquad \nabla^2 f(x_{k-1}) = \nabla f(x_{k-1}) - \nabla f(x_{k-2})$

$\qquad \nabla f(x_{k-1}) = f(x_{k-1}) - f(x_{k-2})$ $\qquad\qquad\qquad \nabla^3 f(x_k) = \nabla^2 f(x_k) - \nabla^2 f(x_{k-1})$

x_{k-1} $f(x_{k-1})$ $\qquad\qquad\qquad \nabla^2 f(x_k) = \nabla f(x_k) - \nabla f(x_{k-1})$

$\qquad \nabla f(x_k) = f(x_k) - f(x_{k-1})$

x_k $\quad f(x_k)$

---

identify all the differences as "starting from" the initial point x_0. In some problems, particularly the solution of ordinary differential equations, it is convenient to identify differences with respect to the last point included. The symbol for such *backward differences* is the inverted delta (∇), and the relationship to divided differences is completely analogous to the forward-difference case. The x's are at equal intervals,

$$
\begin{aligned}
x_k - x_{k-1} &= x_{k-1} - x_{k-2} \\
&= x_{k-2} - x_{k-3} = \cdots \\
&= x_1 - x_0 \\
&= h,
\end{aligned}
$$

and the differences are defined in Table 3.2. One can now define a backward-difference operator which is consistent with this table:

$$ \nabla f(x) = f(x) - f(x - h). $$

With a backward-shift operator,

$$ E^{-1} f(x) = f(x - h), $$

one can make a similar analysis of the form of the high-order differences, starting from the relation

$$ \nabla = (1 - E^{-1}). $$

Since the notation is with respect to x_k, the logical change of variable is

$$ x = x_k + sh, $$

and the fundamental formula in backward-difference notation is

$$ f(x_k + sh) $$
$$ = f(x_k) + s\nabla f(x_k) + s(s + 1)\frac{\nabla^2 f(x_k)}{2!} + s(s + 1)(s + 2)\frac{\nabla^3 f(x_k)}{3!} + \cdots $$

If the last three of the four points of the numerical example are used, one should not expect the same result. The difference table is the same, but the names have changed, and the differences used lie along the lower diagonal as shown by the line in the difference table:

x	f	∇	∇^2	∇^3
2	10			
		20		
3	30		18	
		38		6
4	68		24	
		62		
5	130			

$$ s = \frac{x - x_k}{h} = \frac{3.5 - 5}{1} = -1.5 $$

$$ f(3.5) \approx 130 - 1.5(62) + 1.5(0.5)\tfrac{24}{2} = 46 $$

For the same reasoning as before, the error estimate should give the exact error in this case:

$$ s(s + 1)(s + 2)\frac{\nabla^3 f(x_k)}{3!} = -1.5(-0.5)(0.5)\frac{6}{3!} = 0.375 $$

Central Differences

Neither of the equal-interval difference notations thus far introduced is really well suited for straightforward interpolation. Ideally, in forming an interpolating polynomial, it is desirable to include points symmetrically about some central point (or interval) as terms are added to the polynomial. That is, as additional ordinates are needed for higher-order differences, they should be taken from both sides of the central point rather than consistently forward (as in forward differences) or consistently backward (as in backward differences). The troublesome part of such efforts for notational symmetry is that every other column of a difference table is nonsymmetrical with respect to a point; i.e., if x_0 is a central point, the first difference to be added is either $f(x_1) - f(x_0)$ or $f(x_0) - f(x_1)$, that is, a point is being added unsymmetrically from either one side or the other. One can force symmetry in the notation, however, by referring to "fictional" (i.e., nontabulated) points for all differences that are obtained from an even number of ordinates. For instance, the first difference will include points symmetrically about x_0 if it is defined to be

$$ \delta f(x_0) = f\left(x_0 + \frac{h}{2}\right) - f\left(x_0 - \frac{h}{2}\right), $$

but these designated functional values are not tabulated. However,

$$\delta f\left(x_0 + \frac{h}{2}\right) = f(x_0 + h) - f(x_0) = f(x_1) - f(x_0),$$

which are tabulated values. The second difference, involving three points, is symmetric and is determined from given values:

$$\delta^2 f(x_0) = \delta f\left(x_0 + \frac{h}{2}\right) - \delta f\left(x_0 - \frac{h}{2}\right)$$

$$= (f(x_1) - f(x_0)) - (f(x_0) - f(x_{-1})).$$

The central-difference table then has mid-interval arguments for every other column, as shown in Table 3.3. The equivalence to divided differences has to be written separately for the central differences of even and odd order. This separation of even and odd is often indicated by considering an index $n = 0, 1, 2, \ldots$; then $2n$ represents the even numbers and $2n + 1$ the odd ones:

$$\frac{\delta^{2n} f(x_0)}{(2n)! h^{2n}} = f[x_{2n}, x_{2n-1}, \ldots, x_0],$$

$$\frac{\delta^{2n+1} f(x_0 + h/2)}{(2n + 1)! h^{2n+1}} = f[x_{2n+1}, x_{2n}, \ldots, x_0].$$

It is not possible to select a path through the table which will involve only differences with the x_0-argument. The closest one can come to symmetry about x_0 is to select

$$f(x_0), \quad \delta f\left(x_0 + \frac{h}{2}\right), \quad \delta^2 f(x_0), \quad \delta^3\left(x_0 + \frac{h}{2}\right), \quad \text{etc.,}$$

which is a path that zigzags *forward* from the central point x_0. This path is shown by the lines in Table 3.3. Making the change of variable $x = x_0 + sh$ and writing an interpolation polynomial with the path selected above produce *Gauss' forward formula*,

$$f(x_0 + sh) = f(x_0) + s\, \delta f\left(x_0 + \frac{h}{2}\right) + s(s - 1)\frac{\delta^2 f(x_0)}{2!}$$

$$+ s(s - 1)(s + 1)\frac{\delta^3 f\left(x_0 + \frac{h}{2}\right)}{3!} + \cdots$$

This zigzag path suggests how the desired symmetry can be attained. Add to it a *backward* zigzag-path polynomial (*Gauss' backward formula*) and take one-half of the resulting sum. The resulting interpolating polynomial, called *Stirling's formula*, is symmetric about x_0. New coefficients are formed from combining the forward and backward formulas when paths coincide, and the terms are averages of adjacent

Table 3.3

x	f	δ	δ^2	δ^3
x_{-2}	$f(x_{-2})$			
		$\delta f\left(x_{-1} - \dfrac{h}{2}\right) = f(x_{-1}) - f(x_{-2})$		
x_{-1}	$f(x_{-1})$		$\delta^2 f(x_{-1}) = \delta f\left(x_0 - \dfrac{h}{2}\right) - \delta f\left(x_{-1} - \dfrac{h}{2}\right)$	
		$\delta f\left(x_0 - \dfrac{h}{2}\right) = f(x_0) - f(x_{-1})$		$\delta^3 f\left(x_0 - \dfrac{h}{2}\right) = \delta^2 f(x_0) - \delta^2 f(x_{-1})$
x_0	$f(x_0)$		$\delta^2 f(x_0) = \delta f\left(x_0 + \dfrac{h}{2}\right) - \delta f\left(x_0 - \dfrac{h}{2}\right)$	
		$\delta f\left(x_0 + \dfrac{h}{2}\right) = f(x_1) - f(x_0)$		$\delta^3 f\left(x_0 + \dfrac{h}{2}\right) = \delta^2 f(x_1) - \delta^2 f(x_2)$
x_1	$f(x_1)$		$\delta^2 f(x_1) = \delta f\left(x_1 + \dfrac{h}{2}\right) - \delta f\left(x_0 + \dfrac{h}{2}\right)$	
		$\delta f\left(x_1 + \dfrac{h}{2}\right) = f(x_2) - f(x_1)$		
x_2	$f(x_2)$			

differences when they do not:

$$f(x_0 + sh) = f(x_0) + \frac{s}{2}\left(\delta f\left(x_0 + \frac{h}{2}\right) + \delta f\left(x_0 - \frac{h}{2}\right)\right) + s^2\,\frac{\delta^2 f(x_0)}{2!}$$

$$+ \frac{s(s-1)(s+1)}{2\cdot3!}\left(\delta^3 f\left(x_0 + \frac{h}{2}\right) + \delta^3 f\left(x_0 - \frac{h}{2}\right)\right) + \cdots$$

Similar addition of appropriate zigzag paths can produce a formula which is symmetric about an interval, called *Bessel's formula*.

Even the operator formulation is a little more involved in this case:

$$\delta f(x) = f\left(x + \frac{h}{2}\right) - f\left(x - \frac{h}{2}\right).$$

Half-shift operators are needed to write an equivalent form for δ:

$$E^{1/2}f(x) = f\left(x + \frac{h}{2}\right),$$

$$E^{-1/2}f(x) = f\left(x - \frac{h}{2}\right).$$

Therefore

$$\delta = E^{1/2} - E^{-1/2}.$$

To see that the high-order differences have the same binomial form, observe that

$$\delta^k = (E^{1/2} - E^{-1/2})^k = [E^{-k/2}(E - 1)^k].$$

The same binomial is obtained as in the forward-difference case, but a "halfway" shift of the point of reference of $-k/2$ is also applied.

As an illustration of Stirling's formula, we present the same simple example.

	x	f	$\delta_{1/2}$	δ^2	$\delta^3_{1/2}$
x_{-1}	2	10			
			20		
x_0	3	30		18	
			38		6
x_1	4	68		24	
			62		
x_2	5	130			

$$s = \frac{x - x_0}{h} = \frac{3.5 - 3}{1} = 0.5$$

$$f(3.5) \approx 30 + \frac{0.5}{2}(38 + 20) + (0.5)^2(\tfrac{18}{2}) = 46.75.$$

This result is the same as in the forward-difference case, as it must be since the same three points were used to construct a second-degree polynomial. Note, however, that the first two terms of this approximation add up to 44.5, while the first two terms of the previous forward-difference example are 40. This is an expected result since the central approach means that the first terms will be closer to the final result. If the number of points to be used for an approximation is a variable, it is common sense to adopt the central approach and include first those points which are closest to the given argument.

Instead of writing a program using one of these equal-interval formulas, we shall illustrate the more general divided-difference procedure.

Example Write a program which reads a set of $n + 1$ points with abscissas in ascending order and then computes the divided differences up to order m. After the difference table has been formed, the program reads an argument a and the degree of the interpolation polynomial to be computed, d. Using a, the program searches the list of abscissas to find the appropriate $d + 1$ points from which to construct the polynomial. The interpolation is central in that, as far as possible, the points to be used will be taken equally from either side of the argument. Compute and print the first term neglected as an estimate of the magnitude of the truncation error. If the argument is outside the given tabular values, either the first or last $d + 1$ points are used for extrapolation. The reading of arguments is continued until a prescribed number of sets, NA, has been read. The program should accommodate a table of 100 entries and divided differences up to sixth order. Due to the complexity of the problem statement, it is probably worthwhile to give some description of the procedure before writing a flow chart.

Step 1 Read M, N, NA and

$$x_0 \quad x_1 \quad x_2 \quad x_3 \quad x_4 \quad \cdots \quad x_N$$
$$y_0 \quad y_1 \quad y_2 \quad y_3 \quad y_4 \quad \cdots \quad y_N$$

Step 2 Compute the divided differences. For this purpose the y's can be regarded as the zeroth divided differences and are the zeroth row of a matrix. The first row to be computed is the first divided difference, the second, the divided difference of order two, etc.:

$$
\begin{array}{ccccccc}
y_{0,0} & y_{0,1} & y_{0,2} & y_{0,3} & y_{0,4} & \cdots & y_{0,N} \\
 & y_{1,1} & y_{1,2} & y_{1,3} & y_{1,4} & \cdots & y_{1,N} \\
 & & y_{2,2} & y_{2,3} & y_{2,4} & \cdots & y_{2,N} \\
 & & & y_{M,M} & & \cdots & y_{M,N}
\end{array}
$$

The formula for computing a specific divided difference, $y_{i,j}$, is, in these terms,

$$y_{i,j} = \frac{y_{i-1,j} - y_{i-1,j-1}}{x_j - x_{j-i}},$$

where $1 \leq i \leq M$ and $i \leq j \leq N$. To compute the Mth-order differences, a sufficient number of tabulated points must be available; that is, $N \geq M + 1$.

Step 3 Read the argument a and the degree of interpolation, d, an integer, and search the x's to find the first x, say x_k, for which $a \leq x_k$. If $d < k < N - d/2$, add $d/2$ to k so that the $d + 1$ points ending with x_k are on both sides of a:

$$\overbrace{}^{d + 1 \text{ values}}$$
$$x_0 \qquad x_1 \qquad x_2 \qquad x_3 \qquad x_4 \qquad x_5 \qquad x_6 \quad \cdots$$
$$\uparrow$$
$$a$$

(For this case, $k = 4$, $d = 3$)

It is important to understand here that the result of integer division is truncated to an integer quotient; e.g., $[\frac{1}{2}] = 0$, $[\frac{3}{2}] = 1$.

Step 4 Evaluate the divided-difference polynomial. The relevant divided differences will all be in the kth column so that

$$P(A) = y_{0,k} + (a - x_k)y_{1,k} + (a - x_k)(a - x_{k-1})y_{2,k} + \cdots$$

Step 5 Compute, but do not add into the sum of terms, the $(d + 2)$-term; its value is the estimate of the truncation-error magnitude. With this requirement, it is convenient to compute the polynomial starting from the constant term, rather than using the nested evaluation.

The flow chart and program for this example are presented in Fig. 3.2. The input information and the results from the trial values are printed as output along with the program in Fig. 3.2(b).

The two dimensional y-array is unusual in that the initial element is $y_{0,0}$ instead of $y_{1,1}$. This notation is accommodated by including in the dimension vector for the y-array an appropriate base linear subscript. To allow for an additional row of 100 elements preceding the first row, the linear-subscript equivalent to $y_{1,1}$ is selected to be 101. Since the y-array can have at most 7 rows of 100 elements, the amount to be reserved is 700. A counter, NK, has to be introduced to determine the number of arguments evaluated. When this exceeds the number supplied, which is read in as NA, the computing terminates.

It is important to note that as there are no zero subscripts in FORTRAN programs, the subscripts of X and Y are actually larger by one than the corresponding subscripts on the flow chart. That is x_0 is X(1), and $y_{1,3}$ is Y(2, 4). As a result when a subscript I or J is determined in the program, it is always one larger than that shown on the chart; and instead of reading in values from 0 to N values are read in from 1 to $N + 1$.

(a)

```
C       MASTER DIVDIFF
C       DIVIDED DIFFENCE INTERPOLATION
        INTEGER D,DP2
        DIMENSION X(100),Y(7,100)
        READ(5,10)M,N,NA
        NP1=N+1
        READ(5,11)(X(I),Y(1,I),I=1,NP1)
10      FORMAT(3I9)
11      FORMAT(14F6.2)
        WRITE(6,12)M,N,NA
12      FORMAT(5X,3HM= ,I5,3X,3HN= ,I5,3X,4HNA= I5,/)
        WRITE(6,13)(X(I),I=1,NP1)
        WRITE(6,14)(Y(1,I),I=1,NP1)
13      FORMAT(2X,6HX(0) = ,7F10.6,/)
14      FORMAT(2X,8HY(0,0) = ,7F10.6,/)
        NK=1
        MP1=M+1
        DO 2 I=2,MP1
        DO 2 J=I,NP1
2       Y(I,J)=(Y(I-1,J)-Y(I-1,J-1))/(X(J)-X(J-I+1))
20      READ(5,15)A,D
15      FORMAT(F9.0,I9)
        DO 4 K=1,NP1
        IF(A-X(K))3,3,4
4       CONTINUE
        IF(K.GT.D) GO TO 3
        K=D+2
        GO TO 33
3       IF(K.LE.N+1-D/2) K=K+D/2
33      S=0.
        DP2=D+2
        DO 8 I=1,DP2
        T=Y(I,K)
        J=K
        IF(I.EQ.1) GO TO 6
5       T=T*(A-X(J))
        J=J-1
        IF(J.GE.K+2-I) GO TO 5
        IF(I.GT.D+1) GO TO 8
6       S=S+T
8       CONTINUE
        WRITE(6,16)A,D,S,T
16      FORMAT(3X,3HA= ,F10.5,3X,3HD= ,I4,3X,3HS= ,F9.5,3X,3HT= ,F9.5)
        IF(NK.GE.NA) GO TO 21
        NK=NK+1
        GO TO 20
21      STOP
        END
```

```
        INPUT

6,6,6,
 -.45,1.0,-.15,.866,.15,.5,.45,0.,.75,-.5,.9,-.707,1.35,-1.0,
-.3,2,
0.,2,
.15,3,
.45,3,
1.05,4,
1.20,4,
```

```
        OUTPUT

   M=     6   N=     6   NA=     6
X(0) =  -0.450000 -0.150000  0.150000  0.450000  0.750000  0.900000  1.349999
Y(0,0) =   1.000000  0.866000  0.500000  0.0       -0.500000 -0.707000 -1.000000
   A=     -0.30000   D=     2   S=    0.96200   T=    0.0
   A=      0.0       D=     2   S=    0.69975   T=    0.00612
   A=      0.15000   D=     3   S=    0.50000   T=    0.0
   A=      0.45000   D=     3   S=   -0.00000   T=    0.0
   A=      1.05000   D=     4   S=   -0.86660   T=    0.00094
   A=      1.20000   D=     4   S=   -0.96737   T=    0.00206
```

(b)

Fig. 3.2   Divided differences.   (a) Flow chart.   (b) Program.

Exercises

1 The divided-difference interpolation program does not use the nested evaluation technique because the error-term computation is most easily carried out if each term is computed separately. Alter this program to delete the error-term computation and evaluate the divided-difference polynomial by the more efficient nested method.

2 If only the "upper diagonal" of a difference table is to be used, all differences can be computed without the use of auxiliary storage and stored in the linear array which originally held the ordinate values. For example,

$$\left. \begin{matrix} f(x_0) \\ f(x_1) \\ f(x_2) \\ f(x_3) \end{matrix} \right\} \quad \text{is replaced by} \quad \left\{ \begin{matrix} f[x_0] \\ f[x_0, x_1] \\ f[x_0, x_1, x_2] \\ f[x_0, x_1, x_2, x_3] \end{matrix} \right.$$

Write a program which reads n points and then computes the divided differences in this manner.

3 Lagrangian interpolation can also be computationally simplified when the abscissas are equally spaced. When the substitution $x = x_0 + hs$ is made in the Lagrangian coefficients,

$$c_i = \prod_{i \neq j} \frac{(x - x_j)}{(x_i - x_j)},$$

the interval size, h, cancels out of the numerator and denominator, and the coefficients are functions of the variable s. Obtain the equal-interval Lagrangian formula for the second-degree (i.e., three-point) case and flow-chart an external function which would carry out second-degree interpolation, given as arguments an interpolant x_a and the necessary arrays of ordinates and abscissas.

4 Starting with the assumption that the five "lower diagonal" elements of an equal-interval backward-difference table are stored in a linear array (similar to Problem 1), write a program that computes $f(x_k + h)$ and then replaces the current differences with the new values. Diagrammatically,

$$\left. \begin{matrix} f(x_k) \\ \nabla f(x_k) \\ \nabla^2 f(x_k) \\ \nabla^3 f(x_k) \\ \nabla^4 f(x_k) \end{matrix} \right\} \quad \begin{matrix} \text{Compute} \\ \to f(x_k + h) \to \end{matrix} \quad \left\{ \begin{matrix} f(x_k + h) \\ \nabla f(x_k + h) \\ \nabla^2 f(x_k + h) \\ \nabla^3 f(x_k + h) \\ \nabla^4 f(x_k + h) \end{matrix} \right.$$

5 The function $s(s - 1)(s - 2) \cdots (s - n + 1) = s^{(n)}$, called a *factorial*, appears in the equal-interval forward-difference interpolating polynomial. Using the notation defined on the right above and the forward-difference operator Δ, write this polynomial in concise operational form.

6 a) Using central differences, an interpolating polynomial was developed called the *Gauss forward formula*. Develop the expression for the *Gauss backward formula* for interpolation from the central point x_0 for the variable $x = x_0 + sh$.

b) Combine this expression with the Gauss forward formula to produce Stirling's formula.

7   a) Construct a difference table for the following data:

x	y
1.0	1.0000
1.1	1.5191
1.2	2.0736
1.3	2.6611
1.4	3.2816

b)   Determine the coefficients for an interpolating polynomial of degree three about $x = 1.1$, using Newton's fundamental formula.

8   Using the data of Problem 7 obtain a value for y at $x = 1.14$ with a polynomial of degree three using:

a)   Newton's fundamental formula;

b)   Stirling's formula.

9   a) Use the Lagrangian form to obtain an interpolated value of y at $x = 1.14$ with a polynomial of degree three using the first four points in the table of Problem 7.

b)   Evaluate the error term.

chapter four

the solution of equations

The Taylor series and divided-difference approximations can be used as tools in a familiar context, the solution of equations. Finding the zeros of a function, which is synonymous with solving an equation, is an inverse application of these forms since the value of $f(x)$ is known and what is desired is the value of x that produces the known value $f(x) = 0$. Since the value of x so obtained is only approximate, the methods are iterative, that is, successively applied until the truncation error is acceptably small. (Or, perhaps, until it becomes apparent that the error will not become small!)

Newton's method

The first form of Taylor's series, using only two terms, is the basis for Newton's method (or the Newton-Raphson method) of solving equations:

$$f(x) = f(z) + f'(z)(x - z) + \frac{f''(\xi)}{2}(x - z)^2.$$

Assuming that $f(x) = 0$, one obtains

$$x = z - \frac{f(z)}{f'(z)} + \frac{f''(\xi)}{2f'(z)}(x - z)^2.$$

By dropping the error term $R_2 = [f''(\xi)/2f'(z)](x - z)^2$, we determine an approximate root x_{k+1}, from an estimate z, which will now be named x_k:

$$x_{k+1} = x_k - \frac{f(x_k)}{f'(x_k)}.$$

In order for the results obtained by repeated evaluation of this formula (starting with the initial trial value x_0) to converge to the root x, the sequence of remainders

$$R_2^{(0)}, R_2^{(1)}, \ldots, R_2^{(k)} = \frac{f''(\xi_k)}{2f'(x_k)}(x - x_k)^2$$

must become small. Note that if $(x - x_k)$ does become small, the decrease of these remainders accelerates because this small quantity is squared. Such convergence is called a *second-order* process, or *quadratic* convergence.

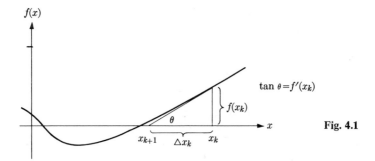

Fig. 4.1

The geometrical interpretation of this process is often used to introduce it. The function $f(x)$ and its first derivative are evaluated at $x = x_k$. The derivative can be regarded as the tangent of the angle which the slope at the point x_k makes with the x-axis. The base of the triangle, Δx_k, can then be determined and subtracted from x_k to produce the new trial value x_{k+1} (Fig. 4.1):

$$f'(x_k) = \frac{f(x_k)}{\Delta x_k}$$

$$\Delta x_k = \frac{f(x_k)}{f'(x_k)}$$

$$x_{k+1} = x_k - \Delta x_k = x_k - \frac{f(x_k)}{f'(x_k)}$$

As intimated earlier, Newton's method does not necessarily lead to convergence. Moreover, unless the initial trial root was close to the root sought, the process may actually yield some other root. Figure 4.2 illustrates these two cases, whose possible occurrence requires that computer programs include some upper bound on the number of iterations to be carried out. Another refinement which is often useful is to limit the minimum magnitude of $f'(x_k)$. If $f'(x_k)$ is zero, division by zero is attempted, and if it is nonzero but very small, Δx_k becomes large and may very well introduce values of x_k that are outside the range of interest. As with all digital algorithms whose operands are real numbers, or more accurately, machine approximations of real numbers, it is rarely possible to compare values for equality since the propagation of round-off error causes variations in what should otherwise be algebraically identical values. One cannot ask, Is $f(x) = 0$?, but only, Is $|f(x)| \leq \epsilon$? where ϵ is small and positive. Occasionally it is difficult to make a good choice of ϵ, and sometimes experimentation is required. The kind of rough reasoning useful for initial choices of ϵ is: A floating-point number is approximately eight decimal digits; hence, if the terms of the expression are fractions in the neighborhood of unity, it would be reasonable to expect that all but the rightmost two digits of two equal numbers would correspond. A first-choice ϵ might then be 1×10^{-6}. The concept of relative (rather than absolute) error is

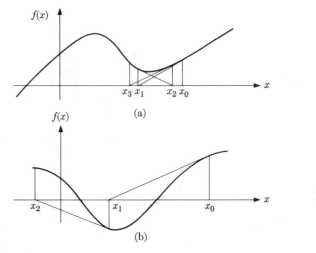

Fig. 4.2

useful in this connection since it avoids the necessity of estimating the magnitudes of the expressions involved. If a and b are to be compared for equality, the question

$$\left| \frac{a - b}{a + b} \right| < \epsilon?$$

asks whether the error $a - b$ is a negligibly small fraction of the sum. Phrased in this manner, the error relative to the magnitude of the operands and $a - b$ could, for instance, be of the order of 10^2 and still be acceptably small, provided that a and b were of the order 10^8.

Before illustrating a complete computer problem utilizing Newton's method, we wish to present an example in which the technique is used to produce an iterative formula for solving the equation $f(x) = x^2 - a = 0$. If this equation can be readily solved by iteration, an effective way of computing square roots $(x = \pm\sqrt{a})$ can be obtained. Applying the iteration formula

$$x_{k+1} = x_k - \frac{f(x_k)}{f'(x_k)},$$

one obtains

$$x_{k+1} = x_k - \frac{x_k^2 - a}{2x_k} = \frac{1}{2}\left(x_k + \frac{a}{x_k}\right).$$

For a simple initial trial, choose $x_0 = a$ so that $x_1 = \frac{1}{2}(a + 1)$. Obviously, this initial choice cannot be made if $a = 0$, since $a/x_0 = 0/0$ cannot be computed; however, for $a = 0$, the root is known to be $x = 0$, and no iteration is required.

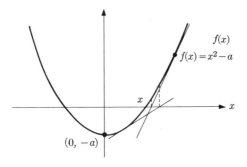

$f(x)$

$f(x) = x^2 - a$

x

x

$(0, -a)$

Fig. 4.3

When the zero case is excluded, the iteration will always converge. Since $f''(x) = 2$, the error term becomes

$$R_2^{(k)} = \frac{(x - x_k)^2}{2x_k}.$$

From the graph of the function $f(x) = x^2 - a$ (Fig. 4.3), it is apparent that after one iterative step the differences

$$(x - x_1), \qquad (x - x_2), \qquad (x - x_3), \qquad \cdots$$

will become steadily smaller. For $a \geq 1$, the initial value is to the right of the positive root (that is, $x_0 = a$), and the tangential projection to the x-axis will always produce a value of x_k that is closer to the root but is still somewhat large. For $a < 1$, the first approximation to x will be to the left of x, but the first iteration will produce a value on the right, and the case described above will apply.

Example Write a computer program to solve the transcendental equation $\cos x - ax^3 = 0$, using Newton's method. Consider only cases where $a \geq 0.04$.

The function is the cosine superimposed on a negative cubic. The restriction on a guarantees that there is only one real root. The method could be applied without this restriction, but a more detailed analysis and a greater number of trials would be required if all roots were to be obtained. Since $\cos x$ has a zero at $\pi/2$, a value between 0 and $\pi/2$, $x_0 = 1.5$, is chosen as an initial approximation to x. The flow chart and the program for the problem are presented in Fig. 4.4. Note that an array of approximations $x_0, x_1, \ldots, x_k$ is not developed in spite of the fact that the iteration formula suggests such a procedure. Only the previous value is needed to obtain the next, and hence one variable, x, is all that is required. Results have been computed for five values of a. The output, shown with the program, gives the roots along with each value of a.

A DO loop has been employed to allow a total of NT values of a to be read. To terminate the loop this value of NT must also be supplied to the program.

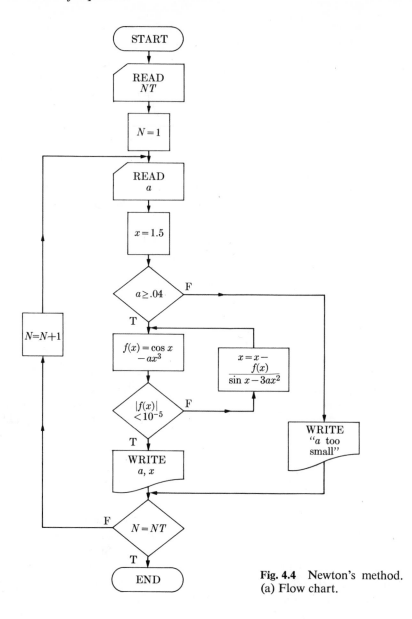

Fig. 4.4 Newton's method.
(a) Flow chart.

Difference methods

One would expect that the first two terms of the divided-difference polynomial could be similarly used to find roots of equations and, indeed, this is the case:

$$f(x) = f[x_0] + (x - x_0)f[x_1, x_0] + (x - x_0)(x - x_1)\frac{f''(\xi)}{2}.$$

```
C          NEWTON'S METHOD
C          FIND THE ROOTS OF COS(X) - A*X**3   (A GREATER OR EQUAL TO .04)
           READ(5,11) NT
           DO 100 III=1,NT
C          NT = NUMBER OF DIFFERENT A'S  TO BE TRIED
           READ(5,12) A
           X=1.5
           F=0.
           IF(A-.04) 1,4,4
1          WRITE(6,2) A
2          FORMAT(//F13.5,16X,24HCOEFFICIENT IS TOO SMALL//)
           GO TO 100
3          X=X+F/( SIN(X)+3.0*A*(X**2))
4          F= COS(X)-A*(X**3)
           IF( ABS(F)-.00001) 5,5,3
5          WRITE(6,13) A,X
6          FORMAT(F13.5,F31.5)
7          FORMAT(19HCOEFFICIENT OF X**3,10X,23HROOT OF COS(X) - A*X**3)
8          FORMAT(19(1H-),10X,23(1H-)//)
11         FORMAT(I5)
12         FORMAT(F10.6)
13         FORMAT(4H A= ,F10.6,5H  X= ,F10.6)
100        CONTINUE
           STOP
           END

              INPUT

     5,
     .04,
     .02,
     .5,
     1.0,
     2.0,

              OUTPUT
     A=   0.040000  X=   1.448841
          0.02000            COEFFICIENT IS TOO SMALL
     A=   0.500000  X=   1.016994
     A=   1.000000  X=   0.865475
     A=   2.000000  X=   0.721406
```

Fig. 4.4   (b) Program.

By writing the divided differences in their expanded form and assuming that $f(x) = 0$, one can derive an expression for the root x:

$$0 = f(x_0) + (x - x_0)\frac{f(x_1) - f(x_0)}{x_1 - x_0} + (x - x_0)(x - x_1)\frac{f''(\xi)}{2},$$

$$x = \frac{-(x_1 - x_0)f(x_0) + x_0(f(x_1) - f(x_0))}{f(x_1) - f(x_0)}$$

$$+ \frac{(x - x_0)(x - x_1)}{f(x_1) - f(x_0)}\frac{f''(\xi)}{2}(x_0 - x_1).$$

Dropping the error term and simplifying, one obtains

$$x \approx \frac{x_0 f(x_1) - x_1 f(x_0)}{f(x_1) - f(x_0)}.$$

Writing this result as an iterative formula yields

$$x_{k+1} = \frac{x_{k-1}f(x_k) - x_k f(x_{k-1})}{f(x_k) - f(x_{k-1})}.$$

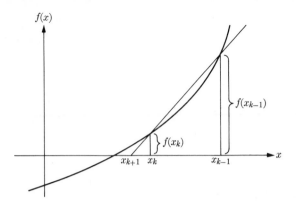

Fig. 4.5

The situation is not quite so clear as it was before, since the new value obtained depends on two previous values, and some rule is needed to determine which of the previous abscissas is to be replaced before the next iteration. Before deriving a rule from geometric considerations, one can extend this observation to the error term. The factors $(x - x_0)(x - x_1)$, which, perhaps, should now be written

$$(x - x_k)(x - x_{k-1}),$$

will not both become small as the iteration converges. Therefore, since only one of these two factors goes to zero, the process is called *first-order* or *linearly* convergent. The geometrical procedure is illustrated in Fig. 4.5. The new approximation to the root x_{k+1} is the abscissa of the intersection of the x-axis and the secant passing through the two given points. One possible replacement rule would be to retain the two abscissas corresponding to the smaller magnitudes of $f(x_{k+1})$, $f(x_k)$, $f(x_{k-1})$.

As outlined, this procedure, called Lin's method or the *secant method*, is fraught with all the convergence difficulties of Newton's method and is only a first-order process. A specialization which makes the method much more useful is to require that $f(x_k)$ and $f(x_{k-1})$ be selected so that they are of opposite sign. In this form the process is called the *method of false position* (*Regula Falsi*) and is guaranteed to converge. Note, however, that the convergence is still linear. In Fig. 4.6, x_{k+1} would replace x_k to preserve the sign change before the procedure was repeated.

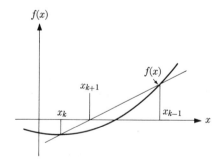

Fig. 4.6

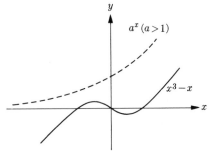

Fig. 4.7

The confusion arising from the change in subscript names is reduced if only two abscissas are considered, x_l for the left endpoint, and x_r for the right endpoint. The iteration formula can then be expressed as follows:

$$x \approx \frac{x_r f(x_l) - x_l f(x_r)}{f(x_l) - f(x_r)}$$

if $f(x)f(x_l) > 0$, $x_l \leftarrow x$; otherwise $x_r \leftarrow x$.

Examining the sign of the product of two numbers is a convenient means of determining whether or not the two are of like sign.

Example Write a computer program to find one root of the equation $f(x) = a^x + x^3 - x = 0$, using the method of false position.

To determine an interval in which the function undergoes a sign change, one sketches two separate graphs for a^x and $x^3 - x$ (Fig. 4.7). Considering only $a \geq 1$, one finds that $f(0) = 1$ and $f(-2) < -5$; so there is a root in the interval $(-2, 0)$. Figure 4.8 shows the flow chart and the program for this example. The trial values of x_r and x_l are saved as AA and BB for later printing. Format statements 13 and 14 have been introduced to print the results in columns under appropriate headings. Results are given for four values of a: 1, 2, 4, and 8.

The method of successive approximations

Another iterative technique for the solution of equations which is particularly useful for hand calculations is the *method of successive approximations*. Iteration is performed directly on an equation, formed by dividing $f(x)$ into two parts:

$$f(x) = g(x) - F(x) = 0.$$

It is usually possible, and always simpler, to choose $g(x) = x$. This leads to the iteration formula,

$$x = F(x).$$

To determine a root, an initial approximation of the root, say x_0, is made, and the right-hand side $F(x_0)$ is evaluated to give a new value x_1. Then $F(x_1)$ is

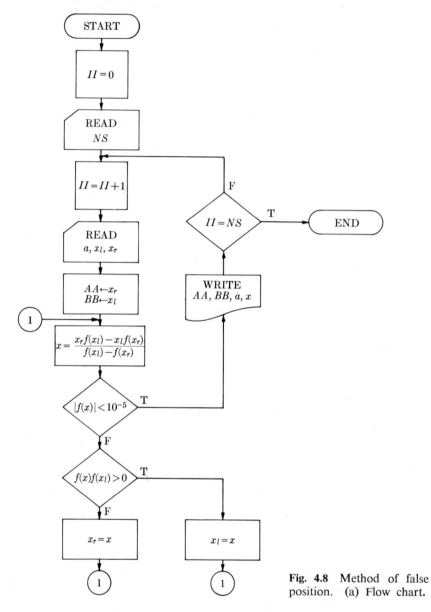

Fig. 4.8 Method of false position. (a) Flow chart.

computed to provide a new approximation to the root, x_2. For the general calculation, the iterative equation is

$$x_{k+1} = F(x_k).$$

Two things must be established:

1. When does the method converge?

```
C       METHOD OF FALSE POSITION
C       TO SOLVE FOR ONE ROOT OF THE EQUATION F(X) = A**X + X**3 - X = 0
C       A IS GREATER THAN OR EQUAL TO 1.0
        F(VAR)=A**VAR+VAR**3-VAR
        WRITE(6,14)
        READ(5,11) NS
C       NS = NUMBER OF SETS OF DATA
        DO 100 II=1,NS
        READ(5,12) XL,XR,A
        AA=XR
        BB=XL
        U=F(XR)
        V=F(XL)
10      X=(XL*U-XR*V)/(U-V)
        W=F(X)
        IF (ABS(W)-.00001) 4,1,1
1       IF (W*V) 3,4,2
2       XL=X
        V=F(XL)
        GO TO 10
3       XR=X
        U=F(XR)
        GO TO 10
4       WRITE (6,13) AA,BB,A,X
11      FORMAT (I3)
12      FORMAT (3F10.2)
13      FORMAT (4HFROM,F5.1,3H TO,F5.1,F13.2,14H**X + X**3 - X,F17.5)
14      FORMAT (16HINTERVAL OF ROOT,5X,23HEQUATION TO BE ANALYZED,5X,4HROOT
       1,//)
100     CONTINUE
        STOP
        END

        INPUT

4,
-2.,0.,1.,
-2.,0.,2.,
-2.,0.,4.,
-2.,0.,8.,

        OUTPUT
```

INTERVAL OF ROOT	EQUATION TO BE ANALYZED	ROOT
FROM 0.0 TO -2.0	1.00**X + X**3 - X	-1.32472
FROM 0.0 TO -2.0	2.00**X + X**3 - X	-1.17373
FROM 0.0 TO -2.0	4.00**X + X**3 - X	-1.09542
FROM 0.0 TO -2.0	8.00**X + X**3 - X	-1.05197

Fig. 4.8   (b) Program.

2. Is the value to which it converges a root of $f(x)$?

If the method does converge, a value x_k, when substituted in $F(x)$, will reproduce itself, or, symbolically,

$$\left| \frac{x_k - x_{k-1}}{x_k} \right| \to 0.$$

When this occurs, then

$$x_k - F(x_{k-1}) = f(x_k) = 0,$$

and x_k is a root. Let us now examine the question of convergence.

If $F(x)$ is continuous and can be differentiated, the differential form of the mean-value theorem can be written for an interval x_{i-1} to x_i as

$$F(x_i) - F(x_{i-1}) = F'(\zeta)(x_i - x_{i-1}),$$

where ζ is some value of x between x_{i-1} and x_i. But, from the iteration equation,

$$F(x_i) = x_{i+1} \quad \text{and} \quad F(x_{i-1}) = x_i.$$

Taking absolute values,

$$|x_{i+1} - x_i| = |F'(\zeta_i)| \, |x_i - x_{i-1}|$$

which, for the preceding steps, is

$$|x_i - x_{i-1}| = |F'(\zeta_{i-1})| \, |x_{i-1} - x_{i-2}|$$

$$\cdots \cdots \cdots \cdots \cdots \cdots \cdots \cdots \cdots \cdots \cdots$$

$$\cdots \cdots \cdots \cdots \cdots \cdots \cdots \cdots \cdots \cdots \cdots$$

$$|x_2 - x_1| = |F'(\zeta_1)| \, |x_1 - x_0|.$$

Assuming that there is an upper bound on the several $|F'(\zeta)|$ which can be expressed as

$$|F'(\zeta)| \le M,$$

then

$$|x_{i+1} - x_i| \le M|x_i - x_{i-1}|$$
$$|x_i - x_{i-1}| \le M|x_{i-1} - x_{i-2}|$$

$$\cdots \cdots \cdots \cdots \cdots \cdots \cdots \cdots \cdots \cdots$$

$$\cdots \cdots \cdots \cdots \cdots \cdots \cdots \cdots \cdots \cdots$$

$$|x_2 - x_1| \le M|x_1 - x_0|.$$

Working forward from the initial approximation we first write

$$|x_2 - x_1| \le M|x_1 - x_0|.$$

Then as

$$|x_3 - x_2| \le M|x_2 - x_1|,$$
$$|x_3 - x_2| \le M^2|x_1 - x_0|.$$

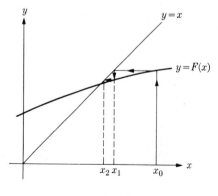

Fig. 4.9

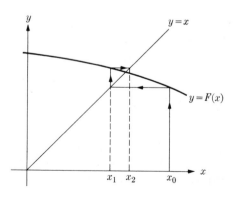

Fig. 4.10

Continuing until a large number of approximations has been made, we can write then for the kth iteration

$$|x_{k+1} - x_k| \leq M^k |x_1 - x_0|.$$

As noted earlier, the condition for convergence is

$$|x_{k+1} - x_k| \rightarrow 0,$$

which requires that $M < 1$. As

$$|F'(x)| \leq M,$$

a sufficient condition for convergence is

$$|F'(x)| < 1.$$

Convergence is more rapid when the slope is small than for cases where the slope is near unity.

This result can be demonstrated graphically. In Figs. 4.9 and 4.10 the conditions are appropriate for convergence. The process starts from the initial value x_0, with the arrows indicating path of the iteration. In Fig. 4.10, the slope $F'(x)$ is negative, but still less than unity in the absolute sense. Examples for which the procedure does not converge are shown in Fig. 4.11, where $F(x)$ has a positive slope, and Fig. 4.12, where the slope is negative.

To illustrate the method of successive approximations, consider the problem of determining the roots of the polynomial

$$f(x) = x^2 - 3x + 2 = 0.$$

It is evident that the roots r_1 and r_2 are 1 and 2 respectively. Selecting

$$F(x) = \tfrac{1}{3}(x^2 + 2),$$

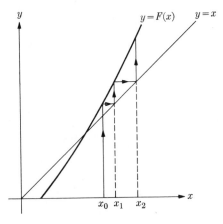

Fig. 4.11

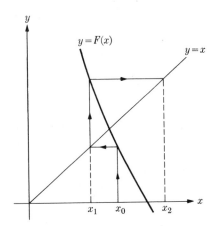

Fig. 4.12

we obtain the iteration formula

$$x = \tfrac{1}{3}(x^2 + 2),$$

and therefore

$$F'(x) = \tfrac{2}{3}x.$$

It is apparent that

$$F'(x) < 1 \quad \text{if} \quad x < 1.5.$$

We can expect the result to converge for r_1 but not for r_2. Starting the process with $x_0 = 0$, the iteration proceeds to $x_1 = 0.6667$, $x_2 = 0.8148$, and, following the tenth step,

$$x_{10} = 0.9916.$$

We shall consider two approaches to computing r_2 with this method. One is to alter the choice of $F(x)$. If the iteration proceeds with the equation

$$x_{k+1} = (3x_k - 2)^{1/2},$$

then

$$F'(x) = \frac{3}{2(3x_k - 2)^{1/2}},$$

and the iteration will converge to the second root for any x_0 greater than 1.41. Although convergence is slow, the approximation after five steps goes to $x_5 = 2.1214$ when the process starts from an initial assumption of $x_0 = 3$.

While it is not always possible to obtain convergence in the manner just described, it is possible to modify the equation to be iterated, by means of a constant which can be selected to make $|F'(x)| < 1$.

Consider again the equation

$$x = F(x).$$

Multiplying both sides by $(1 - p)$, where p is a constant to be selected later, and then subtracting the result from the original expression, we obtain

$$x - (1 - p)x = F(x) - (1 - p)\,F(x),$$

which leads to a new expression

$$x = (1 - p)x + pF(x).$$

If r is a root of the first equation, it is also a root of the second. There is in this new equation a degree of flexibility in that p can be chosen to cause the derivative of the right-hand side of the equation to be less than 1 in the vicinity of the root.

Returning to the example problem, let

$$F_1(x) = (1 - p)x + p(\tfrac{1}{3}(x^2 + 2));$$

then

$$F_1'(x) = (1 - p) + p(\tfrac{2}{3}x).$$

If $p = -\frac{1}{2}$, convergence to r_2 is assured for any initial value of $x > 1.5$, and the new recurrence relation becomes

$$x_{k+1} = \tfrac{3}{2}x_k - \tfrac{1}{6}(x_k^2 + 2).$$

Beginning with $x_0 = 3$, the calculation converges slowly, reaching, after five iterations, an approximation to the root of 2.2206.

Half-interval method

Instead of determining the next approximation to the root by linearly interpolating between the endpoints of the interval in which the sign change occurs, one may proceed by a method called the *bisection* or *Bolzano* algorithm; i.e., the interval is simply divided in half, then the half-interval containing the sign change is *halved*, etc. The iteration formula is more logical than computational but, given that $f(x_l)$ and $f(x_r)$ are of different sign, one step may be described as follows: If

$$f\left(\frac{x_l + x_r}{2}\right) f(x_l) > 0,$$

as is the case in the example illustrated (Fig. 4.13), then

$$x_l \leftarrow \frac{x_l + x_r}{2};$$

otherwise,

$$x_r \leftarrow \frac{x_l + x_r}{2}.$$

Another useful way of viewing these steps is to consider the interval delimited by the abscissas x and $(x + h)$; then

$$\text{if} \quad f\left(x + \frac{h}{2}\right) f(x) > 0, \quad x \leftarrow x + \frac{h}{2}, \quad \text{then} \quad h \leftarrow \frac{h}{2}.$$

The interval-halving step is always performed whether or not the conditional step preceding it is executed. If the initial interval contains one or more roots, this method will always converge.

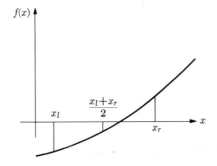

Fig. 4.13

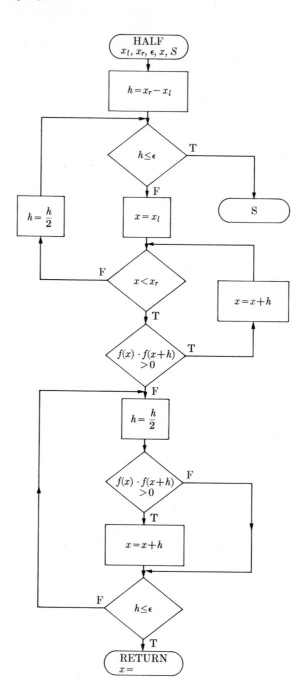

Fig. 4.14   Half-interval search subroutine. (a) Flow chart.

To illustrate the use of this root-finding method, no specific function will be considered; rather the task outlined will be one of writing a function using this process.

Example Write a subroutine for which the following specifications are given:

1. the endpoints of an interval,
2. the name of a function to be examined for roots in the interval,
3. the permissible error in the target root, and
4. the label of a statement which is to be executed if no root is found.

```
C       MASTER HIS
C       HALF-INTERVAL SEARCH MAIN PROGRAM
C       SUB-ROUTINE AND FUNCTION SUB-PROGRAM
C       MAIN PROGRAM
2       READ (5,50) XL,XR,EPS
        CALL HALF (XL,XR,EPS,ROOT,INDS)
        GO TO (3,4),INDS
3       WRITE (6,60) XL,XR
        GO TO 2
4       WRITE (6,70) XL,XR,ROOT
        GO TO 2
50      FORMAT (3F9.0)
60      FORMAT (5X,29HNO ROOT FOUND IN INTERVAL X= ,F10.5,6HTO X= ,F10.5)
70      FORMAT (5X,15HIN INTERVAL X= ,F10.5,6HTO X= ,
        1F10.5,5X,6HROOT= ,F10.5//)
        STOP
        END
        FUNCTION EQUA(X)
        EQUA=EXP (-X)-SIN (1.57296*X)
        RETURN
        END
        SUBROUTINE HALF(XL,XR,EPS,X,INDS)
        H=XR-XL
5       IF (H-EPS) 6,6,7
7       X=XL
11      IF (X.LT.XR) GO TO 8
        H=H/2.
        GO TO 5
8       IF (EQUA (X) *EQUA (X+H) )9,9,10
10      X=X+H
        GO TO 11
9       H=H/2.
        IF (EQUA (X) *EQUA (X+H) ) 13,13,14
14      X=X+H
13      IF (H.GT.EPS) GO TO 9
        INDS=2
        GO TO 18
6       INDS=1
        X=1.
18      RETURN
        END

              INPUT

0.,.25,.0001,
.25,.50,.0001,
.5,1.0,.0001,
1.0,2.0,.0001,
5.0,6.0,.0001,

              OUTPUT

   NO ROOT FOUND IN INTERVAL X=    0.0     TO X=    0.25000
   IN INTERVAL X=    0.25000TO X=    0.50000   ROOT=    0.44312
   NO ROOT FOUND IN INTERVAL X=    0.50000TO X=    1.00000
   IN INTERVAL X=    1.00000TO X=    2.00000   ROOT=    1.90198
   IN INTERVAL X=    5.00000TO X=    6.00000   ROOT=    5.99011
```

Fig. 4.14   (b) Program.

If a root is determined, it is to be directly returned to the calling program. The subroutine is to be named HALF.

These specifications do not guarantee that there is a sign change in the interval given. If the original interval did not indicate a sign change, the function must search for a subinterval containing a sign change. This search may be carried out by bisection; that is, the two intervals formed are examined by evaluating the function at the midpoint of the given interval. If no sign change is observed, halve each of these intervals, test for sign change in the four intervals, etc. When the intervals being examined are less than the root error tolerance, there is no point in proceeding further. This technique has obvious difficulties if there are roots of even multiplicity; and, moreover, it could be extremely time consuming if, in fact, no roots were present in the interval specified. Nonetheless, the inclusion of the interval-search procedure does not seriously reduce the effectiveness of the function in the usual case where the interval designated was known to contain a root. This function illustrates the utility of iteration statements which are not of the repeated product or summation type. The corresponding flow chart and program are presented in Fig. 4.14.

In this example the search terminates when the size of the interval, H, is less than some supplied test value, EPS. As an alternative, it is possible to predict the number of cycles required to reduce the interval to a satisfactory value. A counter in the program can be tested and the search stopped after the proper number of cycles. If h is the size of the first interval, the first half-interval will be $h/2$; the second half-interval will be $h/2^2$; and the nth will be $h/2^n$. The search will be stopped when the interval is less than a specified size, ϵ. This can be expressed as

$$\epsilon \geq \frac{h}{2^n},$$

or in terms of n,

$$n \geq \frac{\log \frac{h}{\epsilon}}{\log 2}.$$

Using this test, the program would be terminated after the number of cycles is one larger than the integer part of $(\log h/\epsilon)/\log 2$.

Included with the subroutine subprogram in Fig. 4.14(b) are the main-line program for calling the subroutine, and the function subprogram for the function to be examined. In this instance, the function called EQUA(X) is

$$e^{-x} - \sin \frac{\pi}{2} x.$$

To determine the roots of another function, appropriate alteration of the function subprogram is all that is necessary.

Roots have been computed for the expression in several intervals. Intervals, and roots, if they exist, are shown in the printed results in Fig. 4.14(b).

In its present form the program will determine only one root in the interval. In the case of multiple roots, that will be the root farthest to the left in the interval. It is not difficult to amend the program to evaluate double roots. An interval containing a double root is first identified. Then after the left-hand root has been found by the present routine, the other half of the interval is searched for the second root. This is left as an exercise.

Multiple roots

A word should be said about obtaining approximate values for multiple roots which are in close proximity to one another, such as the pair of roots r_1 and r_2 in Fig. 4.15. It is necessary to know beforehand that the condition exists. The problem is to obtain distinct values for each root.

One approach is to determine the point on the curve where the slope is zero, and then to approximate the curve with a second-degree polynomial passing through the point. The two roots are then approximated by the zeros of the second-degree polynomial.

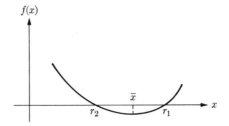

Fig. 4.15

The point of zero slope is found by determining, from one of the methods described, the root of the equation

$$f'(x) = 0$$

in the region of the multiple roots.

Identifying this root as $\bar{x}$, we then expand the function $f(x)$ in a Taylor series about $\bar{x}$:

$$f(x) = f(\bar{x}) + (x - \bar{x})f'(\bar{x}) + \frac{(x - \bar{x})^2}{2!} f''(\bar{x}).$$

Observing that $f'(\bar{x}) = 0$, and denoting the point where the parabola crosses the axis as $\bar{x} + d$, we can evaluate the series at that point to obtain

$$f(\bar{x} + d) = f(\bar{x}) + \frac{d^2}{2} f''(\bar{x}).$$

But $f(\bar{x} + d) = 0$, and therefore

$$d = \pm \sqrt{\frac{-2f(\bar{x})}{f''(\bar{x})}}.$$

Then the approximation to the roots can be expressed as

$$r_1 \approx \bar{x} - d \qquad \text{and} \qquad r_2 \approx \bar{x} + d.$$

Solving polynomial equations

Polynomial equations are of such frequent occurrence that their solution deserves special attention. These equations are of the form $\sum_{i=0}^{n} a_i x^i = 0$ and the n roots, assuming for the moment that they are real, can be obtained by the methods already described. In particular, Newton's method may be used since the derivatives of polynomials are readily obtained. However, the problems of convergence and root separation will be present. For polynomials, one can handle the latter problem by factoring out a root once it is known (thus reducing the degree of the polynomial). It turns out that in applying Newton's method to polynomial equations these tasks of evaluating the polynomial, evaluating the first derivative, and determining the coefficients of the depressed equation (i.e., with determined roots factored out) can be performed quite simply by means of synthetic division.

Synthetic division is merely the division of a polynomial by a linear factor where, for convenience, the variable names are not written and the sign of the divisor is reversed so that the "division" may be carried out by addition instead of subtraction. An example of ordinary division by a linear factor is given below.

$$
\begin{array}{r}
a_n x^{n-1} + (a_n x_0 + a_{n-1}) x^{n-2} + \cdots \\
x - x_0 \overline{\big)\; a_n x^n + a_{n-1} x^{n-1} + \cdots + a_0} \\
\underline{a_n x^n - a_n x_0 x^{n-1}} \\
(a_n x_0 + a_{n-1}) x^{n-1} + a_{n-2} x^{n-2} \\
\underline{(a_n x_0 + a_{n-1}) x^{n-1} - (a_n x_0 + a_{n-1}) x_0 x^{n-2}} \\
((a_n x_0 + a_{n-1}) x_0 + a_{n-2}) x^{n-2} \\
\vdots
\end{array}
$$

Continuing in this manner would produce a last line (a remainder) of

$$(\cdots ((a_n x_0 + a_{n-1}) x_0 + a_{n-2}) x_0 + \cdots + a_0) = p(x_0)$$

where $p(x) = \sum_{i=0}^{n} a_i x^i$. Deleting the x's and reversing the sign of x_0 turns the problem into synthetic division:

$$
\begin{array}{r}
a_n \quad (a_n x_0 + a_{n-1}) \\
x_0 \overline{\big)\; a_n \quad\; a_{n-1} \quad\; a_{n-2} \cdots a_0} \\
+\, a_n x_0 \\
\overline{a_n x_0 + a_{n-1}} \\
+\, (a_n x_0 + a_{n-1}) x_0 \\
\overline{(a_n x_0 + a_{n-1}) x_0 + a_{n-2}}
\end{array}
$$

These expressions are formed according to the following rule: Obtain the expression under the kth coefficient by multiplying the $(k + 1)$-expression by x_0 and adding a_k. Using $b_n, b_{n-1}, \ldots, b_0$ to indicate the expressions, we have

x_0	a_n	a_{n-1}	a_{n-2}	$\ldots$	a_0
		$b_n x_0$	$b_{n-1} x_0$	$\ldots$	$b_1 x_0$
	b_n	b_{n-1}	b_{n-2}	$\ldots$	b_0

and

$$b_n = a_n, \qquad b_j = b_{j+1} x_0 + a_j \quad \text{for} \quad j < n.$$

The remainder is $b_0 = p(x_0)$.

Once it is clear that synthetic division is simply a convenient way of dividing by a linear factor $(x - x_0)$, the fact that the remainder equals $p(x_0)$ can be demonstrated in a simpler way:

$$p(x) = a_n x^n + a_{n-1} x^{n-1} + \cdots + a_0.$$

The division produces a quotient,

$$q(x) = b_n x^{n-1} + b_{n-1} x^{n-2} + \cdots + b_1$$

and a remainder,

$$\frac{R}{x - x_0};$$

that is,

$$\frac{p(x)}{x - x_0} = q(x) + \frac{R}{x - x_0}.$$

This result can be written

$$p(x) = q(x)(x - x_0) + R,$$

and evaluation at $x = x_0$ produces $R = p(x_0)$.

Repeating the process by dividing $q(x)$ by $(x - x_0)$ reveals that the value of the first derivative at $x = x_0$ can be obtained by evaluating $q(x)$:

$$\frac{q(x)}{x - x_0} = q_2(x) + \frac{R_2}{x - x_0},$$

$$q(x) = (x - x_0)q_2(x) + R_2,$$

$$\frac{p(x)}{x - x_0} = (x - x_0)q_2(x) + R_2 + \frac{R}{x - x_0},$$

$$p(x) = (x - x_0)^2 q_2(x) + (x - x_0)R_2 + R,$$

$$p'(x) = (x - x_0)^2 q_2'(x) + 2(x - x_0)q_2(x) + R_2.$$

For $x = x_0$, $p'(x_0) = R_2$. Therefore division by $(x - x_0)$ evaluates the polynomial $p(x)$ at $x = x_0$, and division of $q(x)$, the quotient thus produced, gives the value of the first derivative $p'(x)$ at x_0. These two values are the necessary constituents of Newton's iteration formula, and a value x_1 to be used in the next division cycle is obtained from

$$x_1 = x_0 - \frac{p(x_0)}{p'(x_0)}.$$

Clearly, when x_0 is a root, the remainder R is zero $(R = f(x_0) = 0)$, and the quotient is the polynomial with $(x - x_0)$ factored out.

Before proceeding to a program, we wish to illustrate the steps by a numerical example. Suppose that

$$p(x) = x^4 - 10x^3 + 37x^2 - 60x + 36,$$

and that $x = 1$ is chosen as a first approximation to a root. The polynomial is first evaluated by synthetic division, and then the quotient is similarly evaluated:

1	1	-10	37	-60	36
		1	-9	28	-32
Coefficients of $q(x)$:	1	-9	28	-32	$4 = R = p(1)$
		1	-8	20	
	1	-8	20	$-12 = R_2 = p'(1)$	

$$x_1 = x_0 - \frac{p(x_0)}{p'(x_0)} = 1 - \left(\frac{4}{-12}\right) = 1.33.$$

The new approximation to the root is then used as the divisor in the synthetic division:

1.33	1	-10	37	-60	36
		1.33	-11.5	34	-34.6
	1	-8.67	25.5	-26	$1.4 = R = p(1.33)$
		1.33	-9.76	21.0	
	1	-7.34	15.74	$-5 = R_2 = p'(1.33)$	

$$x_2 = 1.33 - \left(\frac{1.4}{-5}\right) = 1.61.$$

One more cycle brings the approximate value still closer to one of the roots, $x = 2$:

1.61	1	-10	37	-60	36
		1.61	-13.5	37.8	-35.7
	1	-8.39	23.5	-22.2	$0.3 = R = p(1.61)$
		1.61	-10.9	20.3	
	1	-6.78	12.6	$-1.9 = R_2 = p'(1.61)$	

$$x_3 = 1.61 - \left(\frac{0.3}{-1.9}\right) = 1.77.$$

This process is iterative on several levels. The most often repeated, or "inner," cycle is the evaluation of the two polynomials; these evaluations are repeated until a root is found (or until a specified number of trials have been made); the root-improvement cycle must be repeated for each root of the n or less real roots of an nth-degree polynomial.

Example Write a program which reads values for n and for the coefficients of a polynomial of degree $n - 1$, and then finds the real roots of the polynomial until $n - 1$ roots are obtained, or until more than 100 trials of finding a root have been made.

In the absence of any other information, the initial trial value x_0 is chosen to be $-a_1/a_2$, the ratio of the constant and first-order coefficients of the polynomial. This choice is good if the polynomial has a root near zero since in that case the higher powers of x are small and the linear approximation $a_2x + a_1 = 0$ from which the ratio was determined would be reasonably close.

Although the synthetic-division procedure has already been illustrated, it is presented once more to show that the division of the two polynomials $p(x)$ and $q(x)$ can be performed in a single iterative cycle.

$$
\begin{array}{c|ccccccc}
x_0 & a_n & a_{n-1} & a_{n-2} & \cdots & a_j & \cdots & a_1 \\
\hline
& b_n & b_{n-1} & b_{n-2} & \cdots & b_j & \cdots & b_1 = R = p(x_0) \\
& 0 & c_n & c_{n-1} & \cdots & c_{j+1} & \cdots & c_2 = R_2 = p'(x_0)
\end{array}
$$

The displacement of the bottom row is intentional, for it allows the iteration to be described as follows (a single step is considered to be the computation of one column going from left to right):

$$
b_n = a_n,
$$
$$
c_{n+1} = 0,
$$
$$
\left.
\begin{aligned}
b_j &= b_{j+1}x_0 + a_j \\
c_{j+1} &= c_{j+2}x_0 + b_{j+1}
\end{aligned}
\right\}, \quad j = n, n-1, \ldots, 1.
$$

Actually the c's need not be preserved as individual array elements since only the value of c_2 is needed. Because a c-element depends only upon its immediately preceding value, the procedure can be carried out without an array for the c-coefficients:

$$
b_n = a_n,
$$
$$
c = 0,
$$
$$
\left.
\begin{aligned}
b_j &= b_{j+1}x_0 + a_j \\
c &= cx_0 + b_{j+1}
\end{aligned}
\right\}, \quad j = n, \ldots, 1.
$$

The flow chart and the program for this example are presented in Fig. 4.16. The block indicating the substitutions

$$
a_{k-1} = b_k, \qquad a_{k-2} = b_{k-1}, \qquad \ldots, \qquad a_1 = b_2
$$

is actually implemented by an iteration statement, and its purpose is to transfer the coefficients of the depressed equation to the a-array. As indicated, if 100

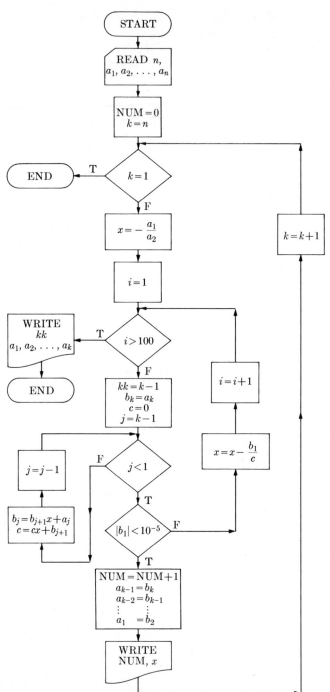

Fig. 4.16 Polynomial solution by synthetic division.   (a) Flow chart.

iterative trials are ever taken, the coefficients of the current polynomial and its degree $k - 1$ are printed before termination of the program.

As it is not possible to decrement an index in FORTRAN programs, dummy indices have been introduced. Thus, for example, when the dummy index KKK is increased by one, the index of importance K is decreased by one. So it is with the dummy index JJJ and the useful index J.

Two polynomials were supplied to the program. The first was fourth degree, and the second was a cubic, $n = 4$. All 4 roots were found for the first polynomial, while only one real root was computed for the second, leaving the residual equa-

```
C       MASTER SYN
C       POLYNOMIAL SOLUTION BY SYNTHETIC DIVISION
        DIMENSION A(10),B(10)
1       FORMAT(I9)
11      FORMAT(10F9.0)
2       FORMAT(5X,22HAFTER 100 ITERATIONS A,I4,25HTH ORDER EQUATION REMAIN
        1S)
3       FORMAT(5X,11HROOT NUMBER,I4,7H EQUALS,F12.4//)
4       FORMAT(//5X,20HITS COEFFICIENTS ARE,//10F8.2)
50      READ(5,1) N
        READ(5,11) (A(I),I=1,N)
C       N=NUMBER OF COEFFICIENTS TO BE READ OR EQUAL TO THE ORDER OF
C       THE POLYNOMIAL PLUS ONE,
C       NUM IS INTRODUCED TO INDICATE ROOT FOUND
        NUM=0
C       DUMMY INDEX KKK INTRODUCED TO ALLOW DECREMENT IN K
        DO 100 KKK=2,N
        X=-A(1)/A(2)
        K=N-KKK+2
        DO 70 I=1,100
        B(K)=A(K)
C       KK IS THE ORDER OF THE RESIDUAL EQUATION
        KK=K-1
        C=0.
C       DUMMY INDEX JJJ INTRODUCED TO ALLOW DECREMENT IN J
        DO 60 JJJ=1,KK
        J=K-JJJ
        B(J)=B(J+1)*X+A(J)
60      C=C*X+B(J+1)
        IF(ABS(B(1)).LT.1.E-5) GO TO 80
70      X=X-B(1)/C
        WRITE(6,2) KK
        WRITE(6,4) (A(L),L=1,K)
        GO TO 50
80      NUM=NUM+1
        DO 90 J=2,K
90      A(J-1)=B(J)
100     WRITE(6,3) NUM,X
        GO TO 50
200     STOP
        END
```

```
        INPUT

5,
36.,-60.,37.,-10.,1.0,
4,
-15.,8.,-4.,1.,
```

```
        OUTPUT

ROOT NUMBER   1 EQUALS      1.9920
ROOT NUMBER   2 EQUALS      2.0082
ROOT NUMBER   3 EQUALS      2.9877
ROOT NUMBER   4 EQUALS      3.0121
ROOT NUMBER   1 EQUALS      3.0000
AFTER 100 ITERATIONS A   2TH ORDER EQUATION REMAINS
ITS COEFFICIENTS ARE
5.00    -1.00    1.00
```

Fig. 4.16 (b) Program.

tion $x^2 - x + 5$. These coefficients were printed in the output. The two other roots, found with the quadratic formula, are $0.5 \pm 2.1794i$.

Complex roots of polynomials

The discussion of Newton's method has been limited to finding real roots. Actually the method is more general, in the sense that if complex arithmetic operations are used, complex roots may be determined by means of this iterative technique. However, a little experience with this method of complex-root determination for equations with real coefficients convinces us that a better technique must be possible. This conviction arises from the fact that the complex roots of a polynomial with real coefficients, if any, occur in conjugate pairs and hence, once one root is known, its conjugate is also known and need not be determined by iteration. The additional complexities of complex arithmetic can be eliminated if the quadratic factor, the product of the linear factors representing the conjugate roots, can be factored from the polynomial by iteration. The division technique can again be applied, but the trial divisor will be of the form $x^2 - px - q$:

$$p(x) = (x^2 - px - q)(b_n x^{n-2} + b_{n-1}x^{n-3} + \cdots + b_2) + (b_1 x + b_0).$$

The parenthesized expressions on the right are the divisor, quotient and remainder, respectively. By deleting the variable names and changing the signs of the divisor to substitute addition for subtraction, one may write these components in synthetic-division form:

p	q	a_n	a_{n-1}	a_{n-2}	a_{n-3}	$\cdots$	a_2	a_1	a_0
			pb_n	qb_n					
				pb_{n-1}	qb_{n-1}				
					pb_{n-2}	$\cdots$	qb_4		
							pb_3	qb_3	
								pb_2	qb_2
		b_n	b_{n-1}	b_{n-2}		$\cdots$	b_2	b_1	b_0

The rules for obtaining the quotient coefficients can be written from this diagram:

$$b_n = a_n,$$
$$b_{n-1} = a_{n-1} + pb_n,$$
$$b_j = a_j + pb_{j+1} + qb_{j+2}, \quad \text{for} \quad j = n-2, n-3, \ldots, 1,$$
$$b_0 = a_0 + qb_2.$$

The divisor is a factor when the coefficients of the remainder, b_1 and b_0, are zero. The basis of the iteration is to select values of p and q that make these coefficients zero:

$$b_0 = a_0 + qb_2,$$
$$b_1 = a_1 + pb_2 + qb_3.$$

These two equations, along with the equations for the remaining b_j, constitute a set of $(n + 1)$ simultaneous, nonlinear equations in $b_n, b_{n-1}, \ldots, b_3, b_2$ in addition to p and q. If p and q were known, the set of $(n - 1)$ equations remaining would be linear and solvable. Trial values of p and q are assumed and the equations solved. The next trial values, p' and q', are therefore

$$q' = -\frac{a_0}{b_2} \quad \text{and} \quad p' = -\left(\frac{a_1 + qb_3}{b_2}\right).$$

Once the process converges, the two roots of the factor can, of course, be computed from $x = (-p \pm \sqrt{p^2 - 4q})/2$. With the factor removed, the depressed equation can again be subjected to iteration or, if it is now of degree 2 or 1, simply solved.

By way of review and as an indication of procedures available for the isolation of quadratic factors, recall that Newton's method was obtained from Taylor's series and Lin's method from the finite-difference approximation. The former was a second-order process and required the evaluation of the derivative, while the latter was of first order and involved only values of the function. Similarly, for polynomials the two approaches are possible with the same qualifications. Specifically, the iterative isolation of quadratic factors, which is not completely described here although it is the most popular approach to polynomial-root finding, is called Lin's method if the finite-difference approach above is used, and Bairstowe's method when the scheme involves the evaluation of the derivative.

Exercises

1   Using the method of successive approximations, calculate the two-real roots of the equation

$$e^x - 4x = 0.$$

Carry out six steps.

2   Using the same trial value as in the previous exercise, calculate the root with Newton's method using six steps.

3   Using Newton's method, write a program to solve $a^x + x = 0$ for values of $a > 1$.

4   In some equations, $f(x) = 0$, the value of the function f may be very small in the neighborhood of a root (e.g., roots of high multiplicity (Fig. 4.17)). The relation $|f(x)| < \epsilon$ permits a large range of x-values. This problem may be eliminated by requiring that the difference between successive trial roots, as well as the value of the function, be small. Alter the statements in Problem 3 to include this additional criterion for convergence.

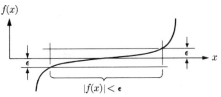

Fig. 4.17

5 Draw the flow chart of an algorithm which employs Newton's method to solve an equation $f(x) = 0$, but in addition, uses the information that there is a sign change in an interval (a, b). One approach is never to permit a trial value outside of the interval and to keep reducing the interval size as function evaluations are made.

6 Modify the half-interval search program so that in the case of a double root in the interval, it computes the right-hand root after determining the one on the left. Use the modified program to determine both roots of the equation

$$x^2 - x + 0.1875 = 0$$

in the interval $x_l = 0$; $x_r = 1$.

7 Modify the half-interval search program to seek successive roots over a large range of values of the variable. Use this program to determine all roots of the equation

$$e^{-x} - \sin \frac{\pi x}{2} = 0$$

in the interval $x = 0$ to $x = 10$.

8 Given the polynomial

$$x^4 - 3x^3 + 2x^2 - 10x + 12 = 0.$$

a) Find the nature of all the roots (i.e., the possible combinations of positive, negative, and imaginary roots) using *Descartes' rule of signs*.

b) Determine all real rational roots by synthetic division.

c) Determine all real irrational roots by the method of false position (*Regula Falsi*).

d) Determine all remaining roots, if any remain. Carry calculations to three decimal places.

9 Write a program which computes the roots of a polynomial by removing quadratic factors from the original polynomial. This program is similar in overall structure to the given program except that the synthetic division is more complex. Also, the roots of the quadratic factors, as well as the root of a linear factor, must be computed if the degree of the original polynomial is odd.

10 An iteration formula was developed for determining the square root of a number, a, by *Newton's method*. By a simple hand calculation, compare the rates of convergence of *Newton's method* for finding square roots, to the *method of false position* and the *half-interval search*. Carry out enough steps with each so the result is within 2% of the correct answer, comparing in each case the number of steps required. Set $a = 5$ and use the following initial values:

Newton's method, $x_0 = 5$.
Method of false position, $x_l = 0$; $x_r = 5$.
Half-interval method, $x_l = 0$; $x_r = 5$.
Note: To four places $\sqrt{5} = 2.2361$.

11 All of the roots of the polynomial

$$x^4 - 3x^3 + 10x^2 - 13x + 15$$

are imaginary. Determine two roots (a conjugate pair) using division by a quadratic factor.

chapter five

numerical integration

One of the reasons for the preeminence of polynomials in numerical methods is that they are readily differentiated and integrated; moreover, the result obtained by differentiating or integrating a polynomial is also a polynomial. Since interpolation polynomials permit one to represent a set of points in functional form, an approximation to the derivative or integral of such a tabular function can be obtained by differentiating or integrating the interpolation polynomial. This procedure is much more effective for integration, since it is a "smoothing" process, than for differentiation, which tends to amplify small variations. This rather vague description may be made somewhat more understandable by the diagram in Fig. 5.1. It is certainly possible that $\int_{x_0}^{x_4} p(x)\, dx$ could be a good approximation to $\int_{x_0}^{x_4} f(x)\, dx$. It looks as if the deviations of $p(x)$ above and below $f(x)$ would about cancel out in the integration. On the other hand, the derivatives of the interpolating polynomial $p(x)$ would differ markedly from those of $f(x)$ since it has much more variation than the relatively smooth $f(x)$. This difference in the degree of approximation of the two processes can be recognized in the corresponding error terms, but since integration has the greater utility it will be emphasized in what follows.

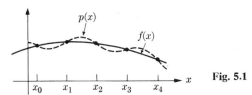

Fig. 5.1

The most frequently used integration formulas are obtained from the equal-interval interpolation polynomials. The simplest example is that of two points where the first-degree forward-difference polynomial is integrated from x_0 to x_1, or by making the substitution $x = x_0 + sh$, where the variable s is varied from 0 to 1. The function, expressed as a polynomial with error term, is

$$f(x) = f(x_0) + s\,\Delta f(x_0) + s(s-1)h^2\frac{f''(\zeta)}{2}, \qquad x_0 \le \zeta \le x_0 + h.$$

From the substitution, $dx = h\, ds$, and the integral becomes

$$\int_{x_0}^{x_1} f(x)\, dx = h \int_0^1 f(x_0 + sh)\, ds$$

$$= h \int_0^1 \left[f(x_0) + s\, \Delta f(x_0) + (s^2 - s) h^2 \frac{f''(\varsigma)}{2} \right] ds$$

$$= h \left[sf(x_0) + \frac{s^2}{2} \Delta f(x_0) + \left(\frac{s^3}{3} - \frac{s^2}{2} \right) h^2 \frac{f''(\varsigma)}{2} \right]_0^1 .$$

When evaluated at the lower limit, all the terms are zero so that the result, with the substitution $f(x_1) - f(x_0) = \Delta f(x_0)$ made, is

$$\int_{x_0}^{x_1} f(x)\, dx = \frac{h}{2} \left[f(x_0) + f(x_1) \right] \boxed{- \frac{h^3}{12} f''(\varsigma).} \;\text{remainder}$$
$$\text{Trapezoidal}$$
$$\text{Rule}$$

Dropping the error term on the right leads to the familiar *trapezoidal rule* which, as the interpolation polynomial required, is simply the area under the straight-line segment connecting the two adjacent points on the function (Fig. 5.2). The next formula, and by far the most popular for numerical integration, is obtained by integrating over two equal intervals. On the assumption that integration over even more intervals will be of interest, a labor-saving method is to obtain the indefinite integral

$$\int_{x_0}^{x} f(x)\, dx = h \int_0^s f(x_0 + hs)\, ds$$

and then truncate the resulting series to the appropriate term:

$$h \int_0^s f(x_0 + hs)\, ds = h \int_0^s \left[f(x_0) + s\, \Delta f(x_0) + s(s - 1) \frac{\Delta^2 f(x_0)}{2!} \right.$$

$$+ s(s - 1)(s - 2) \frac{\Delta^3 f(x_0)}{3!} + \cdots$$

$$\left. + s(s - 1)(s - 2)(s - 3) \frac{\Delta^4 f(x_0)}{4!} + \cdots \right] ds$$

$$= h \left[sf(x_0) + \frac{s^2}{2} \Delta f(x_0) + \left(\frac{s^3}{3} - \frac{s^2}{2} \right) \Delta^2 \frac{f(x_0)}{2} \right.$$

$$+ \left(\frac{s^4}{4} - s^3 + s^2 \right) \frac{\Delta^3 f(x_0)}{6}$$

$$\left. + \left(\frac{s^5}{5} - \frac{6s^4}{4} + \frac{11s^3}{3} - \frac{6s^2}{2} \right) \frac{\Delta^4 f(x_0)}{24} + \cdots \right]_0^s .$$

As before, all the terms are zero when this expression is evaluated at the lower limit $s = 0$. Now to obtain a formula for the two-interval (or three-point) case, substitute $s = 2$ and truncate to the number of terms desired. It would not seem

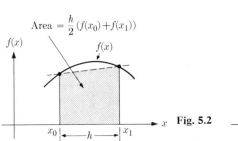

Area $= \dfrac{h}{2}(f(x_0)+f(x_1))$

$f(x)$

$f(x)$

x **Fig. 5.2**

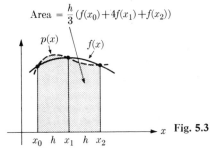

Area $= \dfrac{h}{3}(f(x_0)+4f(x_1)+f(x_2))$

$p(x)$ $f(x)$

x **Fig. 5.3**

reasonable to include terms beyond the second difference since the higher-order differences depend upon more than three points. Actually there is nothing wrong with using many points to obtain the interpolating polynomial and then integrating this approximation to the function over a smaller interval. In the two-interval case under consideration, including the third difference yields an unexpected dividend:

$$h\int_0^2 f(x_0 + hs)\,ds = h\left[2f(x_0) + 2\,\Delta f(x_0) + \frac{2}{3}\frac{\Delta^2 f(x_0)}{2}\right.$$
$$\left. + 0\,\frac{\Delta^3 f(x_0)}{6} - \frac{4}{15}\frac{h^4 f^{(4)}(\zeta)}{24}\right].$$

Again, the rightmost term is the remainder obtained by substituting

$$\frac{h^4 f^{(4)}}{4!} \qquad \text{for} \qquad \frac{\Delta^4 f(x_0)}{4!}.$$

Note that since the coefficient of the third difference is zero, a fourth-order remainder term is applicable although only three points are required to compute the nonzero terms. By substituting

$$f(x_1) - f(x_0) = \Delta f(x_0) \qquad \text{and} \qquad f(x_2) - 2f(x_1) + f(x_0) = \Delta^2 f(x_0)$$

for the differences, one obtains the usual form of *Simpson's rule:*

$$\int_{x_0}^{x_2} f(x)\,dx = \frac{h}{3}\left[f(x_0) + 4f(x_1) + f(x_2)\right]\boxed{-\frac{h^5 f^{(4)}(\zeta)}{90}}.$$

remainder of Simpson's Rule

Without the error term, evaluation of this formula amounts to computing the area under a third-degree polynomial passing through three points with equally spaced abscissas (Fig. 5.3). Comparison of the error terms for the trapezoidal and Simpson's rules shows why the latter is so popular. In the two-interval case, the computation of the formula is only slightly more complicated than in the one-interval case, and yet the error term is proportional to the fourth derivative instead of, as might be expected, the third.

A similar jump in the order of the error term is encountered for all even-interval (or odd-point) formulas and hence they are preferred to the others. The next two formulas are:

$$\int_{x_0}^{x_3} f(x)\,dx = \frac{3h}{8}\left[f(x_0) + 3f(x_1) + 3f(x_2) + f(x_3)\right] - \frac{3h^5}{80}f^{(4)}(\varsigma),$$

$$\int_{x_0}^{x_4} f(x)\,dx = \frac{2h}{45}\left[7f(x_0) + 32f(x_1) + 12f(x_2) + 32f(x_3) + 7f(x_4)\right]$$
$$- \frac{8h^7}{945}f^{(6)}(\varsigma).$$

These formulas, of which Simpson's rule is a special case, are called the *Newton-Cotes formulas*. When these formulas are repeatedly applied they can be written in a combined form over the entire interval of application. Thus, if the trapezoidal rule is used in the intervals (x_0, x_1) and (x_1, x_2), then

$$\int_{x_1}^{x_2} f(x)\,dx \approx \frac{h}{2}[f(x_0) + f(x_1)] + \frac{h}{2}[f(x_1) + f(x_2)]$$
$$\approx \frac{h}{2}[f(x_0) + 2f(x_1) + f(x_2)],$$

and in general, for $x_0, x_1, \ldots, x_n$,

$$\int_{x_0}^{x_n} f(x)\,dx \approx \frac{h}{2}[f(x_0) + 2f(x_1) + 2f(x_2) + \cdots + 2f(x_{n-1}) + f(x_n)].$$

The repeated application of Simpson's rule for the total interval (x_0, x_{2n}) is expressed in the same manner:

$$\int_{x_0}^{x_{2n}} f(x)\,dx \approx \frac{h}{3}[f(x_0) + 4f(x_1) + 2f(x_2) + 4f(x_3) + 2f(x_4) + \cdots$$
$$+ 2f(x_{2n-2}) + 4f(x_{2n-1}) + f(x_{2n})].$$

The terminal subscript is expressed as $2n$ to indicate that the subscript must be even. This requirement makes the total number of points odd since there is a zeroth point.

To illustrate the relative effectiveness of the two methods developed, a familiar definite integral, $\int_0^\pi \sin\theta\,d\theta = 2$, is approximated by numerical integration in several ways. The given values are:

x	$f(x) = \sin x$
0	0
$\pi/4$	0.7071
$\pi/2$	1.0
$3\pi/4$	0.7071
π	0

1. Trapezoidal rule applied twice ($h = \pi/2$):

$$\int_0^\pi \sin x \, dx \approx \frac{\pi}{4}\left[\sin 0 + 2 \sin \frac{\pi}{2} + \sin \pi\right] = 1.5708$$

$$\left|\frac{h^3}{12} \sin (\zeta)\right| \leq \frac{h^3}{12} \max \sin \zeta, \qquad 0 \leq \zeta \leq \frac{\pi}{2},$$

$$\leq \frac{(1.5708)^3}{12} 1 = 0.32.$$

Since there are two similar intervals involved, the error bound is $2 \times 0.32 = 0.64$.

2. Trapezoidal rule applied four times ($h = \pi/4$):

$$\int_0^\pi \sin x \, dx \approx \frac{\pi}{8}\left[\sin 0 + 2 \sin \frac{\pi}{4} + 2 \sin \frac{\pi}{2} + 2 \sin \frac{3\pi}{4} + \sin \pi\right] = 1.895.$$

It is necessary to obtain the sum of the maximum error from two different intervals, $(0, \pi/4)$ and $(\pi/4, \pi/2)$, and, due to the symmetry, then to double this sum to obtain the error bound.

$$\left|\frac{h^3}{12} \sin (\zeta)\right| \leq \left(\frac{\pi}{4}\right)^3 \frac{1}{12} \max \sin \zeta = 0.0287, \qquad 0 \leq \zeta \leq \frac{\pi}{4}$$

$$= 0.0405, \qquad \frac{\pi}{4} < \zeta \leq \frac{\pi}{2}.$$

The error bound is $2(0.0287 + 0.0405) = 0.1384$.

3. Simpson's rule applied once ($h = \pi/2$):

$$\int_0^\pi \sin x \, dx \approx \frac{\pi}{6}\left[\sin 0 + 4 \sin \frac{\pi}{2} + \sin \pi\right] = 2.0943,$$

$$\left|\frac{h^5}{90} \cos \zeta\right| \leq \left(\frac{\pi}{2}\right)^5 \times \frac{1}{90} \max \cos (\zeta) = 0.1006.$$

4. Simpson's rule applied twice ($h = \pi/4$):

$$\int_0^\pi \sin x \, dx \approx \frac{\pi}{12}\left[\sin 0 + 4 \sin \frac{\pi}{4} + 2 \sin \frac{\pi}{2} + 4 \sin \frac{3\pi}{4} + \sin \pi\right] = 2.0043.$$

As before, the error bound can be determined from one interval and, due to symmetry, simply doubled,

$$\left|\frac{h^5}{90} \cos \zeta\right| \leq \left(\frac{\pi}{4}\right)^5 \frac{1}{90} \max \cos \zeta = .00331,$$

so that the error bound is $2(0.00331) = 0.00662$.

Gaussian quadrature

Several observations can be made at this point to motivate an alternative approach to numerical integration, or *quadrature* as it is often called.

1. Applying Simpson's rule, one is able to integrate a third-degree polynomial *exactly*, using only three points. One would at first glance expect four to be necessary.

2. The formulas for integration are of the form $\sum_{i=0}^{n} a_i f(x_i)$. Since $n + 1$ constants, as well as $n + 1$ ordinates, are given in this formula, the total of $2n + 2$ parameters suggests that sufficient information is available to permit specification of a polynomial of degree $2n + 1$.

3. In the equal-interval formulas derived, the ordinates could not be freely chosen in the range of interest but were limited by the equal-interval requirement. It might be possible, then, to realize in greater measure the benefits of Simpson's rule, i.e., integrate exactly a $(2n + 1)$-degree polynomial using only an nth-degree polynomial approximation, by appropriately selecting the abscissas as well as the constant coefficients.

The Lagrangian interpolation polynomial is a logical starting point in determining the constants and abscissas since the ordinates appear explicitly and there is no restriction on the selection of x's:

$$f(x) = \sum_{i=0}^{n} \bar{c}_i(x) f(x_i) + \prod_{i=0}^{n} (x - x_i) \frac{f^{(n+1)}(\zeta)}{(n + 1)!}, \qquad a \le \zeta \le b$$

where

$$\bar{c}_i = \prod_{\substack{j=0 \\ j \ne i}}^{n} \left(\frac{x - x_j}{x_i - x_j} \right).$$

It will be convenient to make a change of variable so that the interval containing the abscissas,

$$a \le x_0, \quad x_1, \quad \ldots, \quad x_n \le b,$$

is transformed to the interval $(-1, 1)$:

$$t = \frac{2x - (a + b)}{b - a}.$$

Then

$$f(x) = f\left(\frac{(b - a)t + a + b}{2} \right) = F(t),$$

and

$$F(t) = \sum_{i=0}^{n} c_i(t) F(t_i) + \prod_{i=0}^{n} (t - t_i) \frac{F^{(n+1)}(\bar{\zeta})}{(n + 1)!},$$

$$c_i = \prod_{\substack{j=0 \\ j \ne i}}^{n} \left(\frac{t - t_j}{t_i - t_j} \right), \qquad -1 \le \bar{\zeta} \le 1.$$

Now if $F(t)$ is assumed to be a polynomial of degree $2n + 1$, the derivative term can be replaced by a polynomial of degree n, $p_n(t)$:

$$F(t) = \sum_{i=0}^{n} c_i(t)F(t_i) + \prod_{i=0}^{n} (t - t_i)p_n(t).$$

The rationale for this replacement is simply that there is a $(2n + 1)$-degree polynomial on the left, and hence the right must represent the same polynomial. The first expression on the right, $\sum_{i=0}^{n} c_i(t)F(t_i)$, is an nth-degree polynomial, and therefore the remaining term on the right must be of degree $2n + 1$. But the product, $\prod_{i=0}^{n} (t - t_i)$, is a polynomial in factored form of degree $n + 1$, and hence the remaining polynomial factor $p_n(t)$ must be of nth degree.

Integrating both sides, one obtains

$$\int_{-1}^{1} F(t)\, dt = \int_{-1}^{1} \left[\sum_{i=0}^{n} c_i(t)F(t_i) \right] dt + \int_{-1}^{1} \prod_{i=0}^{n} (t - t_i)p_n(t)\, dt$$

$$= \sum_{i=0}^{n} \left(\int_{-1}^{1} c_i(t)\, dt \right) F(t_i) + \int_{-1}^{1} \prod_{i=0}^{n} (t - t_i)p_n(t)\, dt$$

$$= \sum_{i=0}^{n} a_i F(t_i) + \int_{-1}^{1} \prod_{i=0}^{n} (t - t_i)p_n(t)\, dt.$$

To achieve the desired goal the remainder term, the integral on the right, must be zero. The question now is: Are there values t_i such that the integral of the polynomial product $\prod_{i=0}^{n} (t - t_i)p_n(t)$ in the interval $(-1, 1)$ is zero? The factor $p_n(t)$ is arbitrary but one degree less than the other factor. Two polynomials meeting this integration requirement are said to be *orthogonal*. There are families of polynomials $p_0(x)$, $p_1(x)$, $p_2(x)$, $\ldots$, $p_n(x)$ which have this property; i.e.,

$$\int_{-1}^{1} p_i(x)p_j(x)\, dx = 0 \qquad \text{for} \qquad i \neq j.$$

But the most common set of polynomials,

$$P_0(x) = 1$$
$$P_1(x) = x$$
$$P_2(x) = x^2$$
$$\vdots$$
$$P_n(x) = x^n$$

does not. There is a set of polynomials, called *Legendre* polynomials, which do have this orthogonal property:

$$L_0(x) = 1$$
$$L_1(x) = x$$
$$L_2(x) = \tfrac{1}{2}(3x^2 - 1)$$
$$L_3(x) = \tfrac{1}{2}(5x^3 - 3x).$$
$$\vdots$$

The name given to this family arises from the fact that one way to derive them is as solutions of Legendre's equation

$$(1 - x^2)\frac{d^2y}{dx^2} - 2x\frac{dy}{dx} + n(n + 1)y = 0,$$

where the subscript for the polynomial corresponds to the value of n in the differential equation.

It is worth a short digression at this point to present a geometrical meaning for the term "orthogonality" which is, in fact, a generalization of "perpendicularity." Two directed line segments (vectors) which have no collinear tendency are perpendicular (Fig. 5.4a). Segment 2 has no component which is collinear with 1. When the segments are not perpendicular (Fig. 5.4b and c), then segment 2 can be decomposed into two components (the dashed lines), one of which is collinear with, and the other perpendicular to, segment 1. Note that the condition for perpendicularity is that the slope of one line is the negative reciprocal of the other, i.e.,

$$\frac{y_1}{x_1} = -\frac{x_2}{y_2},$$

or, as the condition is more often stated,

$$x_1x_2 + y_1y_2 = 0.$$

When the coordinates are regarded as endpoints of a vector, this sum of products is called the *scalar product*. The condition is readily extended to three-dimensional vectors whose termini are (x_1, y_1, z_1) and (x_2, y_2, z_2):

$$x_1x_2 + y_1y_2 + z_1z_2 = 0.$$

It is difficult to picture how this condition is extended beyond three dimensions, but formally it is simply a matter of including more coordinate values. If n values are named, that is, $x_1, x_2, x_3, \ldots, x_n$, instead of $(x_1, y_1, z_1, \ldots)$, and $y_1, y_2, \ldots,$

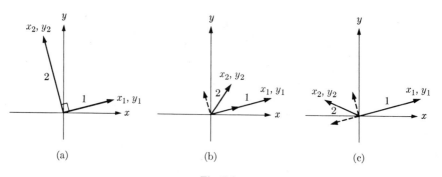

(a)   (b)   (c)

Fig. 5.4

y_n, instead of $(x_2, y_2, z_2, \ldots)$, the condition for orthogonality is

$$x_1y_1 + x_2y_2 + x_3y_3 + \cdots + x_ny_n = \sum_{i=1}^{n} x_iy_i = 0.$$

As n becomes large, and presuming that the x- and y-values can be represented as a function of an independent variable t, one sees that this sum tends to approach the integral

$$\int_a^b x(t)y(t)\, dt = 0.$$

This is the definition for orthogonality over the interval (a, b) when functions are involved. Although "orthogonal" is a two-valued term (lines are perpendicular or they are not), the value of the integral and sums of products above are measures of the "similar tendency" of two line segments or functions; or, from a slightly different point of view, it is a measure of how much of a function can be represented by another function. In other contexts this measure is called *correlation*.

An example from each of the families of polynomials shown serves to illustrate this idea.

$$\int_{-1}^{1} P_1(x)P_3(x)\, dx = \int_{-1}^{1} x \cdot x^3\, dx = \tfrac{2}{5}$$

Plotting these functions on the same graph (Fig. 5.5) shows that they do indeed have a similar tendency in $(-1, 1)$. Both functions continually increase as x goes from -1 to 1. As a matter of fact, the best lower-degree approximation for x^3 is $\tfrac{3}{4}x$. On the other hand,

$$\int_{-1}^{1} L_1(x)L_3(x)\, dx = \int_{-1}^{1} x \cdot \tfrac{1}{2}(5x^3 - 3x)\, dx = \tfrac{1}{2}[x^5 - x^3]_{-1}^{1} = 0.$$

In this case the graph (Fig. 5.6) shows that in the center portion of the interval the functions have a tendency opposite to that of the remaining portion of the interval. The resultant orthogonality implies that $L_3(x)$ has no part that is representable by $L_1(x)$. Just as it is possible to represent arbitrary functions by weighted sums of

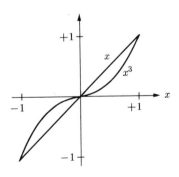

Fig. 5.5

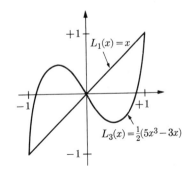

Fig. 5.6

$P_0(x)$, $P_1(x)$, $P_2(x)$, for example,

$$q(x) = d_0 + d_1 x + d_2 x^2 + \cdots + d_n x^n$$
$$= d_0 P_0(x) + d_1 P_1(x) + \cdots + d_n P_n(x),$$

it is also possible to represent the function in terms of the orthogonal family. This ability to write

$$q(x) = a_0 L_0(x) + a_1 L_1(x) + a_2 L_2(x) + \cdots + a_n L_n(x)$$

permits a solution to the problem of finding the Gaussian integration formulas.

Returning now to the problem of eliminating the integral term, one finds that the expansion of the two polynomial factors in terms of Legendre polynomials provides a method of determining the t_i's which will make this error term zero:

$$p_n(t) = a_0 L_0(t) + a_1 L_1(t) + \cdots + a_n L_n(t) = \sum_{i=0}^{n} a_i L_i(t),$$

$$\prod_{i=0}^{n} (t - t_i) = b_0 L_0(t) + b_1 L_1(t) + \cdots + b_{n+1} L_{n+1}(t) = \sum_{j=0}^{n+1} b_j L_j(t).$$

The product of these two expanded polynomials will be a sum of terms of the form $a_i b_j L_i(t) L_j(t)$. Integrating this sum from -1 to 1 causes all the terms where $i \neq j$ to become zero since

$$\int_{-1}^{1} L_i(t) L_j(t)\, dt = 0 \qquad \text{when} \qquad i \neq j.$$

The remaining terms are now

$$\int_{-1}^{1} \prod_{i=0}^{n} (t - t_i) p_n(t)\, dt = \int_{-1}^{1} a_0 b_0 (L_0(t))^2\, dt + \int_{-1}^{1} a_1 b_1 (L_1(t))^2\, dt + \cdots$$

$$+ \int_{-1}^{1} a_n b_n (L_n(t))^2\, dt$$

$$= \int_{-1}^{1} \left[\sum_{i=0}^{n} a_i b_i (L_i(t))^2 \right] dt.$$

These terms can be made zero by choosing $b_0 = b_1 = b_2 = \cdots = b_n = 0$. With this choice the error term will be zero, but it is now necessary to select b_{n+1} and the t_i so that

$$\prod_{i=0}^{n} (t - t_i) = b_{n+1} L_{n+1}(t).$$

As the polynomial on the left is in factor form, the equation will be satisfied if the t_i are roots of the polynomial on the right. This means that the abscissas to be used in forming the interpolating polynomial, which in turn is the basis of the integration formula, are roots of the $(n + 1)$th Legendre polynomial, and that the coefficient, b_{n+1}, is chosen to cause the coefficient of the t^{n+1} term on the right-hand side of the equation to be unity. Once these roots, $t_0, t_1, t_2 \ldots t_n$ are determined, then the coefficients and the ordinates a_i in the integration approximation can be obtained by integration.

$$a_i = \int_{-1}^{1} c_i(t)\, dt = \int_{-1}^{1} \prod_{\substack{j=0 \\ i \neq j}}^{n} \frac{t - t_j}{t_i - t_j}\, dt$$

For example, when $n = 1$, the Legendre polynomial is

$$L_2(t) = \tfrac{1}{2}(3t^2 - 1),$$

having roots

$$t_0 = \frac{-\sqrt{3}}{3} \qquad \text{and} \qquad t_1 = \frac{\sqrt{3}}{3}.$$

As a matter of interest

$$b_2 = \tfrac{2}{3}.$$

The coefficients, a_i, are found from the integration

$$a_0 = \int_{-1}^{1} \frac{t - t_1}{t_0 - t_1}\, dt = \frac{1}{t_0 - t_1}\left[\frac{t^2}{2} - t_1 t\right]_{-1}^{1}$$

$$= -\frac{1}{(2\sqrt{3})/3}\left[\frac{1}{2} - \frac{\sqrt{3}}{3} - \frac{1}{2} - \frac{\sqrt{3}}{3}\right] = 1.$$

Repeating the procedure for the second coefficient, we obtain

$$a_1 = 1.$$

The roots and coefficients for $n = 0, 1, 2$ are:

n	t_0	t_1	t_2	a_0	a_1	a_2
0	0			2		
1	$-1/\sqrt{3}$	$1/\sqrt{3}$		1	1	
2	$-\sqrt{\tfrac{3}{5}}$	0	$\sqrt{\tfrac{3}{5}}$	$\tfrac{5}{9}$	$\tfrac{8}{9}$	$\tfrac{5}{9}$

(Additional values of roots and coefficients are given in Appendix B.) Thus in the interval $(-1, 1)$, the integral of a first-degree polynomial $p_1(x)$ is, without error, $2p_1(0)$. Similarly,

$$\int_{-1}^{1} p_3(x)\, dx = p_3\left(-\frac{1}{\sqrt{3}}\right) + p_3\left(\frac{1}{\sqrt{3}}\right)$$

and

$$\int_{-1}^{1} p_5(x)\, dx = \frac{5}{9} p_5\left(-\sqrt{\tfrac{3}{5}}\right) + \frac{8}{9} p_5(0) + \frac{5}{9} p_5\left(\sqrt{\tfrac{3}{5}}\right).$$

For comparison, the two- and three-point Gaussian formulas are applied to the problem used as an equal-interval example:

x	$\sin x$
0	0
$\pi/2$	1.0
π	0

1. Double application of two-point Gaussian formula. From the symmetry one sees that

$$\int_0^\pi \sin x \, dx = 2 \int_0^{\pi/2} \sin x \, dx.$$

For $a = 0$ and $b = \pi/2$, the transformations

$$x = \frac{(b-a)t + a + b}{2}, \qquad dx = \frac{b-a}{2} dt$$

become

$$x = \frac{\pi}{4}(t+1), \qquad dx = \frac{\pi}{4} dt,$$

and

$$\int_0^\pi \sin x \, dx = 2 \left(\frac{\pi}{4}\right) \int_{-1}^1 \sin\left[\frac{\pi}{4}(t+1)\right] dt$$

$$\approx \frac{\pi}{2}\left[\sin \frac{\pi}{4}\left(-\frac{1}{\sqrt{3}}+1\right) + \sin \frac{\pi}{4}\left(\frac{1}{\sqrt{3}}+1\right)\right] = 1.9968.$$

2. Single application of three-point Gaussian formula:

$$x = \frac{\pi}{2}(t+1), \qquad dx = \frac{\pi}{2} dt$$

$$\int_0^\pi \sin x \, dx = \frac{\pi}{2} \int_{-1}^1 \sin \frac{\pi}{2}(t+1) \, dt$$

$$\approx \frac{\pi}{2}\left[\frac{5}{9} \sin \frac{\pi}{2}\left(-\sqrt{\frac{3}{5}}+1\right) + \frac{8}{9} \sin \frac{\pi}{2}(0+1)\right.$$

$$\left. + \frac{5}{9} \sin\left(\sqrt{\frac{3}{5}}+1\right)\right] = 2.0014.$$

Both approximations are better than the best equal-interval result, and yet fewer evaluations of the function were required. Gaussian integration should be used when there is freedom to select the abscissas.

There are other families of orthogonal polynomials which are used when the interval of interest cannot be transformed to $(1, -1)$. Using the roots of these polynomials produces Gaussian formulas which are similar to the ones developed here.

To illustrate the programming of numerical integration, two subroutine functions are included. Both functions are fairly general in that the calling program specifies (a, b), the total interval; n, the number of subintervals n to be integrated separately (i.e., the number of applications of the formula selected); m, the number of points to be used in the integration over one subinterval; and f, the name of the function to be integrated. In both instances it is assumed that the function can be evaluated; it is not tabulated. Equal-interval formulas are used for the first routine, and the Gaussian formula is used for the second.

Example Write a subroutine called EQUINT, which has dummy variables a, b, m, n defined as above and a function subprogram for f. The numerical integration of f over the interval (a, b) is to be accomplished by n applications of an equal-interval m-point formula $(2 \leq m \leq 5)$.

The flow chart (Fig. 5.7a) is simple enough to be readily followed if the following two points are understood.

1. The ordinate which corresponds to the division between subintervals is used twice, once as the last ordinate in a formula and again as the first ordinate in the next application of the formula. The algorithm is constructed so that this ordinate is computed only once.

2. The constants which are multipliers of the ordinates are arranged in a linear array. Moreover, the constants are arranged so that h is always a multiplier of the entire sum of weighted ordinates.

Number of points	Formula	Constants				
2	$(h/2)(f(x_0) + f(x_1))$	$\frac{1}{2},$	$\frac{1}{2}$			
3	$(h/3)(f(x_0) + 4f(x_1) + f(x_2))$	$\frac{1}{3},$	$\frac{4}{3},$	$\frac{1}{3}$		
4	$(3h/8)(f(x_0) + 3f(x_1) + 3f(x_2) + f(x_3))$	$\frac{3}{8},$	$\frac{9}{8},$	$\frac{9}{8},$	$\frac{3}{8}$	
5	$(2h/45)(7f(x_0) + 32f(x_1) + 12f(x_2)$ $\quad + 32f(x_3) + 7f(x_4))$	$\frac{14}{45},$	$\frac{64}{45},$	$\frac{24}{45},$	$\frac{64}{45},$	$\frac{14}{45}$

The elements of the linear array (A) are then:

A_1	A_2	A_3	A_4	A_5	A_6	A_7	A_8	A_9	A_{10}	A_{11}	A_{12}	A_{13}	A_{14}
$\frac{1}{2}$	$\frac{1}{2}$	$\frac{1}{3}$	$\frac{4}{3}$	$\frac{1}{3}$	$\frac{3}{8}$	$\frac{9}{8}$	$\frac{9}{8}$	$\frac{3}{8}$	$\frac{14}{45}$	$\frac{64}{45}$	$\frac{24}{45}$	$\frac{64}{45}$	$\frac{14}{45}$

2 points   3 points   4 points   5 points

Note that if the first constant in an m-point formula is selected, the linear subscript, k, is $k = m(m - 1)/2$. Then if $m = 2$, $k = 1$; $m = 3$, $k = 3$; $m = 4$, $k = 6$; $m = 5$, $k = 10$.

The program for this example is presented in Fig. 5.7b. Note that to avoid confusion with the interval endpoint a (here A), the array designated before A is represented by **ARRAY**.

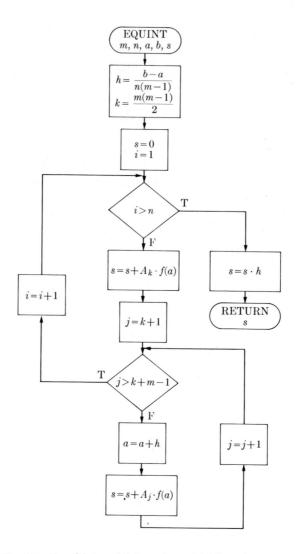

Fig. 5.7 Equal-interval integration. (a) Flow chart.

A function subprogram is included in Fig. 5.7b along with the main program. In this example the integral $\int_a^b \sin x\, dx$ has been integrated for several sets of m and n. It is a simple matter to change the function to be integrated.

Example Write a subroutine called GAUINT, which has dummy variables a, b, f, m, n as before. The numerical integration is to be accomplished by n applications of a Gaussian m-point formula ($2 \leq m \leq 4$).

```
C MASTER EQUINT
C MAINLINE
      READ(5,4)NSETS
4     FORMAT(I5)
      DO 100 LL=1,NSETS
      READ(5,1)A,B,M,N
      AA=A
1     FORMAT(2F9.5,2I5)
      CALL EQUINT(M,N,A,B,S)
100   WRITE(6,2)AA,B,S
2     FORMAT(5X,13HINTEGRAL FROM,F7.2,4H TO,F7.2,3H =,F10.7)
      STOP
      END

      SUBROUTINE EQUINT(M,N,A,B,S)
C EQUAL INTERVAL SUBROUTINE
C (A,B) IS THE INTERVAL OF INTEGRATION
C N IS THE NUMBER OF THESE INTERVALS
C M IS THE NUMBER OF POINTS PER INTERVAL
C EQUA(X) MUST BE ENTERED AS A FUNCTION SUBPROGRAM
C S IS THE VALUE OF THE INTEGRAL WHICH IS RETURNED
      DIMENSION ARRAY(14)
      DATA ARRAY/.5,.5,.333333,1.333333,.333333,.375,1.125,
     11.125,.375,.311111,1.422222,.533333,1.422222,.311111/
C ABOVE ARE EQUAL INTERVAL COEFFICIENTS
      Q=N
      R=M
      H=(B-A)/(Q*(R-1.))
      K=(M*(M-1))/2
      S=0.
      DO 10 I=1,N
      S=S+ARRAY(K)*EQUA(A)
      KP1=K+1
      KPMM1=K+M-1
      DO 10 J=KP1,KPMM1
      A=A+H
10    S=S+ARRAY(J)*EQUA(A)
      S=S*H
      RETURN
      END

      FUNCTION EQUA(X)
      EQUA=SIN(X)
      RETURN
      END

      INPUT

5,
0.,3.1416,2,5,
0.,3.1416,3,5,
0.,3.1416,4,5,
0.,3.1416,5,5,
0.,3.1416,3,10,

      OUTPUT

INTEGRAL FROM    0.0    TO    3.14    = 1.9337635
INTEGRAL FROM    0.0    TO    3.14    = 2.0001078
INTEGRAL FROM    0.0    TO    3.14    = 2.0000515
INTEGRAL FROM    0.0    TO    3.14    = 1.9999990
INTEGRAL FROM    0.0    TO    3.14    = 2.0000048
```

Fig. 5.7   (b) Program.

Since the Gaussian formulas do not involve ordinates at the endpoints of the intervals, the steps preventing the reevaluation of the function f at the same point are not relevant. There is in this case, however, the additional complication of having to transform each subinterval to the interval $(-1, 1)$. Also, in addition to the linear array (*a*) of weighting coefficients, a linear array (*t*) is required for the roots of the appropriate Legendre polynomials. The formula for the linear subscript for the first relevant item in the m-point formula is

$$k = m(m - 1)/2.$$

Number of points	Ordinate weights	Legendre polynomial roots
2	1, 1	$-\dfrac{1}{\sqrt{3}}, \dfrac{1}{\sqrt{3}}$
3	$\frac{5}{9}, \frac{8}{9}, \frac{5}{9}$	$-\sqrt{\frac{3}{5}}, 0, \sqrt{\frac{3}{5}}$
4	$\dfrac{18 - \sqrt{30}}{36}, \dfrac{18 + \sqrt{30}}{36},$	$-(\frac{3}{7} + \frac{2}{7}\sqrt{\frac{6}{5}})^{1/2}, \quad -(\frac{3}{7} - \frac{2}{7}\sqrt{\frac{6}{5}})^{1/2}$
	$\dfrac{18 + \sqrt{30}}{36}, \dfrac{18 - \sqrt{30}}{36},$	$(\frac{3}{7} - \frac{2}{7}\sqrt{\frac{6}{5}})^{1/2}, \quad (\frac{3}{7} + \frac{2}{7}\sqrt{\frac{6}{5}})^{1/2}$

The transformation to the t variables is

$$x = \frac{(b - a)}{2} t + \frac{(b + a)}{2}$$

and

$$dx = \frac{b - a}{2} dt.$$

If $h = b - a$, $x = (h/2)t + \frac{1}{2}(2a + h)$. As a further simplification, let $h' = h/2$ and $g = (2a + h)/2$; then $x = h't + g$.

The flow chart and program are presented in Fig. 5.8. An example was run evaluating the integral $\int_a^b \sin x \, dx$ over the interval $a = 0$ to $b = \pi$. The cases were: $m = 2, n = 2$; $m = 3, n = 2$; $m = 4, n = 2$; and $m = 4, n = 6$.

Numerical differentiation

The shortcomings of numerical differentiation were mentioned earlier, but they are verified by examination of the error terms after differentiation. Starting with the first-order forward-difference interpolating polynomial

$$f(x) = f(x_0) + (x - x_0)\frac{\Delta f(x_0)}{h} + (x - x_0)(x - x_1)\frac{f''(\zeta)}{2}$$

and differentiating with respect to x, one arrives at

$$f'(x) = \frac{\Delta f(x_0)}{h} + [(x - x_0) + (x - x_1)]\frac{f''(\zeta)}{2}.$$

Evaluating the derivative at x_0, and remembering that $x_0 - x_1 = -h$, one

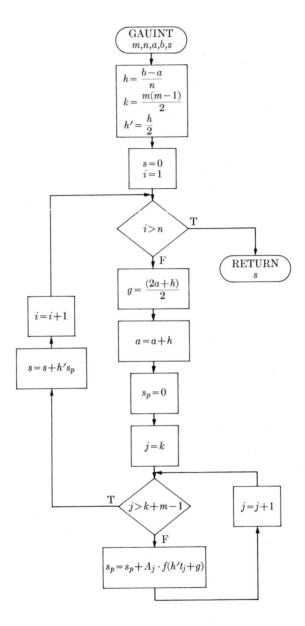

Fig. 5.8 Gaussian *M*-point integration. (a) Flow chart.

obtains a familiar approximation for the first derivative:

$$f'(x_0) = \frac{\Delta f(x_0)}{h} - \frac{h}{2} f''(\zeta), \qquad f'(x_0) \approx \frac{\Delta f(x_0)}{h} = \frac{f(x_1) - f(x_0)}{h}.$$

```
C MASTER GAUINT
C MAINLINE
        READ (5,1) NSETS
1       FORMAT (I2)
        DO 10 I=1,NSETS
        READ(5,2) A,B,M,N
2       FORMAT (2F9.5,2I3)
        AA=A
        CALL GAUINT(M,N,A,B,S)
10      WRITE(6,3) AA,B,S
3       FORMAT(5X,13HINTEGRAL FROM,F7.2,4H  TO,F7.2,3H  =,F10.7)
        STOP
        END

        SUBROUTINE GAUINT(M,N,A,B,S)
        DIMENSION ARRAY(9),T(9)
        DATA ARRAY/1.,1.,.555556,.888889,.555556,.347855,.652145,.652145,.
        1347855/
        DATA T/-.577350,.577350,-.774597,0.,.774597,-.861136,-.339981,.339
        1981,.861136/
        R=N
        H=(B-A)/R
        K=(M*(M-1))/2
        HP=H/2.
        S=0.
        DO 100 I=1,N
        G=(2.*A+H)/2.0
        A=A+H
        SP=0.
        MPKM1=M+K-1
        DO 50 J=K,MPKM1
        X=HP*T(J)+G
50      SP=SP+ARRAY(J)*EQUA(X)
100     S=S+HP*SP
        RETURN
        END

        FUNCTION EQUA(X)
        EQUA=SIN(X)
        RETURN
        END

        INPUT

4,
0.,3.1416,2,2,
0.,3.1416,3,2,
0.,3.1416,4,2,
0.,3.1416,4,6,

        OUTPUT

    INTEGRAL FROM   0.0   TO   3.14  = 1.9969454
    INTEGRAL FROM   0.0   TO   3.14  = 2.0000162
    INTEGRAL FROM   0.0   TO   3.14  = 1.9999981
    INTEGRAL FROM   0.0   TO   3.14  = 1.9999981
```

Fig. 5.8 (b) Program.

In the same way, a second-order approximation can be derived:

$$f(x) = f(x_0) + (x - x_0)\frac{\Delta f(x_0)}{h} + (x - x_0)(x - x_1)\frac{\Delta^2 f(x_0)}{2h^2}$$
$$+ (x - x_0)(x - x_1)(x - x_2)\frac{f'''(\zeta)}{3!},$$

$$f'(x) = \frac{\Delta f(x_0)}{h} + [(x - x_0) + (x - x_1)]\frac{\Delta^2 f(x_0)}{2h^2}$$
$$+ [(x - x_0)(x - x_1) + (x - x_0)(x - x_2)$$
$$+ (x - x_1)(x - x_2)]\frac{f'''(\zeta)}{6}.$$

At $x = x_0$,
$$f'(x_0) = \frac{\Delta f(x_0)}{h} - \frac{\Delta^2 f(x_0)}{2h} + 2h^2 \frac{f'''(\zeta)}{6}.$$

By expanding the differences one obtains
$$f'(x_0) \approx \frac{2f(x_1) - 2f(x_0) - f(x_2) + 2f(x_1) - f(x_0)}{2h}$$

$$\approx \frac{-f(x_2) + 4f(x_1) - 3f(x_0)}{2h}.$$

However, at $x = x_1$,
$$f'(x_1) = \frac{\Delta f(x_0)}{h} + \frac{\Delta^2 f(x_0)}{2h} - \frac{h^2 f'''(\zeta)}{6}$$

and

$$f'(x) \approx \frac{2f(x_1) - 2f(x_0) + f(x_2) - 2f(x_1) + f(x_0)}{2h} = \frac{f(x_2) - f(x_0)}{2h}.$$

When the derivative was evaluated at the center point, a smaller error was made. The odd-point formulas are preferred due to the symmetry about a central point, which produces a corresponding reduction in the error term. The five-point expression, evaluated at the central point x_2, which could be obtained in the same manner as those above, is

$$f'(x_2) = \frac{f(x_0) + 8f(x_1) + 8f(x_3) + f(x_4)}{12h} + \frac{h^4}{30} f^{(5)}(\zeta).$$

Exercises

1 a) Evaluate
$$I = \int_0^1 (x^2 - 2x)\, dx$$

by Simpson's rule, using 2, 4, and 10 intervals.

b) Plot the error versus the width of the interval and, from these results, obtain a relation between the interval and the order of magnitude of error.

2 Write a program which reads a set of values tabulated at equal intervals of the argument and the number of entries in the table   Using Simpson's rule, find the definite integral of this tabular function. If there is an odd number of intervals, use one application of the trapezoidal rule to complete the numerical integration.

3 If the tabular values are not given at equal intervals, the Newton-Cotes formulas do not directly apply. Integrate the second-degree Lagrangian interpolating polynomial over two intervals to obtain a formula which could be used for this unequal-interval case.

4 Write and execute a computer program to evaluate the integral
$$I = \int_0^b n \sin m\pi x\, dx$$

by Simpson's rule for an interval $h = b/p$. Run the program for $n = 2$, $b = \frac{1}{2}$, $m = 1$, for $p = 2, 4, 10, 50, 100, 500, 1000$. Plot error versus h. The program should be written for the general case. You can incorporate in your program a calculation for error, and supply the particular value of I_{exact} in the data.

5 Consider the integral $\int_{-1}^{1} (1 + x + x^2 + x^3)\, dx$. By a hand calculation,

a) evaluate the integral by Simpson's rule for $h = 1$;

b) evaluate by Gauss-Legendre quadrature using the two-point formula;

c) compare the errors in these two cases and comment on the result.

6 a) By a hand calculation, evaluate the integral

$$\int_{0}^{1/2} 2 \sin \pi x \, dx$$

by five-point Gaussian quadrature.

b) Compute the error.

7 Write a program which integrates a function F over the interval (A, B) by applying the three-point Gaussian formula N times. Read the values of A, B, and N. What changes and additions must be made to turn this program into an external subroutine which directly returns the definite integral value?

8 It is useful if the "degree of orthogonality" of two vectors is scaled to vary from 0 to 1; that is, the degree of 0 if the vectors are orthogonal, and 1 if they are collinear. If each element of a vector x_1, x_2, x_3 is divided by the length of the vector $\sqrt{x_1^2 + x_2^2 + x_3^2}$, the new vector is of length 1.

$$\left\{ \left(\frac{x_1}{\sqrt{x_1^2 + x_2^2 + x_3^2}} \right)^2 + \left(\frac{x_2}{\sqrt{x_1^2 + x_2^2 + x_3^2}} \right)^2 + \left(\frac{x_3}{\sqrt{x_1^2 + x_2^2 + x_3^2}} \right)^2 \right\}^{1/2} = 1.$$

Flow chart the algorithm for computing this normalized measure of orthogonality between two vectors $x_1, x_2, \ldots, x_n$ and $y_1, y_2, \ldots, y_n$.

9 Write a computer program that obtains a derivative from discrete values of a function prescribed at equal intervals of the argument. The method is to consist of fitting a polynomial to an appropriate number of points and evaluating the derivative of the polynomial at the proper value of the argument.

Function values will be supplied along with the corresponding equally-spaced arguments. A value is read in for the argument at which the derivative is to be computed, along with an integer number to indicate the degree of the polynomial to be fitted.

The program will seek a sufficient number of values in the vicinity of the argument through which the interpolating polynomial will be fit. Coefficients will then be found and the derivative evaluated.

To test the method we wish to evaluate the error produced by this method as a function of the spacing of the data and the degree of the polynomial. This is to be done by obtaining approximations to the derivative of $\sin x$ at $x = \pi/6$ and $x = \pi/3$. Data values can be generated by the trigonometric functions of the computer. Compute derivatives for intervals of $\pi/12$, 0.1 and 0.001, and for polynomials of degree 1, 2, 3, 4, and 5. Present the results for the error in graphical form.

chapter six

systems of equations

The solution of simultaneous linear equations is probably the most frequently occurring problem in numerical computation. One class of problems in which sets of simultaneous equations arise is in the analysis of systems having a finite number of degrees of freedom (called lumped-parameter systems) which are in equilibrium or steady-state conditions. These problems occur in the analysis of elastic bodies, of dc and ac networks, of heat transfer in steady state, of chemical equilibrium (usually nonlinear), and of hydraulic networks (usually nonlinear), to name a few applications. Solution of equilibrium problems in continuous systems (bending of beams, heat-conduction) reduces to sets of simultaneous linear equations when the method of finite differences is employed. These will be treated in a later chapter. We have seen, in the section on complex roots (Chapter 4), an example which leads to simultaneous equations, and we shall see another in the method of least squares (Chapter 7).

Most people who have had courses in algebra know a method of solving such systems of equations (at least in principle), but not all can describe unambiguously a procedure for doing so. The description of such algorithms is the principal concern of this chapter.

As an example of how such problems arise from a physical system, consider the dc network shown in Fig. 6.1. The system has a finite number of resistances, and is called a lumped-parameter system. To formulate the mathematical problem, Ohm's law is applied and the sum of the voltages in each circuit is set to zero. This leads to three simultaneous equations, in the three unknown currents, I_1, I_2, and I_3, in terms of a voltage E and a resistance R. Proceeding then to circuit 1,

$$E - RI_1 - 4R(I_1 - I_2) - 2R(I_1 - I_3) = 0;$$

for circuit 2,

$$-3RI_2 - R(I_2 - I_3) - 4R(I_2 - I_1) = 0;$$

and for circuit 3

$$3E - 2R(I_3 - I_1) - R(I_3 - I_2) - 5RI_3 = 0.$$

It is convenient to put the equations in dimensionless form. This can be done by

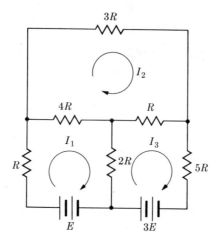

3R

I_2

4R

R

I_1

I_3

R

2R

5R

E

3E

Fig. 6.1

defining a dimensionless current, x_i, as

$$x_i = \frac{I_i}{(E/R)}.$$

Introducing this, the set of equations becomes,

$$7x_1 - 4x_2 - 2x_3 = 1,$$
$$-4x_1 + 8x_2 - x_3 = 0,$$
$$-2x_1 - x_2 + 8x_3 = 3.$$

This linear set of simultaneous equations is typical of the mathematical formula-tion of problems in a variety of systems. In general, these systems of n equations in n unknowns can be represented as follows:

$$\sum_{j=1}^{n} a_{ij}x_j = b_i \quad \text{for} \quad i = 1, 2, 3, \ldots, n.$$

In the expanded form, the equations are written

$$a_{11}x_1 + a_{12}x_2 + a_{13}x_3 + \cdots + a_{1n}x_n = b_1$$
$$a_{21}x_1 + a_{22}x_2 + \cdots \qquad\qquad + a_{2n}x_n = b_2$$
$$\vdots$$
$$a_{n1}x_1 + a_{n2}x_2 + \cdots \qquad\qquad + a_{nn}x_n = b_n.$$

Methods of solving simultaneous linear equations can be grouped under three headings: (1) by determinants, (2) by elimination, and (3) by iterative techniques. Actually, from a computational point of view, the first is not an effective method, and determinants will be discussed only to the extent required to show that they may be computed by the elimination methods.

Properties of systems

In the context above the word "linear" is usually taken to mean that a variable appears to the first power in every term of the function. Thus

$$f(x_1, x_2) = a_1 x_1 + a_2 x_2 \quad \text{and} \quad g(x_1, x_2) = b_1 x_1 + b_2 x_2$$

are linear. Two properties of linear functions that are extensively used in the solution of simultaneous equations are:

$$f(x_1, x_2) + f(x_3, x_4) = f(x_1 + x_3, x_2 + x_4); \quad cf(x_1, x_2) = f(cx_1, cx_2).$$

In the latter, c is an arbitrary constant. The nonlinear function $h(x_1, x_2) = a_1 x_1^2 + a_2 x_1 x_2$ quite obviously does not have either of these properties. In the first property above, the same linear function is applied to two different x-vectors, first to x_1, x_2 and then to x_3, x_4. In simultaneous equations the opposite point of view is taken. There is only one solution vector, but there are several different linear functions (the left-hand parts of the equations which are applied). Since the linear functions are symmetric, that is, both the coefficients and the unknowns appear to the first power in every term, one may define another function, say g, which represents the x's operating on various coefficients:

$$g(a_{11}, a_{12}) = x_1 a_{11} + x_2 a_{12},$$
$$g(a_{21}, a_{22}) = x_1 a_{21} + x_2 a_{22}.$$

The linear operations are:

$$g(a_{11} + a_{21}, a_{12} + a_{22}) = g(a_{11}, a_{12}) + g(a_{21}, a_{22}),$$
$$g(ca_{11}, ca_{12}) = cg(a_{11}, a_{22}).$$

In prose, the first equation states that if one adds the corresponding coefficients of two linear equations before multiplying each by a value of x, the result is the same as if the multiplication by the x's were carried out for each equation separately and then the resultant sums (the right-hand sides) were added. The second equation states that if the coefficients are multiplied by a constant c and then used as multipliers of x_1 and x_2, the results are the same as those obtained by multiplying by x_1, x_2 first and then multiplying the sum (the right-hand side) by the constant c. The repeated application of these two properties is the basis for the elimination methods of solving simultaneous linear equations. "Equations," multiplied by a constant, are added in such a way that a simpler set of equations is produced.

Let us take a simple example to illustrate how these properties are employed to obtain a solution. Consider the set of equations

$$a_{11} x_1 + a_{12} x_2 = b_1,$$
$$a_{21} x_1 + a_{22} x_2 = b_2.$$

Multiplying the first equation by a_{22} and the second by a_{12}, and then subtracting one from the other, we can eliminate x_2. Then, solving for x_1, we obtain

$$x_1 = \frac{b_1 a_{22} - b_2 a_{12}}{a_{11} a_{22} - a_{21} a_{12}}.$$

This suggests that there are certain combinations of coefficients for which no solution exists, or for which an accurate solution might be difficult to obtain. For example, in the case

$$\left. \begin{aligned} 2x_1 - x_2 &= 2 \\ 6x_1 - 3x_2 &= 3 \end{aligned} \right\}$$

no solution is possible, since

$$a_{11} a_{22} - a_{21} a_{22} = 0.$$

These are parallel lines on a graph of x_1 versus x_2 (Fig. 6.2), with no intersection or solution. The set

$$\left. \begin{aligned} 2x_1 - x_2 &= 2 \\ 6x_1 - 3x_2 &= 6 \end{aligned} \right\}$$

has an infinity of solutions, as both numerator and denominator in the expression for x_1 are zero. The lines for the two expressions are coincident. Each of these is a clearcut situation. It is possible to have systems of equations for which an accurate numerical solution is difficult to obtain. Such systems are said to be *ill-conditioned* or *near-singular*. An example of a near-singular system was given in Chapter 1, with a graph of the two equations shown in Fig. 1.1. A well-behaved system of equations in two unknowns is illustrated in Fig. 6.3. Here the lines intersect at a large angle, θ, with the intersection clearly defined, in contrast to the graph for the ill-conditioned system of Fig. 1.1. Referring to the set of equations, and the angles defined in Fig. 6.3, it is evident that

$$\tan \alpha = -\frac{a_{11}}{a_{12}} \quad \text{and} \quad \tan \beta = -\frac{a_{21}}{a_{22}}$$

since

$$\tan \theta = \tan (\alpha - \beta) \quad \text{and} \quad \tan \theta = \frac{a_{21} a_{12} - a_{11} a_{22}}{a_{11} a_{21} + a_{12} a_{22}}.$$

Thus, to ensure that the angle θ be large,

$$|a_{11} a_{22} - a_{21} a_{12}|$$

must be large. Notice that this expression is the denominator in the equation for the calculation of x_1 (or x_2), and is also the determinant of the coefficients of the governing equations.

While it is not easy to extend these arguments to systems having many degrees of freedom, it is sufficient to say here that the same principles apply, and further, that large values of the diagonal elements in the coefficient matrix are associated with well-behaved systems of equations. Further discussion of near-singular sets of equations is presented in the next chapter.

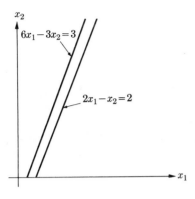

Fig. 6.2

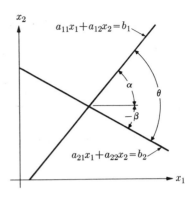

Fig. 6.3

Matrix operations

Before returning to a discussion of the solution of systems of equations, let us examine some properties of matrices, and summarize some elements of matrix algebra, concentrating on topics that will be used in this and later chapters. It is convenient to employ matrix notation and matrix algebra in the discussions that follow.

A matrix is a rectangular array of elements arranged in rows and columns. Using a capital letter to denote a matrix, the element can be located with the first subscript representing the row and the second subscript the column. A matrix A, with m rows and n columns, is said to be an $m \times n$ matrix. It can be written as

$$A = \begin{bmatrix} a_{11} & a_{12} & \cdots & a_{1n} \\ a_{21} & a_{22} & \cdots & a_{2n} \\ \vdots & & & \\ a_{m1} & a_{m2} & \cdots & a_{mn} \end{bmatrix}.$$

Matrices having the same dimensions can be added or subtracted by forming elements of the new matrix as the sum or difference of corresponding elements in the two matrices. To form the sum C of two matrices A and B,

$$A + B = C,$$

elements are added in the form

$$a_{ij} + b_{ij} = c_{ij}.$$

For the difference

$$A - B = C,$$

elements are subtracted

$$a_{ij} - b_{ij} = c_{ij}.$$

Multiplication of matrices,

$$A \times B = C,$$

is described in terms of elements as

$$c_{ij} = \sum_{k=1}^{n} a_{ik}b_{kj}.$$

This requires that there be n columns in matrix A, and n rows in matrix B. If A has the dimensions $m \times n$, and B has the dimensions $n \times b$, the matrix C will have the dimensions $m \times b$.

Here are a few examples: A 2×3 matrix multiplied by a 3×2 matrix produces a 2×2 matrix, as follows:

$$\begin{bmatrix} 2 & 2 & -1 \\ -1 & 2 & 0 \end{bmatrix} \times \begin{bmatrix} 2 & 3 \\ -1 & 0 \\ 2 & 1 \end{bmatrix} = \begin{bmatrix} 0 & 5 \\ -4 & -3 \end{bmatrix}.$$

A row matrix multiplied by a square matrix produces a row matrix:

$$\begin{bmatrix} 1 & 2 & 3 \end{bmatrix} \times \begin{bmatrix} 2 & -1 & 0 \\ 1 & 2 & 1 \\ 1 & 2 & 1 \end{bmatrix} = \begin{bmatrix} 7 & 9 & 5 \end{bmatrix}.$$

A square matrix times a square matrix produces a square matrix:

$$\begin{bmatrix} 3 & 2 \\ -2 & 0 \end{bmatrix} \times \begin{bmatrix} 2 & 3 \\ 2 & -1 \end{bmatrix} = \begin{bmatrix} 10 & 7 \\ -4 & -6 \end{bmatrix};$$

and a row matrix times a column matrix produces a scalar:

$$\begin{bmatrix} 1 & 2 & -3 \end{bmatrix} \times \begin{bmatrix} 2 \\ -1 \\ -2 \end{bmatrix} = 6.$$

A matrix having a single row or column is called a *vector* or *column vector*.

It can be shown that matrix products are associative,

$$(AB)C = A(BC),$$

but, in general, are not commutative; that is,

$$AB \neq BA.$$

A *symmetric matrix* is a matrix in which the elements satisfy the condition

$$a_{ij} = a_{ji}.$$

If $\hat{a}_{ij}$ is the *complex conjugate* of a_{ij}, then the matrix $\hat{A}$, formed by replacing the elements a_{ij} of matrix A with the elements $\hat{a}_{ij}$, is the conjugate of A. The *transpose* of matrix A is designated A^T, and is formed by replacing the a_{ij} with elements b_{ij} such that

$$b_{ji} = a_{ij}.$$

To demonstrate these last two definitions, if

$$A = \begin{bmatrix} 1 + i & -2 - i \\ + i & 2 - 2i \end{bmatrix},$$

then

$$\hat{A} = \begin{bmatrix} 1 - i & -2 + i \\ -i & 2 + 2i \end{bmatrix} \quad \text{and} \quad A^T = \begin{bmatrix} 1 + i & i \\ -2 - i & 2 - 2i \end{bmatrix}.$$

A square matrix is *triangular* if all elements above or below the main diagonal are zero. Matrices can be *upper* triangular or *lower* triangular. The matrix

$$\begin{bmatrix} 1 & 2 & 4 \\ 0 & 2 & 3 \\ 0 & 0 & -1 \end{bmatrix}$$

is upper triangular.

A square matrix having all elements equal to zero except those along the main diagonal is called a *scalar* matrix or a *diagonal* matrix. When all the elements of the main diagonal of a scalar matrix are unity, it is called a *unit* or *identity* matrix. It will be denoted by the symbol I.

For a square matrix A, having a nonvanishing determinant, there is a matrix A^{-1} such that

$$AA^{-1} = I.$$

This new matrix A^{-1} is called the *inverse* of A. We shall examine it more closely later in this chapter. Forming the inverse is equivalent in matrix algebra to division. It is evident that

$$AA^{-1} = A^{-1}A = I.$$

Several comments should be made regarding the special matrices called vectors. If $\bar{x}$ and $\bar{y}$ are vectors,

$$\bar{x} = [x_1, x_2] \quad \text{and} \quad \bar{y} = [y_1, y_2],$$

then their scalar product is

$$\bar{x}^T \bar{y} = [x_1 \ x_2] \times \begin{bmatrix} y_1 \\ y_2 \end{bmatrix} = x_1 y_1 + x_2 y_2.$$

For the square matrix

$$B = \begin{bmatrix} 2 & 1 \\ 1 & 3 \end{bmatrix},$$

the product

$$\bar{x}^T B \bar{y} = 2x_1 y_1 + x_1 y_2 + x_2 y_1 + 3x_2 y_2.$$

When B is *symmetric* as it is here, then

$$\bar{x}^T B \bar{y} = \bar{y}^T B \bar{x}.$$

If the product

$$\bar{x}^T B \bar{y} = 0,$$

the vectors $\bar{x}$ and $\bar{y}$ are said to be *orthogonal* with respect to B. If $\bar{x}^T B \bar{x} = 1$, the vector $\bar{x}$ is said to be normalized with respect to B.

Referring again to the system of linear equations under consideration, in matrix notation it is simply

$$A\bar{x} = \bar{b}.$$

Written out, the matrix equation is:

$$
\begin{bmatrix}
a_{11} & a_{12} & a_{13} & \cdots & a_{1n} \\
a_{21} & a_{22} & \cdots & \cdots & a_{2n} \\
\vdots & & & & \\
a_{n1} & a_{n2} & & \cdots & a_{nn}
\end{bmatrix}
\begin{bmatrix}
x_1 \\
x_2 \\
\vdots \\
x_n
\end{bmatrix}
=
\begin{bmatrix}
b_1 \\
b_2 \\
\vdots \\
b_n
\end{bmatrix}.
$$

The coefficients when written as a two-dimensional array without the variable names form the *coefficient matrix*. Its dimensions are $n \times n$. The right-hand coefficients are regarded as a vector, as are the variables. When the right-hand sides are appended on the right of the coefficient matrix, the combined array is called the *augmented matrix:*

$$
\begin{bmatrix}
a_{11} & a_{12} & a_{13} & \cdots & a_{1n} & b_1 \\
a_{21} & a_{22} & \cdots & \cdots & a_{2n} & b_2 \\
\vdots & & & & & \\
a_{n1} & a_{n2} & & \cdots & a_{nn} & b_n
\end{bmatrix}.
$$

$$\underbrace{}_{\substack{\text{Matrix of} \\ \text{coefficients}}} \quad \underbrace{}_{\substack{\text{Vector of} \\ \text{right-hand sides}}}$$

In computing it is often desirable to treat the augmented matrix as a single array having more columns than rows. The right-hand sides are designated as elements of the same matrix rather than by a separate name. With such notation, the n equations are

$$\sum_{j=1}^{n} a_{ij} x_j = a_{i,n+1}, \qquad i = 1, 2, \ldots, n,$$

and the augmented matrix is written

$$\begin{bmatrix} a_{11} & a_{12} & \cdots & a_{1,n+1} \\ a_{21} & a_{22} & \cdots & a_{2,n+1} \\ \vdots & & & \\ a_{n1} & a_{n2} & \cdots & a_{n,n+1} \end{bmatrix}.$$

It is possible to append more than one vector on the right-hand side, in which case the index of the right-hand column would be greater than $n + 1$.

Elimination or reduction

To illustrate an elimination (or reduction) procedure, we shall use the following steps to solve the set of equations given below.

$$2x_1 - 7x_2 + 4x_3 = 9$$
$$x_1 + 9x_2 - 6x_3 = 1$$
$$-3x_1 + 8x_2 + 5x_3 = 6$$

The ordinary (human) approach is to look for simple relations between coefficients. One could observe, for instance, that multiplying the second equation by 2 would yield $2x_1$ as the leading term, and therefore subtracting the first equation from the multiplied second equation would eliminate x_1 from the second equation:

$$2x_1 - 7x_2 + 4x_3 = 9$$
$$25x_2 - 16x_3 = -7$$
$$-3x_1 + 8x_2 + 5x_3 = 6$$

Only the coefficients and right-hand sides entered into this elimination step; the variable names did not. Hence only the augmented matrix need be written. The step could be written

$$\begin{bmatrix} 2 & -7 & 4 & 9 \\ 1 & 9 & -6 & 1 \\ -3 & 8 & 5 & 6 \end{bmatrix} \rightarrow \begin{array}{c} 2 \times \text{row } 2 \\ - \\ 1 \times \text{row } 1 \\ \\ = \\ \text{New row } 2 \end{array} \rightarrow \begin{bmatrix} 2 & -7 & 4 & 9 \\ 0 & 25 & -16 & -7 \\ -3 & 8 & 5 & 6 \end{bmatrix}$$

As a next step, multiplying the first row by 4 and adding the second would eliminate

x_3 from the first equation, i.e., make the coefficient of x_3 zero:

$$\begin{bmatrix} 2 & -7 & 4 & 9 \\ 0 & 25 & -16 & -7 \\ -3 & 8 & 5 & 6 \end{bmatrix} \rightarrow \begin{matrix} 4 \times \text{row 1} \\ + \\ 1 \times \text{row 2} \\ = \\ \text{New row 1} \end{matrix} \rightarrow \begin{bmatrix} 8 & -3 & 0 & 29 \\ 0 & 25 & -16 & -7 \\ -3 & 8 & 5 & 6 \end{bmatrix}$$

Two more steps permit the solution for one of the variables, after which substitution in the other equations determines the remaining two.

$$\begin{bmatrix} 8 & -3 & 0 & 29 \\ 0 & 25 & -16 & -7 \\ -3 & 8 & 5 & 6 \end{bmatrix} \rightarrow \begin{matrix} 8 \times \text{row 3} \\ + \\ 3 \times \text{row 1} \\ = \\ \text{New row 3} \end{matrix} \rightarrow \begin{bmatrix} 8 & -3 & 0 & 29 \\ 0 & 25 & -16 & -7 \\ 0 & 55 & 40 & 135 \end{bmatrix}$$

$$\begin{bmatrix} 8 & -3 & 0 & 29 \\ 0 & 25 & -16 & 7 \\ 0 & 55 & 40 & 135 \end{bmatrix} \rightarrow \begin{matrix} 2 \times \text{row 3} \\ + \\ 5 \times \text{row 2} \\ = \\ \text{New row 3} \end{matrix} \rightarrow \begin{bmatrix} 8 & -3 & 0 & 29 \\ 0 & 25 & -16 & -7 \\ 0 & 235 & 0 & 235 \end{bmatrix}$$

The result is:

$$235x_2 = 235 \qquad \therefore \quad x_2 = 1$$
$$8x_1 - 3x_2 = 29 \qquad \therefore \quad x_1 = 4$$
$$25x_2 - 16x_3 = -7 \qquad \therefore \quad x_3 = 2$$

The procedure is then to add and subtract the proper multiples of rows until the equations are reduced to a simpler form that is readily solved. The simplest form is the system of independent equations:

$$a_{11}x_1 = b_1$$
$$a_{22}x_2 = b_2$$
$$a_{33}x_3 = b_3$$

However, a system in which one equation involves only one variable, another two variables, another three, etc., is also readily solved by systematic substitution (or *back solution* as this procedure is often called):

$$a_{11}x_1 = b_1, \qquad a_{21}x_1 + a_{22}x_2 = b_2, \qquad a_{31}x_1 + a_{32}x_2 + a_{33}x_3 = b_3.$$

The coefficient matrices corresponding to these two readily solvable cases are *diagonal* and *triangular*, respectively. In the diagonal matrix the equations in matrix form are

$$\begin{bmatrix} a_{11} & 0 & 0 \\ 0 & a_{22} & 0 \\ 0 & 0 & a_{33} \end{bmatrix} \begin{bmatrix} x_1 \\ x_2 \\ x_3 \end{bmatrix} = \begin{bmatrix} b_1 \\ b_2 \\ b_3 \end{bmatrix}.$$

In the triangular form the matrix equation is

$$\begin{bmatrix} a_{11} & 0 & 0 \\ a_{21} & a_{22} & 0 \\ a_{31} & a_{32} & a_{33} \end{bmatrix} \begin{bmatrix} x_1 \\ x_2 \\ x_3 \end{bmatrix} = \begin{bmatrix} b_1 \\ b_2 \\ b_3 \end{bmatrix}.$$

The coefficient matrix is said to be in lower triangular form.

Assuming for the moment that the elimination should proceed until the independent equations are obtained, one can describe the process of solving a system of equations as follows. Given an augmented matrix, add (or subtract) multiples of the other rows to (or from) a multiple of a particular row until the coefficient matrix has been reduced to diagonal form. What is now needed is a systematic description of the order in which the rows are modified. In the example, the order was determined by considerations of arithmetic simplicity. Elements were set to zero when the relations between the coefficients made such a step convenient. If a computer is to do the computation, such considerations are not important, but what is needed is a systematic sequence of steps to accomplish the diagonalization.

Assuming that $a_{11} \neq 0$, one may use multiples of the first row to set to zero the other elements in the first column. The row selected to simplify the other rows is called the *pivot row;* the element in the pivot row which is in the column to be set to zero is called the *pivot element.*

$$\begin{bmatrix} a_{11} & a_{12} & a_{13} & a_{14} \\ a_{21} & a_{22} & a_{23} & a_{24} \\ a_{31} & a_{32} & a_{33} & a_{34} \end{bmatrix} \rightarrow \begin{matrix} a_{11} \text{ row } 2 \\ - \\ a_{21} \text{ row } 1 \\ = \\ \text{New row 2} \end{matrix} \rightarrow \begin{bmatrix} a_{11} & a_{12} & a_{13} & a_{14} \\ 0 & a'_{22} & a'_{23} & a'_{24} \\ a_{31} & a_{32} & a_{33} & a_{34} \end{bmatrix} \rightarrow \begin{matrix} a_{11} \text{ row } 3 \\ - \\ a_{31} \text{ row } 1 \\ = \\ \text{New row 3} \end{matrix}$$

$$\rightarrow \begin{bmatrix} a_{11} & a_{12} & a_{13} & a_{14} \\ 0 & a'_{22} & a'_{23} & a'_{24} \\ 0 & a'_{32} & a'_{33} & a'_{34} \end{bmatrix}$$

In this sequence of steps, *row 1 was a pivot row* (a_{11} the pivot element) and *rows 2 and 3 were reduced*. The same steps for the previously considered numerical example are:

$$
\begin{bmatrix} 2 & -7 & 4 & 9 \\ 1 & 9 & -6 & 1 \\ -3 & 8 & 5 & 6 \end{bmatrix} \rightarrow \begin{bmatrix} 2 & -7 & 4 & 9 \\ 0 & 25 & -16 & -7 \\ -3 & 8 & 5 & 6 \end{bmatrix}
$$

$$
\rightarrow \begin{bmatrix} 2 & -7 & 4 & 9 \\ 0 & 25 & -16 & -7 \\ 0 & -5 & 22 & 39 \end{bmatrix}
$$

One can continue the process, using multiples of row 2 to alter the other rows so that the other elements in column 2 are set to zero. Note that row 3 could be used for the same purpose, but not row 1, since adding multiples of it to the other rows would destroy the zeros already created.

$$
\begin{bmatrix} a_{11} & a_{12} & a_{13} & a_{14} \\ 0 & a'_{22} & a'_{23} & a'_{24} \\ 0 & a'_{32} & a'_{33} & a'_{34} \end{bmatrix} \begin{matrix} a'_{22} \text{ row 1} \\ - \\ a_{12} \text{ row 2} \\ = \\ \text{New row 1} \end{matrix} \rightarrow \begin{bmatrix} a'_{11} & 0 & a'_{13} & a'_{14} \\ 0 & a'_{22} & a'_{23} & a'_{24} \\ 0 & a'_{32} & a'_{33} & a'_{34} \end{bmatrix} \begin{matrix} a'_{22} \text{ row 3} \\ - \\ a'_{32} \text{ row 2} \\ = \\ \text{New row 3} \end{matrix}
$$

$$
\rightarrow \begin{bmatrix} a'_{11} & 0 & a'_{13} & a'_{14} \\ 0 & a'_{22} & a'_{23} & a'_{24} \\ 0 & 0 & a''_{33} & a''_{34} \end{bmatrix}
$$

Row 2 was a pivot row, and *rows 1 and 3 were reduced*. Note that, except for the diagonal elements, the computation did not need to be carried out on the elements to the left of the pivot column (column 2), since the results are zero. In fact, the computations of elements *in* the pivot column can also be omitted since the method is designed to make these zero. The same steps for the example are:

$$
\begin{bmatrix} 2 & -7 & 4 & 9 \\ 0 & 25 & -16 & -7 \\ 0 & -5 & 22 & 39 \end{bmatrix} \rightarrow \begin{bmatrix} 50 & 0 & -12 & 176 \\ 0 & 25 & -16 & -7 \\ 0 & -5 & 22 & 39 \end{bmatrix}
$$

$$
\rightarrow \begin{bmatrix} 50 & 0 & -12 & 176 \\ 0 & 25 & -16 & -7 \\ 0 & 0 & 470 & 940 \end{bmatrix}
$$

There is one independent equation at this point ($470x_3 = 940$), but one more cycle is required to obtain the diagonal matrix.

At this point, only row 3 may be used as a pivot since the use of any other would destroy existing zeros.

$$
\begin{bmatrix}
a'_{11} & 0 & a'_{13} & a'_{14} \\
0 & a'_{22} & a'_{23} & a'_{24} \\
0 & 0 & a''_{33} & a''_{34}
\end{bmatrix}
\begin{array}{l} a''_{33} \text{ row 1} \\ \overline{} \\ a'_{13} \text{ row 3} \\ \overline{\overline{}} \\ \text{New row 1} \end{array}
\rightarrow
\begin{bmatrix}
a''_{11} & 0 & 0 & a''_{14} \\
0 & a'_{22} & a'_{23} & a'_{24} \\
0 & 0 & a''_{33} & a''_{34}
\end{bmatrix}
\begin{array}{l} a''_{33} \text{ row 2} \\ \overline{} \\ a'_{23} \text{ row 3} \\ \overline{\overline{}} \\ \text{New row 2} \end{array}
$$

$$
\rightarrow
\begin{bmatrix}
a''_{11} & 0 & 0 & a''_{14} \\
0 & a''_{22} & 0 & a''_{24} \\
0 & 0 & a''_{33} & a''_{34}
\end{bmatrix}
$$

Row 3 was used as a pivot row to reduce rows 1 and 2. The matrix of coefficients is now in the desired form. The analogous steps bring the example to diagonal form also.

$$
\begin{bmatrix}
50 & 0 & -12 & 176 \\
0 & 25 & -16 & -7 \\
0 & 0 & 470 & 940
\end{bmatrix}
\rightarrow
\begin{bmatrix}
23500 & 0 & 0 & 94000 \\
0 & 25 & -16 & -7 \\
0 & 0 & 470 & 940
\end{bmatrix}
\rightarrow
\begin{bmatrix}
23500 & 0 & 0 & 94000 \\
0 & 11750 & 0 & 11750 \\
0 & 0 & 470 & 940
\end{bmatrix}
$$

From the above, one obtains

$$
\begin{aligned}
23500x_1 &= 94000 & \therefore \quad x_1 &= 4 \\
11750x_2 &= 11750 & \therefore \quad x_2 &= 1 \\
470x_3 &= 940 & \therefore \quad x_3 &= 2
\end{aligned}
$$

In this process three indices are employed. One index designates the pivot row (k), another the row being reduced (i), and the third determines the specific element being changed within the row being reduced (j). To develop a formula for the reduction, it may be helpful to write out a specific row reduction and then identify the elements with the subscripts i, j, k. With the first row used as a pivot ($k = 1$), the modification of the third row ($i = 3$) to set $a_{31} = 0$ is:

$$
\text{New row 3} = (a_{11} \text{ row 3} - a_{31} \text{ row 1}) =
\begin{cases}
a'_{31} = a_{11}a_{31} - a_{31}a_{11} = 0 \\
a'_{32} = a_{11}a_{32} - a_{31}a_{12} \\
a'_{33} = a_{11}a_{33} - a_{31}a_{13} \\
a'_{34} = a_{11}a_{34} - a_{31}a_{14}
\end{cases}
$$

The formulas on the right can be replaced by

$$
a'_{3j} = a_{11}a_{3j} - a_{31}a_{1j} \qquad \text{for} \qquad j = 1, 2, 3, 4.
$$

By observing that the pivot row (row 1) used in this specific case was also employed in the reduction of row 2, the formula can be generalized to include this reduction also:

$$a'_{ij} = a_{11}a_{ij} - a_{i1}a_{1j} \quad \text{for} \quad i = 2, j = 1, 2, 3, 4, \quad \text{and}$$
$$\text{for} \quad i = 3, j = 1, 2, 3, 4.$$

Since every row was used once as a pivot row, the inclusion of the pivot index k permits the statement of the entire reduction by a single formula, with some accompanying statements to quantify the indices:

$$a'_{ij} = a_{kk}a_{ij} - a_{ik}a_{kj}$$

The ranges of i, j, k are

$$k = 1, 2, 3$$

and for each value of k, $i = 1, 2, 3$ (excluding $i = k$) and for each value of i, $j = 1, 2, 3, 4$. This method of reduction, which is called the Gauss-Jordan method, produces a diagonal matrix and can be extended to n equations by simply replacing the 3 by n. In such highly iterative procedures, it is useful to introduce a little more formalism in the description of the algorithms so that such statements as ". . . and for each value of . . ." need not be included. In addition, the occasionally confusing mathematical practice of using the same index in a variety of independent expressions can be clarified by such notation. The convention to be used in algorithms of this repetitive type is that a quantifying statement (for example, $i = 1, 2, \ldots, n$) applies to all the statements below it that are indented to the right. In this form, the Gauss-Jordan algorithm is:

$$k = 1, 2, \ldots, n$$
$$i = 1, 2, \ldots, n; \quad i \neq k$$
$$j = 1, 2, \ldots, n + 1$$
$$a'_{ij} = a_{kk}a_{ij} - a_{ik}a_{kj}$$

The method described has some shortcomings, but the discussion of these will be postponed until this algorithm has been presented as a computer program. A fairly subtle problem is encountered in translating these steps into a computing procedure. As has been shown, the alteration of the ith row, using pivot row k, is written

$$a'_{ij} = a_{kk}a_{ij} - a_{ik}a_{kj}, \quad j = 1, 2, 3, \ldots, n + 1.$$

The prime is used to designate that a new array distinct from the original is produced, and in this mathematical statement questions about the dynamics of computation, such as, Does a'_{ij} replace a_{ij}?, simply do not arise. Since in the computer formulation only one array is involved, the formula, in effect, is

$$a_{ij} \leftarrow a_{kk}a_{ij} - a_{ik}a_{kj}, \quad j = 1, 2, \ldots, n + 1.$$

The reduction of row i causes all the elements in that row to be given new values, and, once given a new value, the old value is no longer available. The element

a_{ij} is used just once and is immediately replaced by its new value. However, a_{ik} is used in the computation of every element in the row (i.e., it does not have a j-index), and at some point ($j = k$) it will be given a new value. In the remaining computation in the row, the new value of a_{ik} will be used in the formula, and this was not intended! There are several ways of overcoming this difficulty. One solution is to postpone the computation of a_{ik} until the last step of a row; another is to save the value of a_{ik} before the row reduction commences. A better solution, because it reduces the amount of computation as well, is to compute only those elements which are used in the later stages of reduction. This selective computation is illustrated by the following partially reduced matrix with five rows:

Pivot column

$$\downarrow$$

$$
\begin{matrix}
a_{11} & 0 & 0 & a_{14} & a_{15} \\
0 & a_{22} & 0 & a_{24} & a_{25} \\
0 & 0 & a_{33} & a_{34} & a_{35} \\
\text{Pivot row} \rightarrow 0 & 0 & 0 & a_{44} & a_{45} \\
0 & 0 & 0 & a_{45} & a_{55}
\end{matrix}
$$

Row 4 is the pivot row. To see the problem described above, consider the left-to-right reduction of row 2. The new elements are:

$$(a_{44} \cdot 0 - a_{24} \cdot 0) \quad (a_{44} \cdot a_{22} - a_{24} \cdot 0) \quad (a_{44} \cdot 0 - a_{24} \cdot 0)$$
$$a_{21} \leftarrow 0 \qquad\qquad a_{22} \leftarrow a_{44} \cdot a_{22} \qquad\qquad a_{23} \leftarrow 0$$

$$(a_{44} \cdot a_{24} - a_{24} \cdot a_{kk}) \quad (a_{44} \cdot a_{25} - 0 \cdot a_{45})$$
$$a_{24} \leftarrow 0 \qquad\qquad \text{The zero above should be the original } a_{24} \text{ element!}$$

Note that the correct values would be obtained by proceeding from right to left, since to the left of the pivot column the multiplier of a_{24} is zero anyway. However, with the exception of the diagonal element (a_{22} in row 2), every computation to the left of and including the pivot column results in a zero and need not be carried out. In other words, compute from left to right starting at column $k + 1$. To handle the diagonal elements to the left of the kth column (this refers only to those rows preceding the pivot row, that is, $i < k$) the additional step,

$$a_{ii} \leftarrow a_{kk} a_{ii} \qquad \text{for} \qquad i < k$$

must be included. With these revisions, the process is:

$$k = 1, 2, \ldots, n$$
$$i = 1, 2, \ldots, n; \quad i \neq k$$
$$\text{whenever} \quad i < k, \quad a'_{ii} = a_{kk} a_{ii}$$
$$j = k + 1, \quad k + 2, \ldots, n + 1$$
$$a'_{ij} = a_{kk} a_{ij} - a_{ik} a_{kj}.$$

This description is almost directly translatable into a program, but the flow chart is helpful in understanding the nested iterations.

Example Write a subroutine which solves by the Gauss-Jordan method a system of n simultaneous linear equations. The solutions obtained are to replace the right-hand sides of the given augmented matrix, a. The calling sequence is SLEQ(N,M,A), where A is the name of the augmented matrix and N the number of equations. A dummy variable, $M = N + 1$, has been carried to facilitate a flexible dimension statement. Assume that the pivot elements are nonzero. The flow chart and program are given in Fig. 6.4. The last iteration divides the right-hand side by the diagonal element to produce the solutions and replaces the right side with these values. The conditional skipping of the scope of an iteration when $i = k$ illustrates the need for the CONTINUE statement. It is necessary to skip the scope of the iteration but not leave the iteration; i.e., the index i must be incremented from its present value and the computation continued.

Results have been computed for the following set of equations:

$$x_1 + 2x_2 - 2x_3 = -3,$$
$$2x_1 - 4x_2 + 4x_3 = 0,$$
$$8x_1 - 6x_2 + 2x_3 = 4.$$

This routine has two serious deficiencies:

1. Since the elements are computed by multiplication, the numbers increase in magnitude quite rapidly if the original elements are greater than unity. Correspondingly, the computed elements diminish rapidly if the original elements are less than unity. The numerical example illustrates this growth. The numbers in the augmented matrix are all integers less than ten and yet, in spite of the fact that there are only three equations, the final diagonal elements are as large as 23,500. With such growth it is not uncommon for systems of as few as six equations to exceed the floating-point capacity of the machine in the reduction. What is needed is some normalizing step in the process to scale the elements to reasonable magnitudes.

2. The assumption that the pivot element must be nonzero leads to the second deficiency. Certainly it is possible to have zero coefficients in the given set of equations, and although the pivot elements after the first reduction are not the given coefficients, it is conceivable that some of these could be zero and yet have a solvable system of equations. The solution to this problem is fairly obvious, but its exact statement is fairly complex. The procedure is to rearrange the remaining rows and columns of a partially reduced matrix to guarantee that the pivot element will be nonzero. If such an arrangement is not possible, the system does not have a unique solution. Both shortcomings will be remedied in subsequent programs.

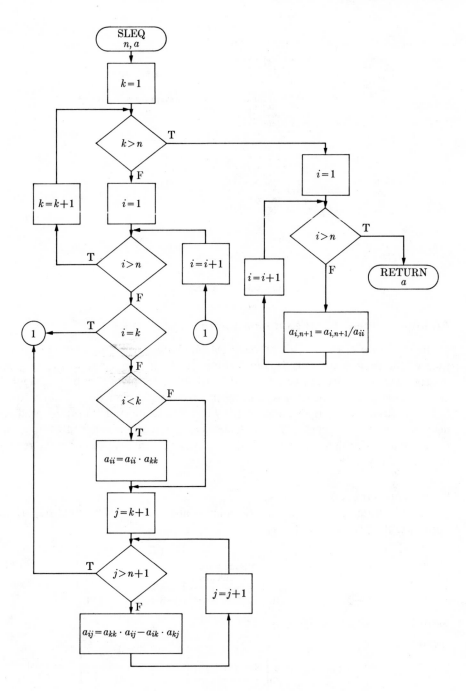

Fig. 6.4 Gauss-Jordan subroutine. (a) Flow chart.

```
        DIMENSION A(3,4)
        DATA A/1.,2.,8.,2.,-4.,-6.,-2.,4.,2.,-3.,0.,4./
        CALL SLEC(3,4,A)
        WRITE(6,15)  A(1,4),A(2,4),A(3,4)
15      FORMAT(5H  X1=F10.5,5H  X2=F10.5,5H  X3=F10.5)
        END

C       GAUSS-JORDAN REDUCTION OF SIMULTANEOUS EQUATIONS
        SUBROUTINE SLEC(N,M,A)
        DIMENSION A(N,M)
C       M IS A DUMMY VARIABLE.  M=N+1
        DO 10 K=1,N
        DO 10 I=1,N
        IF (I.EQ.K) GO TO 10
        IF (I.LT.K)  A(I,I)=A(I,I)*A(K,K)
        KP1=K+1
        DO 5 J=KP1,M
5       A(I,J)=A(K,K)*A(I,J)-A(I,K)*A(K,J)
10      CONTINUE
        DO 15 I=1,N
15      A(I,N+1)=A(I,N+1)/A(I,I)
        RETURN
        END

          OUTPUT

     X1= -1.50000   X2=  -3.62500   X3=  -2.87500
```

Fig. 6.4 (b) Program.

An improvement for the elimination method

In the reduction of the sample problem by the Gauss-Jordan method, the magnitude of the matrix elements becomes extremely large compared to the original matrix. This difficulty can be alleviated by dividing the pivot row by its diagonal element (the pivot element) before using this row in the reduction of other rows. Stated another way, the elements of the kth row are divided by a_{kk} to produce new elements a'_{kj}. A new diagonal element, $a'_{kk} = 1$, is a result of this division. (It is still assumed that $a_{kk} \neq 0$.) With the pivot element always unity, the reduction formula is

$$a'_{ij} = a_{ij} - a_{ik}a_{kj}$$

for the same ranges of i, j, and k.

To illustrate, consider again the numerical example. The first step is to normalize the pivot element a_{11}:

$$\begin{bmatrix} 2 & -7 & 4 & 9 \\ 1 & 9 & -6 & 1 \\ -3 & 8 & 5 & 6 \end{bmatrix} \xrightarrow[\text{pivot element}]{\text{Normalize the}} \begin{bmatrix} 1 & -\frac{7}{2} & 2 & \frac{9}{2} \\ 1 & 9 & -6 & 1 \\ -3 & 8 & 5 & 6 \end{bmatrix}$$

Next the first column is reduced:

$$\begin{bmatrix} 1 & -\frac{7}{2} & 2 & \frac{9}{2} \\ 0 & \frac{25}{2} & -8 & \frac{7}{2} \\ 0 & -\frac{5}{2} & 11 & \frac{39}{2} \end{bmatrix} \xrightarrow{\text{Normalize } a_{22}} \begin{bmatrix} 1 & -\frac{7}{2} & 2 & \frac{9}{2} \\ 0 & 1 & -\frac{16}{25} & -\frac{7}{25} \\ 0 & -\frac{5}{2} & 11 & \frac{39}{2} \end{bmatrix}$$

Reduce the second column:

$$\begin{bmatrix} 1 & 0 & -\frac{6}{25} & \frac{76}{25} \\ 0 & 1 & -\frac{16}{25} & -\frac{7}{25} \\ 0 & 0 & \frac{47}{5} & \frac{94}{5} \end{bmatrix} \xrightarrow{\text{Normalize } a_{33}} \begin{bmatrix} 1 & 0 & -\frac{6}{25} & \frac{76}{25} \\ 0 & 1 & -\frac{16}{25} & -\frac{7}{25} \\ 0 & 0 & 1 & 2 \end{bmatrix}$$

and reduce the third column:

$$\begin{bmatrix} 1 & 0 & 0 & 4 \\ 0 & 1 & 0 & 1 \\ 0 & 0 & 1 & 2 \end{bmatrix} \qquad \begin{array}{l} \therefore \; x_1 = 4 \\ \therefore \; x_2 = 1 \\ \therefore \; x_1 = 2 \end{array}$$

Normalizing the pivot element has resulted in control of the size of the matrix elements, and has produced the values of the x's directly by transforming the coefficient matrix to a unit matrix.

Determinants

Every square matrix has associated with it a number called a determinant, given as

$$A = \begin{vmatrix} a_{11} & a_{12} & \cdots & a_{1n} \\ a_{21} & a_{22} & \cdots & a_{2n} \\ \vdots & & & \vdots \\ a_{n1} & a_{n2} & \cdots & a_{nn} \end{vmatrix}.$$

It is assumed that the student is familiar with the evaluation of the determinant for small n.

The solution of simultaneous linear equations by determinants, or Cramer's rule, is a useful theoretical tool and a practical procedure when the number of equations is small (say, four or less). But determinants do not provide an effective computing procedure when n is large, because the number of arithmetic steps becomes excessively large. Using the number of multiplications as a measure of efficiency of the solution procedure, one finds that solving n equations by determinants requires approximately $2(n + 1)!$ multiplications, while the Gauss-Jordan method previously described requires approximately n^3 multiplications. (Specifically, the method is accomplished with $[n(n + 1)(n - 1) + n(n - 1)/2]$ multiplications; the scaling modification alluded to would reduce this number to $[n(n + 1)(n - 1)/2 + n(n - 1)]$.) For $n = 10$, comparison shows that the order of magnitude is 70,000,000 multiplications for determinants versus 1000 for elimination. Nonetheless, it is important to be able to calculate the value of determinants. One of the most important reasons for being able to compute such values is that if the determinant of a system is zero (or, in a computational problem, is very small) the equations are not independent. Perhaps before one examines some specific properties of determinants, a review of Cramer's rule is in order.

Given n simultaneous linear equations in n unknowns,

$$\sum_{j=1}^{n} a_{ij}x_j = b_i, \qquad i = 1, 2, \ldots, n,$$

the determinant of the coefficients is

$$\Delta = \begin{vmatrix} a_{11} & a_{12} & \cdots & a_{1n} \\ a_{21} & a_{22} & \cdots & a_{2n} \\ \vdots & & & \vdots \\ a_{n1} & a_{n2} & & a_{nn} \end{vmatrix}.$$

If $\Delta \neq 0$, the solution is

$$x_1 = \frac{\begin{vmatrix} b_1 & a_{12} & \cdots & a_{1n} \\ b_2 & a_{22} & \cdots & a_{2n} \\ \vdots & & & \vdots \\ b_n & a_{n2} & \cdots & a_{nn} \end{vmatrix}}{\Delta},$$

$$x_2 = \frac{\begin{vmatrix} a_{11} & b_1 & \cdots & a_{1n} \\ a_{21} & b_2 & \cdots & a_{2n} \\ \vdots & & & \vdots \\ a_{n1} & b_n & \cdots & a_{nn} \end{vmatrix}}{\Delta},$$

$$\vdots \qquad x_n = \frac{\begin{vmatrix} a_{11} & a_{12} & \cdots & b_1 \\ a_{21} & a_{22} & \cdots & b_2 \\ \vdots & & & \vdots \\ a_{n1} & a_{n2} & \cdots & b_n \end{vmatrix}}{\Delta}.$$

To evaluate such determinants by the methods of elimination, one must understand the evaluation of determinants by minors and determine the effect that the linear operations on the rows of the determinant have on the determinant's value.

Evaluation by Minors

This technique is developed here by induction. The determinant of a single number is itself,

$$|a_{11}| = a_{11}.$$

The rule for computing a second-order determinant can be found by solving the system

$$a_{11}x_1 + a_{12}x_2 = b_1, \qquad a_{21}x_1 + a_{22}x_2 = b_2,$$

for x_1 by substitution and identifying the resultant denominator as the determinant:

$$x_1 = \frac{b_1a_{22} - b_2a_{12}}{a_{11}a_{22} - a_{12}a_{21}}.$$

The denominator is obtained by the familiar "criss-cross" rule:

$$\begin{vmatrix} a_{11} & a_{12} \\ a_{21} & a_{22} \end{vmatrix} = a_{11}a_{22} - a_{12}a_{21} = a_{11}|a_{22}| - a_{12}|a_{21}|.$$

In the rightmost form, the determinant is expanded by taking the element and multiplying it by the determinant that results when the row and column containing the first row element are deleted. The signs of the terms alternate.

The same method can be used to compute a third-order determinant:

$$\begin{vmatrix} a_{11} & a_{12} & a_{13} \\ a_{21} & a_{22} & a_{23} \\ a_{31} & a_{32} & a_{33} \end{vmatrix} = a_{11}a_{22}a_{33} + a_{31}a_{12}a_{23} + a_{13}a_{21}a_{32}$$
$$- a_{13}a_{22}a_{31} - a_{23}a_{32}a_{11} - a_{33}a_{21}a_{12}$$

$$= a_{11}\begin{vmatrix} a_{22} & a_{23} \\ a_{32} & a_{33} \end{vmatrix} - a_{12}\begin{vmatrix} a_{21} & a_{23} \\ a_{31} & a_{33} \end{vmatrix} + a_{13}\begin{vmatrix} a_{21} & a_{22} \\ a_{31} & a_{33} \end{vmatrix}$$

The same description applies. The elements of a given row are multiplied by the determinants obtained by deleting the row and column which contain the element. This subdeterminant is called a *minor*, and when the sign is attached to the determinant, it is called a *cofactor*. Using the latter form one can write the third-order determinant:

$$\begin{vmatrix} a_{11} & a_{12} & a_{13} \\ a_{21} & a_{22} & a_{23} \\ a_{31} & a_{32} & a_{33} \end{vmatrix} = a_{11}\Delta_{11} + a_{12}\Delta_{12} + a_{13}\Delta_{13}$$

where

$$\Delta_{ij} = (-1)^{i+j} \times \begin{vmatrix} a_{11} & \cdots & a_{1,j-1}, & a_{1,j+1} & \cdots & a_{1n} \\ \vdots & & & & & \vdots \\ a_{i-1,1} & & & & & \\ a_{i+1,1} & & & & & \\ \vdots & & & & & \\ a_{n,1} & & & & \cdots & a_{nn} \end{vmatrix}$$

The ith row and the jth column are deleted from the determinant to form the cofactor of a_{ij}. If the quantity $i + j$ is even, the cofactor is positive; otherwise it is negative. In general, then,

$$\Delta = \sum_{j=1}^{n} a_{ij}\Delta_{ij}.$$

Notice that if elements in either row i or column j are zero, the determinant is zero. As i and j are arbitrary, we conclude that a square matrix in which all elements in a column or row are zero has a determinant of zero.

Each determinant Δ_{ij} can, in turn, be expanded in terms of cofactors. This expansion process can continue recursively until the cofactors are single elements. To illustrate numerically, the now familiar third-order matrix is used.

$$\begin{vmatrix} 2 & -7 & 4 \\ 1 & 9 & -6 \\ -3 & 8 & 5 \end{vmatrix} = 2 \begin{vmatrix} 9 & -6 \\ 8 & 5 \end{vmatrix} + 7 \begin{vmatrix} 1 & -6 \\ -3 & 5 \end{vmatrix} + 4 \begin{vmatrix} 1 & 9 \\ -3 & 8 \end{vmatrix}$$

$$= 2(93) + 7(-13) + 4(35) = 235$$

Consider now the determinants of two special cases, a diagonal matrix and a triangular matrix. For the diagonal matrix,

$$\begin{vmatrix} a_{11} & 0 & 0 \\ 0 & a_{22} & 0 \\ 0 & 0 & a_{33} \end{vmatrix} = a_{11} \begin{vmatrix} a_{22} & 0 \\ 0 & a_{33} \end{vmatrix} + 0 \begin{vmatrix} 0 & 0 \\ 0 & a_{33} \end{vmatrix} + 0 \begin{vmatrix} 0 & a_{22} \\ 0 & 0 \end{vmatrix}$$

$$= a_{11} a_{22} a_{33}$$

The determinant is simply the product of the diagonal elements.

In the triangular case,

$$\begin{vmatrix} a_{11} & 0 & 0 \\ a_{21} & a_{22} & 0 \\ a_{31} & a_{32} & a_{33} \end{vmatrix} = a_{11} \begin{vmatrix} a_{22} & 0 \\ a_{32} & a_{33} \end{vmatrix} + 0 \begin{vmatrix} a_{21} & 0 \\ a_{31} & a_{33} \end{vmatrix} + 0 \begin{vmatrix} a_{21} & a_{22} \\ a_{31} & a_{32} \end{vmatrix}$$

$$= a_{11} a_{22} a_{33}$$

In both cases, the determinant is the product of the diagonal elements. These matrix forms are the ones produced by the elimination methods, but in the process, rows were multiplied by constants and added together. The next step is to determine how these operations affect the value of a determinant.

Multiplication of a Row by a Constant

The effect of multiplication by a constant is relatively easy to determine once expansion by cofactors is understood. The procedure is simply to expand about the multiplied row and observe that the multiplier may be factored out of the determinant. In the third-order case,

$$\begin{vmatrix} a_{11} & a_{12} & a_{13} \\ ca_{21} & ca_{22} & ca_{23} \\ a_{31} & a_{32} & a_{33} \end{vmatrix} = ca_{21} \Delta_{21} + ca_{22} \Delta_{22} + ca_{23} \Delta_{23}$$

$$= c(a_{21} \Delta_{21} + a_{22} \Delta_{22} + a_{23} \Delta_{23}) = c\Delta,$$

where Δ was the determinant before the multiplication by c. Thus one sees that multiplying a row (or a column) by a constant modifies the determinant by the same factor. If more than one row multiplier is used, the original determinant is modified by the product of the multipliers.

Addition of Rows

If a certain fact is known, it is readily seen that the addition of rows of a determinant does not affect the value of the determinant. This fact is that if the cofactors of a given row are multiplied by the elements of another row, the resultant sum of products is zero. One can easily demonstrate this result for the third-order determinant. By taking the cofactors of the first row and multiplying these by second-row elements one sees that the result is zero.

$$a_{21}\Delta_{11} + a_{22}\Delta_{12} + a_{23}\Delta_{13}$$

$$= a_{21}\begin{vmatrix} a_{22} & a_{23} \\ a_{32} & a_{33} \end{vmatrix} - a_{22}\begin{vmatrix} a_{21} & a_{23} \\ a_{31} & a_{33} \end{vmatrix} + a_{23}\begin{vmatrix} a_{21} & a_{22} \\ a_{31} & a_{32} \end{vmatrix}$$

$$= a_{21}a_{22}a_{33} - a_{21}a_{23}a_{32} - a_{22}a_{21}a_{33}$$

$$+ a_{22}a_{23}a_{31} + a_{23}a_{21}a_{32} - a_{23}a_{22}a_{31} = 0$$

This property is to be expected if the matrix elements are considered to be coefficients of linear equations, because it is the result that would be obtained if one row of the matrix were duplicated, e.g.,

$$\begin{vmatrix} a_{21} & a_{22} & a_{23} \\ a_{21} & a_{22} & a_{23} \\ a_{31} & a_{32} & a_{33} \end{vmatrix}.$$

But from the point of view of linear equations this duplication is equivalent to including the same equation twice in the system of equations. Obviously, in this case the equations are not independent (two are identical!) and the determinant of the system must therefore be zero. With this information, the lack of effect of adding multiples of rows can be seen by expanding the determinant about the augmented row.

$$\begin{vmatrix} a_{11} + ca_{21} & a_{12} + ca_{22} & a_{13} + ca_{23} \\ a_{21} & a_{22} & a_{23} \\ a_{31} & a_{32} & a_{33} \end{vmatrix}$$

$$= (a_{11} + ca_{21})\Delta_{11} + (a_{21} + ca_{22})\Delta_{12} + (a_{13} + ca_{23})\Delta_{13}$$

$$= (a_{11}\Delta_{11} + a_{12}\Delta_{12} + a_{13}\Delta_{13}) + c(a_{21}\Delta_{11} + a_{22}\Delta_{12} + a_{23}\Delta_{13})$$

$$= \Delta + 0$$

These observations demonstrate that the determinant of a matrix can be computed by an elimination procedure which reduces the given matrix to triangular or diagonal form, provided the product of the row multipliers used to produce these forms is recorded. This product is divided into the determinant of the triangular (or diagonal) form to obtain the true determinant.

To illustrate, the coefficients of the numerical example are presumed to be reduced to triangular form by the elimination process. The necessary multipliers for this reduction are recorded adjacent to the multiplied row.

$$\begin{vmatrix} 2 & -7 & 4 \\ 1 & 9 & -6 \\ -3 & 8 & 5 \end{vmatrix} \rightarrow \begin{vmatrix} 2 & -7 & 4 \\ 0 & 25 & -16 \\ 0 & -5 & 22 \end{vmatrix} \begin{matrix} \\ \times 2 \\ \times 2 \end{matrix} \rightarrow \begin{vmatrix} 2 & -7 & 4 \\ 0 & 25 & -16 \\ 0 & 0 & 470 \end{vmatrix} \begin{matrix} \\ \times 2 \\ \times 2 \times 25, \end{matrix}$$

$$\Delta = \frac{2 \cdot 25 \cdot 470}{2 \cdot 2 \cdot 25} = 235.$$

It will be seen that the incorporation of the scaling into the elimination process, which was mentioned previously, will simplify the task of keeping track of the multiplicative constants.

In summary then, it has been shown that

1. A determinant can be evaluated in terms of its cofactors.

2. If all elements in a row or column of a matrix are zero, its determinant is zero.

3. The addition of a scalar multiple of one row to another does not alter the determinant.

4. If all elements of a row or column are multiplied by a constant, the determinant is multiplied by the same factor.

5. If a matrix is diagonal or triangular, the determinant is simply the product of the elements on the main diagonal

$$|A| = a_{11} \quad a_{22} \quad a_{33} \quad \cdots \quad a_{nn}.$$

In addition, it is simple to demonstrate the following:

6. Interchange of two rows or two columns changes the sign of the determinant, but not its magnitude.

7. If a matrix has two identical rows or columns, its determinant is zero. Referring to item 4, this can be extended to rows or columns which are scalar multiples.

From Cramer's rule it is evident that if the determinant of the coefficients is zero, there is no solution for the set of equations. The matrix is called singular. This can occur when two equations are identical or when their coefficients are multiples of one another. There can be an infinity of solutions, if in addition to $\Delta = 0$, the numerators in Cramer's rule are zero. This would result if all the elements on the right-hand side, $\bar{b}$, were zero.

Reduction with the Evaluation of the Determinant

Although one would not expect to employ Cramer's rule for solving large sets of equations, there are occasions when it is necessary to evaluate the determinant. In the Gauss-Jordan method, as the matrix is reduced it becomes triangular, then diagonal, which suggests that the determinant can be calculated during the reduction process. This is particularly direct when the pivot elements have been normalized.

From the results obtained in the discussion of determinants it follows that the value of the determinant is not changed by the reduction of the ith row. It is true, however, that the division of the pivot row by the diagonal element does affect the determinant value. This step is tantamount to multiplying by the constant $1/a_{kk}$. Therefore the value of the determinant is the determinant of the reduced matrix, Δ_r (a diagonal or triangular matrix), divided by the product of all the row multipliers:

$$\frac{1}{a_{11}} \cdot \frac{1}{a'_{22}} \cdot \frac{1}{a''_{33}} \cdots \frac{1}{a_{nn}^{(n-i)}} \cdot$$

The primes indicate that these diagonal elements are not the original matrix but are altered in the reduction process. More concisely,

$$\Delta = \frac{\Delta_r}{\prod_{i=1}^{n} [1/a_{ii}^{(i-1)}]} = \Delta_r \prod_{i=1}^{n} a_{ii}^{(i-1)}$$

But as a result of the division of the pivot rows all the diagonal elements will be unity, and the product of these diagonal elements, Δ_r, is also unity. Hence $\Delta = \prod_{i=1}^{n} a_{ii}^{(i-1)}$. In prose, once again, the determinant of the system is simply the product of the pivot-row divisors used in reducing the matrix of coefficients to diagonal or triangular form. If the primary goal is the computation of the determinant, reduction to triangular form is superior since it can be carried out in fewer operations than are required to obtain the diagonal form. As observed earlier, the determinant is the product of the diagonal elements.

In the process of solving the equations by reduction of the coefficient matrix to a unity matrix, the diagonal terms stand alone and it is possible to form the product of the pivot elements. This must be done after a column has been reduced, but before the pivot element has been normalized. In the numerical example the determinant is simply

$$(2)(\tfrac{25}{2})(\tfrac{47}{5}) = 235.$$

In the program for this calculation, the product must be initialized, then updated at the proper place in the calculation loop, just before the pivot element is normalized. If the determinant is to be called DELTA, then the product is initialized as

$$\text{DELTA} = 1.0,$$

and updated in each loop as

$$\text{DELTA} = \text{DELTA}*A(K,K)$$

In the example program to follow, subroutines are written to evaluate the determinant and to solve the set of equations. The matrix is first reduced to a triangular matrix with unit diagonal elements. To illustrate this, consider a matrix which is partially reduced to triangular form. In such a matrix, one can determine the range of the indices.

$$\begin{matrix} 1 & a_{12} & a_{13} & a_{14} \\ 0 & 1 & a_{23} & a_{24} \\ 0 & 0 & a_{33} & a_{34} \\ 0 & 0 & a_{43} & a_{44} \end{matrix}$$

The next step is to select row 3 as a pivot row (row 4 could be used, but that is a complication which is postponed until later) and divide the remaining elements in row 3 by a_{33}.

$$\begin{matrix} 1 & a_{12} & a_{13} & a_{14} \\ 0 & 1 & a_{23} & a_{24} \\ 0 & 0 & 1 & a'_{34} \\ 0 & 0 & a_{43} & a_{44} \end{matrix}$$

Subtracting $a_{43} \cdot$ (row 3) from row 4 will make

$$a'_{43} = a_{43} - a_{43} \cdot 1 = 0$$

as desired. As observed earlier, this computation need not be carried out since the result must be zero. Also, the zero elements to the left of a_{43} will not be changed by the subtraction of row 3, and hence the computation of new elements in a reduced row can start with the first element to the right of the pivot column, i.e., the $(k + 1)$-column.

The description of a procedure for reducing a matrix to triangular form with 1's on the diagonal can be described in terms of the pivot index, k, the reduced row index, i, and the column index, j, as follows:

$$k = 1, 2, \ldots, n$$
$$j = k + 1, k + 2, \ldots, n + 1$$
$$a'_{kj} = \frac{a_{kj}}{a_{kk}}$$
$$i = k + 1, k + 2, \ldots, n$$
$$j = k + 1, k + 2, \ldots, n + 1$$
$$a'_{ij} = a_{ij} - a_{ik}a_{kj}$$

After these steps have been completed, the reduction of the triangular matrix to diagonal form must be carried out before the solution to the linear equations is obtained explicitly. The triangular form for three equations is:

$$x_1 + a_{12}x_2 + a_{13}x_3 = a_{14}$$
$$x_2 + a_{23}x_3 = a_{24}$$
$$x_3 = a_{34}$$

Putting this result in matrix form suggests a familiar approach to the diagonalization, namely, finishing the pivoting process. Use row 3 to put zeros in column 3 (rows 1 and 2); use row 2 to put a zero in column 2 (row 1). This procedure is merely the "triangularization" of the upper triangle, but this time there is no need for division since the diagonal elements are already unity.

$$
\begin{array}{cccc}
1 & a_{12} & a_{13} & a_{14} \\
0 & 1 & a_{23} & a_{24} \\
0 & 0 & 1 & a_{34}
\end{array}
$$

To reduce, use the triangularization algorithm without division.

$$k = 2, 3, \ldots, n$$
$$i = 1, 2, \ldots, k - 1$$
$$j = k + 1, k + 2, \ldots, n + 1$$
$$a'_{ij} = a_{ij} - a_{ik}a_{kj}$$

Although this approach to the back solution is familiar, it is a little more complicated than it needs to be. There are two alternative approaches: one is to substitute the known value x_3 into all the other equations and then x_2 into all the remaining equations, etc.; the other is to do the same thing but in a different order. That is, one substitutes x_3 in the preceding equation to obtain x_2, then substitutes x_2 and x_3 in the preceding equation, etc. The difference is simply whether the row index or the column index is the "inner" iteration. Using the first approach, one selects row n as a pivot, and only the pivot column and $(n + 1)$-column are altered. Since zeros are produced in the pivot column, there is no need to explicitly carry out the computation. When column $n - 1$ is used as a pivot, only the $(n + 1)$-column need be changed.

$$k = n, n - 1, \ldots, 2$$
$$i = k - 1, k - 2, \ldots, 1$$
$$a_{i,n+1} = a_{i,n+1} - a_{ik}a_{k,n+1}$$

When this process is completed, the roots will be in the positions originally occupied by the right-hand sides. The entire procedure is now complete enough to be programmed.

Example Write a subroutine called SLEQ1 which, by an elimination method, computes the determinant of the matrix of coefficients of a given augmented matrix. The function should have two entries: DET = 1 if only the value of the determinant is desired, and DET = 2 if, in addition, the solution of the system is to replace the right-hand sides. The parameters should be n, the number of equations, and a, the name of the augmented matrix. The flow chart and program are presented in Figs. 6.5a and 6.5b respectively.

The section from the entry points to ⎣1⎦ computes the number of columns, m, involved in the reduction: $m = n$ if only the determinant is desired; otherwise, $m = n + 1$ on the assumption that there is only one vector of right-hand sides in

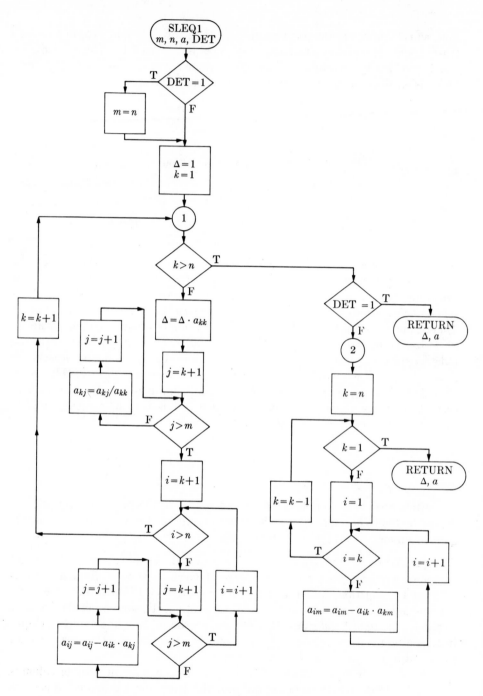

Fig. 6.5   Gauss-Jordan and Det subroutine.   (a) Flow chart.

```
      DIMENSION A(3,4)
      DATA A/1.,2.,8.,2.,-4.,-6.,-2.,4.,2.,-3.,0.,4./
      CALL SLEQ1(3,2,A,D)
      WRITE(6,15)D,(A(I,4),I=1,3)
15    FORMAT(4HDET=F10.5,5H  X1=F10.5,5H  X2=F10.5,5H  X3=F10.5)
      END


      SUBROUTINE SLEQ1(N,DET,A,DELTA)
C     SUBROUTINE FOR GAUSS-JORDAN REDUCTION AND
C     DETERMINANT EVALUATION
C     N EQUALS THE NUMBER OF EQUATIONS. M EQUALS N,
C     AND DET IS 1 IF ONLY THE THE DETERMINANT
C     IS TO BE RETURNED.OTHERWISE DET IS 2 AND M=N+1
      INTEGER DET
      DIMENSION A(3,4)
      M=N+1
      GO TO (3,6),DET
3     M=N
C     INITIALIZE THE DETERMINANT
6     DELTA=1.
      DO 12 K=1,N
      DELTA=DELTA*A(K,K)
      KP1=K+1
      DO 9 J=KP1,M
9     A(K,J)=A(K,J)/A(K,K)
      IF(KP1.GT.N) GO TO 13
      DO 12 I=KP1,N
      DO 12 J=KP1,M
12    A(I,J)=A(I,J)-A(I,K)*A(K,J)
13    IF(DET.EQ.1) GO TO 18
      NM1=N-1
      DO 15 IND=1,NM1
      K=N+1-IND
      KM1=K-1
      DO 15 I=1,KM1
15    A(I,M)=A(I,M)-A(I,K)*A(K,M)
18    RETURN
      END


      OUTPUT

DET=  32.00000  X1=  -1.50000  X2=  -3.62500  X3=  -2.87500
```

Fig. 6.5 (b) Program.

the augmented matrix. From (1) to (2) the triangular matrix with unit diagonal elements is computed. Every pivot-row divisor is multiplied by a partial product to form the determinant. At the conclusion of this section, if $m = n$, the process is completed, and the value of the determinant is the direct result. The last section, from (2) to the end, carries out the back-solution of the reduced matrix.

The program has been used to solve the same set of equations that was used for the subroutine SLEQ, namely:

$$\begin{bmatrix} 1. & 2. & -2. \\ 2. & -4. & 4. \\ 8. & -6. & 2. \end{bmatrix} \begin{bmatrix} x_1 \\ x_2 \\ x_3 \end{bmatrix} = \begin{bmatrix} -3. \\ 0. \\ 4. \end{bmatrix}$$

In addition, DET was set at 2 so the determinant is returned.

Matrix inversion

Let us return for the moment to the equation $A\bar{x} = \bar{b}$. If the range of these variables were real numbers instead of arrays, solving for $\bar{x}$ would be a simple task:

$$A\bar{x} = \bar{b}$$

$$\frac{1}{A} A\bar{x} = \frac{1}{A} \bar{b}$$

$$1 \cdot \bar{x} = A^{-1}\bar{b}$$

Thus $1 \cdot \bar{x} = \bar{x}$ would be obtained by multiplying the reciprocal (or inverse) of A and $\bar{b}$. In the domain of real numbers a unit element (1) and an inverse for any nonzero number x exist, and the latter is designated x^{-1}. The defining properties of these two entities are:

$$\left. \begin{array}{l} 1 \cdot x = x \\ x \cdot x^{-1} = 1 \end{array} \right\} \quad \text{for any real number } x.$$

The question is whether analogous entities exist for arrays, and the answer is that in some cases they do. A square diagonal matrix with ones on the diagonal is called an identity matrix (I) and fulfills the first condition:

$$\begin{bmatrix} 1 & 0 & 0 \\ 0 & 1 & 0 \\ 0 & 0 & 1 \end{bmatrix} \begin{bmatrix} a_{11} & a_{12} & a_{13} & a_{14} \\ a_{21} & a_{22} & a_{23} & a_{24} \\ a_{31} & a_{32} & a_{33} & a_{34} \end{bmatrix} = \begin{bmatrix} a_{11} & a_{12} & a_{13} & a_{14} \\ a_{21} & a_{22} & a_{23} & a_{24} \\ a_{31} & a_{32} & a_{33} & a_{34} \end{bmatrix}$$

$$IA = A$$

Note that in this case, AI is not defined since the rows and columns do not match, and hence the matrix multiplication formula cannot be applied. The search for inverses is limited to square matrices and reduces to the question of whether there is a matrix $\mathcal{A}$ such that

$$A\mathcal{A} = I$$

Writing out this matrix equation in array form illustrates that $\mathcal{A}$, if it exists, can be found by solving a system of n equations with n different right-hand side vectors.

$$\begin{bmatrix} a_{11} & a_{12} & a_{13} \\ a_{21} & a_{22} & a_{23} \\ a_{31} & a_{32} & a_{33} \end{bmatrix} \begin{bmatrix} \alpha_{11} & \alpha_{12} & \alpha_{13} \\ \alpha_{21} & \alpha_{22} & \alpha_{23} \\ \alpha_{31} & \alpha_{32} & \alpha_{33} \end{bmatrix} = \begin{bmatrix} 1 & 0 & 0 \\ 0 & 1 & 0 \\ 0 & 0 & 1 \end{bmatrix}$$

There are three different unknown vectors (the columns of $\mathcal{A}$) and three corresponding right-hand side vectors (the columns of I). To find the elements of matrix $\mathcal{A}$, the combined system must be solved. This can be done most easily by separating the three vector equations into the form:

$$\begin{bmatrix} a_{11} & a_{12} & a_{13} \\ a_{21} & a_{22} & a_{23} \\ a_{31} & a_{32} & a_{33} \end{bmatrix} \begin{bmatrix} \alpha_{11} \\ \alpha_{21} \\ \alpha_{31} \end{bmatrix} = \begin{bmatrix} 1 \\ 0 \\ 0 \end{bmatrix},$$

$$\begin{bmatrix} a_{11} & a_{12} & a_{13} \\ a_{21} & a_{22} & a_{23} \\ a_{31} & a_{32} & a_{33} \end{bmatrix} \begin{bmatrix} \alpha_{12} \\ \alpha_{22} \\ \alpha_{32} \end{bmatrix} = \begin{bmatrix} 0 \\ 1 \\ 0 \end{bmatrix},$$

$$\begin{bmatrix} a_{11} & a_{12} & a_{13} \\ a_{21} & a_{22} & a_{23} \\ a_{31} & a_{32} & a_{33} \end{bmatrix} \begin{bmatrix} \alpha_{13} \\ \alpha_{23} \\ \alpha_{33} \end{bmatrix} = \begin{bmatrix} 0 \\ 0 \\ 1 \end{bmatrix}.$$

For this system to be solvable, and for α to exist, the coefficient matrix, A, must be nonsingular. Furthermore, the matrix α is the inverse of A, designated A^{-1}, and satisfies the relation

$$AA^{-1} = I = A^{-1}A.$$

Returning to the set of simultaneous equations, it is evident that a solution for $\bar{x}$ can be obtained by premultiplying both sides by the inverse of the coefficient matrix. In matrix representation that is

$$A^{-1}A\bar{x} = A^{-1}\bar{b},$$

which leads directly to a solution for $\bar{x}$ as

$$I\bar{x} = \bar{x} = A^{-1}\bar{b}.$$

This is deceptively simple in appearance, for it is first necessary to find A^{-1}. To indicate the complexity of solving directly for A^{-1}, return to the matrix equations in the elements of α. Applying Cramer's rule, the element α_{23} can be found from the third set of equations as

$$\alpha_{23} = \frac{\begin{vmatrix} a_{11} & 0 & a_{13} \\ a_{21} & 0 & a_{23} \\ a_{31} & 1 & a_{33} \end{vmatrix}}{\Delta}.$$

Notice that the determinant in the numerator is simply the minor of the element a_3. Therefore, in terms of the cofactor Δ_{32},

$$\alpha_{23} = \frac{\Delta_{32}}{\Delta},$$

or for any element α_{ij} of the inverse matrix,

$$\alpha_{ij} = \frac{\Delta_{ji}}{\Delta}.$$

The method obtains for any size matrix, with each element calculated in this way. For large matrices it is evident that this will be a long computation.

It is simple to show that the Gauss-Jordan reduction does produce the inverse matrix of the coefficients. The procedure begins with the matrix equation $A\bar{x} = \bar{b}$, and proceeds until the system has been reduced to the form $I\bar{x} = \bar{c}$, where $\bar{c}$ is the solution vector.

All of the steps in the reduction can be represented as a single matrix, having the same dimensions as the coefficient matrix, A. When the matrix A is pre-multiplied by this matrix the identity matrix results. This then is A^{-1}. As

$$A^{-1}I = A^{-1},$$

it is a simple matter to keep account of the inverse, by performing all the operations of the Gauss-Jordan reduction on a unit matrix. One may carry this out by operating on a separately stored copy of the unit matrix or, alternatively, by using the zeros and diagonal elements of the coefficient matrix, as they become available, for storing the modified unit matrix. Algorithms using this latter storage economization are called "inversion in place."

An alternative way of reaching the same conclusion is to solve the sets of simultaneous equations for the elements of the inverse α_{ij}, using the Gauss-Jordan reduction procedure. Combining the right-hand sides into a single matrix, we can write the augmented matrix as follows:

Using the numerical example, the augmented matrix is:

$$\begin{bmatrix} 2 & -7 & 4 & 9 & 1 & 0 & 0 \\ 1 & 9 & -6 & 1 & 0 & 1 & 0 \\ -3 & 8 & 5 & 6 & 0 & 0 & 1 \end{bmatrix}.$$

The first row is normalized and the first column reduced:

$$\begin{bmatrix} 1 & -\frac{7}{2} & 2 & \frac{9}{2} & \frac{1}{2} & 0 & 0 \\ 0 & \frac{25}{2} & -8 & -\frac{7}{2} & -\frac{1}{2} & 1 & 0 \\ 0 & -\frac{5}{2} & 11 & \frac{39}{2} & \frac{3}{2} & 0 & 1 \end{bmatrix}.$$

Normalize the second row and reduce the second column:

$$\begin{bmatrix} 1 & -\frac{7}{2} & 2 & \frac{9}{2} & \frac{1}{2} & 0 & 0 \\ 0 & 1 & -\frac{16}{25} & -\frac{7}{25} & -\frac{1}{25} & \frac{2}{25} & 0 \\ 0 & -\frac{5}{2} & 11 & \frac{32}{2} & \frac{3}{2} & 0 & 1 \end{bmatrix} \rightarrow \begin{bmatrix} 1 & 0 & \frac{6}{25} & \frac{76}{25} & \frac{18}{50} & \frac{14}{50} & 0 \\ 0 & 1 & -\frac{16}{25} & -\frac{7}{25} & -\frac{1}{25} & \frac{2}{25} & 0 \\ 0 & 0 & \frac{47}{5} & \frac{94}{5} & \frac{7}{5} & \frac{1}{5} & 1 \end{bmatrix},$$

and for the third pivot:

$$\begin{bmatrix} 1 & 0 & -\frac{6}{25} & \frac{76}{25} & \frac{18}{50} & \frac{14}{50} & 0 \\ 0 & 1 & -\frac{16}{25} & -\frac{7}{25} & -\frac{1}{25} & \frac{2}{25} & 0 \\ 0 & 0 & 1 & 2 & \frac{7}{47} & \frac{1}{47} & \frac{5}{47} \end{bmatrix} \rightarrow \begin{bmatrix} 1 & 0 & 0 & 4 & \frac{93}{235} & \frac{67}{235} & \frac{6}{235} \\ 0 & 1 & 0 & 1 & \frac{13}{235} & \frac{22}{235} & \frac{16}{235} \\ 0 & 0 & 1 & 2 & \frac{7}{47} & \frac{1}{47} & \frac{5}{47} \end{bmatrix}$$

Unit   Solution   Inverse
matrix   vector   matrix

As a check, we can write

$$AA^{-1} = \begin{bmatrix} 2 & -7 & 4 \\ 1 & 9 & 6 \\ -3 & 8 & 5 \end{bmatrix} \begin{bmatrix} \frac{93}{235} & \frac{67}{235} & \frac{6}{235} \\ \frac{13}{235} & \frac{22}{235} & \frac{16}{235} \\ \frac{7}{47} & \frac{1}{47} & \frac{5}{47} \end{bmatrix} = \begin{bmatrix} 1 & 0 & 0 \\ 0 & 1 & 0 \\ 0 & 0 & 1 \end{bmatrix}.$$

It is also possible with the inverse to obtain the solution vector directly as:

$$A^{-1}\bar{b} = \begin{bmatrix} \frac{93}{235} & \frac{67}{235} & \frac{6}{235} \\ \frac{13}{235} & \frac{22}{235} & \frac{16}{235} \\ \frac{7}{47} & \frac{1}{47} & \frac{5}{47} \end{bmatrix} \begin{bmatrix} 9 \\ 1 \\ 6 \end{bmatrix} = \begin{bmatrix} 4 \\ 1 \\ 2 \end{bmatrix} = \bar{x}.$$

Example Write an external function INV(N,A) which replaces the $N \times N$ matrix A with its inverse. Since the algorithm is so similar to the Gauss-Jordan method, only the program is presented.

```
      DIMENSION A(3,3)
      DATA A/1.,2.,8.,2.,-4.,-6.,-2.,4.,2./
      CALL INV(3,A)
      WRITE(6,15) ((A(I,J),J=1,3),I=1,3)
   15 FORMAT(3F10.5/)
      END

      SUBROUTINE INV(N,A)
      DIMENSION A(N,N)
      DO 1 K=1,N
      DO 2 J=1,N
    2 IF(J.NE.K)A(K,J)=A(K,J)/A(K,K)
      A(K,K)=1./A(K,K)
      DO 1 I=1,N
      IF(I.EQ.K) GO TO 1
      DO 3 J=1,N
    3 IF(J.NE.K)A(I,J)=A(I,J)-A(I,K)*A(K,J)
      A(I,K)=A(I,K)*A(K,K)
    1 CONTINUE
      RETURN
      END

            OUTPUT

      1.50000   -0.25000   0.0
     -0.87500   -0.81250   0.25000
     -0.62500   -0.68750  -0.25000
```

Elementary row operations

The various reduction steps on augmented matrices that have been considered thus far could be posed as matrix multiplications where the applied or "reducing" matrices are of particularly simple forms. Specifically, the multiplications can be limited to the accomplishment of the three following row operations and still be sufficient to describe the algorithms.

1. The interchange of any two rows can be effected by multiplying by a *permutation matrix* (i.e., a matrix of zeros and ones where there is exactly one nonzero element in any row or column) which has only two unit elements not on the main diagonal. As an example, multiplication of a four-row matrix on the left by

$$\begin{bmatrix} 1 & 0 & 0 & 0 \\ 0 & 0 & 1 & 0 \\ 0 & 1 & 0 & 0 \\ 0 & 0 & 0 & 1 \end{bmatrix}$$

would cause the interchange of the second and third rows. Notice in passing that a permutation matrix can be encoded as a vector of integers where each vector element corresponds to a row (or column) and gives the position of the unit element in the row or column. With this symmetric example both row and column correspondence produce the same vector $(1, 3, 2, 4)$.

2. Multiplication of any row by a nonzero constant is accomplished by a multiplier matrix the same as the identity matrix with the exception of one diagonal element where the constant appears; e.g.,

$$\begin{bmatrix} 1 & 0 & 0 & 0 \\ 0 & 5 & 0 & 0 \\ 0 & 0 & 1 & 0 \\ 0 & 0 & 0 & 1 \end{bmatrix}$$

would cause the second row to be multiplied by 5 with the other rows unchanged.

3. The addition of any multiple of one row to another row requires the multiplier to appear as an off-diagonal element in what would otherwise be the identity matrix. The elementary matrix

$$\begin{bmatrix} 1 & 0 & 0 & 0 \\ 0 & 1 & 0 & 0 \\ 0 & 0 & 1 & 0 \\ 0 & -4 & 0 & 1 \end{bmatrix},$$

when used as a left factor, would result in four times the second row subtracted from the fourth row.

Actually, only the last two row operations have been used so far in reduction algorithms. If E_i is used to represent one of these latter operations, the Gauss-Jordan method of determining the inverse can be described as

$$E_n \ldots E_3 E_2 E_1 A = I.$$

That is, the application of elementary row operations on A until it is reduced to the identity matrix. It is clear that the product $E_n \ldots E_3 E_2 E_1 = A^{-1}$ and that, for inversion in place, as A is reduced, the terms of this product are stored in its place.

Factoring

When the reduction is divided into a distinct forward and back solution, it is possible to store, in place, a slight variant of the forward solution operations as well as the matrix resulting from their application to obtain a pair of matrices that are often more useful than the inverse. The procedure is essentially that of factoring the given matrix A into a lower triangular factor, L, and an upper triangular factor, U. The production of U is nearly identical to the forward solution previously described, while the determination of L amounts to a systematic recording of the row multipliers used. That the product of the two is A is more difficult to see.

As a preliminary, observe that the product of two elementary row operations of the third type simply results in the inclusion of both off-diagonal constants. That is,

$$
\begin{bmatrix} 1 & 0 & 0 & 0 \\ -b_{21} & 1 & 0 & 0 \\ 0 & 0 & 1 & 0 \\ 0 & 0 & 0 & 1 \end{bmatrix}
\begin{bmatrix} 1 & 0 & 0 & 0 \\ 0 & 1 & 0 & 0 \\ -b_{31} & 0 & 1 & 0 \\ 0 & 0 & 0 & 1 \end{bmatrix}
=
\begin{bmatrix} 1 & 0 & 0 & 0 \\ -b_{21} & 1 & 0 & 0 \\ -b_{31} & 0 & 1 & 0 \\ 0 & 0 & 0 & 1 \end{bmatrix}.
$$

This means that the reduction of the first column to zero (after the first element) could be accomplished by multiplying by

$$
B = \begin{bmatrix} 1 & 0 & 0 & 0 \\ -b_{21} & 1 & 0 & 0 \\ -b_{31} & 0 & 1 & 0 \\ -b_{41} & 0 & 0 & 1 \end{bmatrix},
$$

where $b_{i,1} = a_{i1}/a_{11}$ and the elements of A are a_{ij}.

Similarly, the second column of the now reduced matrix A' could be set to zero below the diagonal by applying

$$
B' = \begin{bmatrix} 1 & 0 & 0 & 0 \\ 0 & 1 & 0 & 0 \\ 0 & -b'_{32} & 1 & 0 \\ 0 & -b'_{42} & 0 & 1 \end{bmatrix},
$$

where $b'_{32} = a'_{32}/a'_{22}$. The resulting A'' could be completely reduced to upper triangular form, U, by multiplying by

$$B'' = \begin{bmatrix} 1 & 0 & 0 & 0 \\ 0 & 1 & 0 & 0 \\ 0 & 0 & 1 & 0 \\ 0 & 0 & -b_{43} & 1 \end{bmatrix},$$

with $b''_{43} = a''_{43}/a'_{33}$. In summary, the successive applications of B, B', and B'' reduce A to an upper triangular matrix U; that is,

$$B''B'BA = U \quad \text{or} \quad \overline{B}A = U.$$

If the inverse of $\overline{B}$ can be determined, then A will be expressible as a product with factors $\overline{B}^{-1}$ and U since

$$\overline{B}^{-1}\overline{B}A = \overline{B}^{-1}U.$$

The inverse of an elementary row transformation of type 3 is the same matrix with the sign of the off-diagonal element reversed. (That is, the multiples of other rows are added and then the same multiples are subtracted.) The desired inverse can therefore be expressed as the reversed product of the individual inverses:

$$\underbrace{B^{-1}B'^{-1}B''^{-1}}_{\overline{B}^{-1}}\underbrace{B''B'B}_{\overline{B}}$$

This inverse is lower triangular and is usually designated L:

$$L = \begin{bmatrix} 1 & 0 & 0 & 0 \\ b_{21} & 1 & 0 & 0 \\ b_{31} & 0 & 1 & 0 \\ b_{41} & 0 & 0 & 1 \end{bmatrix}\begin{bmatrix} 1 & 0 & 0 & 0 \\ 0 & 1 & 0 & 0 \\ 0 & b_{32} & 1 & 0 \\ 0 & b_{42} & 0 & 1 \end{bmatrix}\begin{bmatrix} 1 & 0 & 0 & 0 \\ 0 & 1 & 0 & 0 \\ 0 & 0 & 1 & 0 \\ 0 & 0 & b_{43} & 1 \end{bmatrix} = \begin{bmatrix} 1 & 0 & 0 & 0 \\ b_{21} & 1 & 0 & 0 \\ b_{31} & b_{32} & 1 & 0 \\ b_{41} & b_{42} & b_{43} & 1 \end{bmatrix}.$$

It can be seen that the steps to obtain the factoring $LU = A$ are essentially those of Gaussian elimination or reduction where distinct forward and backward solutions are used. The upper matrix U is the result obtained at the end of the forward solution except that division of the pivot rows has not been done so the diagonal elements are not unity. The lower matrix can be obtained by simply recording, in place, the pivot row multipliers used in the forward solution except that the unit diagonal elements are not actually stored. As before the determinant is the product of the diagonal elements of U.

An interesting result is that the factors obtained are usually more useful than the inverse. If more solutions to a system of equations are desired, two simple triangular systems can be solved in succession to obtain them. From $A\overline{x} = LU\overline{x} = \overline{b}$ it can be seen that if $U\overline{x} = \overline{y}$, an intermediate vector, then finding first $\overline{y}$ by solving $L\overline{y} = \overline{b}$ and, secondly, solving $U\overline{x} = \overline{y}$ the desired solution $\overline{x}$ is obtained. To illustrate the factorization of a matrix and the use of the factors to obtain a solution an example three-dimensional system will be used.

$$BA = \begin{bmatrix} 1 & 0 & 0 \\ -\frac{2}{1} & 1 & 0 \\ \frac{2}{1} & 0 & 1 \end{bmatrix} \begin{bmatrix} 1 & 2 & 8 \\ 2 & -4 & -6 \\ -2 & 4 & 2 \end{bmatrix} = \begin{bmatrix} 1 & 2 & 8 \\ 0 & -8 & -22 \\ 0 & 8 & 18 \end{bmatrix},$$

$$B'BA = \begin{bmatrix} 1 & 0 & 0 \\ 0 & 1 & 0 \\ 0 & \frac{8}{8} & 1 \end{bmatrix} \begin{bmatrix} 1 & 2 & 8 \\ 0 & -8 & -22 \\ 0 & 8 & 18 \end{bmatrix} = \begin{bmatrix} 1 & 2 & 8 \\ 0 & -8 & -22 \\ 0 & 0 & -4 \end{bmatrix} = U,$$

and

$$LU = \begin{bmatrix} 1 & 0 & 0 \\ 2 & 1 & 0 \\ -2 & -1 & 1 \end{bmatrix} \begin{bmatrix} 1 & 2 & 8 \\ 0 & -8 & -22 \\ 0 & 0 & -4 \end{bmatrix} = \begin{bmatrix} 1 & 2 & 8 \\ 2 & -4 & -6 \\ -2 & 4 & 2 \end{bmatrix} = A.$$

The determinant is the product of the diagonal elements of U and is equal to 32. To obtain solutions for a set of right-side vectors, say

$$B = \begin{bmatrix} -3 & 1 & 0 & 0 \\ 0 & 0 & 1 & 0 \\ 4 & 0 & 0 & 1 \end{bmatrix},$$

the solution of $LY = B$ is first obtained and then the equations represented by $UX = Y$ are solved:

$$\begin{bmatrix} 1 & 0 & 0 \\ 2 & 1 & 0 \\ -2 & -1 & 1 \end{bmatrix} Y = \begin{bmatrix} -3 & 1 & 0 & 0 \\ 0 & 0 & 1 & 0 \\ 4 & 0 & 0 & 1 \end{bmatrix},$$

$$Y = \begin{bmatrix} -3 & 1 & 0 & 0 \\ 6 & -2 & 1 & 0 \\ 4 & 0 & 1 & 1 \end{bmatrix};$$

$$\begin{bmatrix} 1 & 2 & 8 \\ 0 & -8 & -22 \\ 0 & 0 & -4 \end{bmatrix} X = \begin{bmatrix} -3 & 1 & 0 & 0 \\ 6 & -2 & 1 & 0 \\ 4 & 0 & 1 & 1 \end{bmatrix},$$

$$X = \begin{bmatrix} 1 & \frac{1}{2} & \frac{7}{8} & \frac{5}{8} \\ 2 & \frac{1}{4} & \frac{9}{16} & \frac{11}{16} \\ -1 & 0 & -\frac{1}{4} & -\frac{1}{4} \end{bmatrix}.$$

The first column is an example solution while the remaining three columns are the inverse. The algorithm for the solution of an upper triangular system has been discussed before. Only a slight modification is required for the solution of the lower triangular system. A surprising fact is that there are fewer multiplications required to solve for m solutions (i.e., m right-hand side vectors) by the factor method than by obtaining the inverse and multiplying. The comparison (see Isaacson and Keller, *The Analysis of Numerical Method*) is

$$\frac{n^3}{3} - \frac{n}{3} + mn^2$$

for the *LU* approach and

$$n^3 + mn^2$$

for the multiplication by an inverse. Even for a single solution it is more effective.

Pivot element selection

The algorithms considered thus far are ineffective if the next pivot element happens to be zero—even though the matrix of coefficients is now singular. This difficulty can be removed by augmenting the reduction algorithms so that the next pivot is not constrained to the diagonal element in the next row but rather may be selected from any of the rows and columns which do not contain a pivot element. With this freedom of selection, if the only choices are zero elements then the matrix is truly singular.

One criterion for the selection of a pivot element is that it be largest in magnitude of all of the remaining candidates. The rationale for this selection comes from considerations of the error produced in division operations. Since divisions are usually approximate operations it would be well to postpone them as long as possible in the execution of an algorithm, so that the resulting propagation of error would be reduced. The original Gauss-Jordan scheme (without division) is good from this point of view, but practically, the divisions cannot be postponed. A differential analysis of the division operation (similar to the approximation-error number analysis) shows the advantage of picking maximum-magnitude divisors when division must be carried out. If $f = y/x$,

$$df = \frac{x\,dy - y\,dx}{x^2}.$$

That is, a large divisor produces a large denominator in the expression for error above.

An interchange algorithm

The complications introduced by searching for the maximal element in the remaining submatrix, whose rows and columns do not include previous pivot elements, do not seem warranted in practice. It is sufficient to search only the current pivot column. This simpler procedure still has the virtue that finding no alternative to a zero pivot element implies that the matrix being reduced is singular. This fact is clear if one considers that the inability to find a nonzero pivot element in a column means that the reduced triangular matrix will have a zero on the diagonal, with the result that the determinant will be zero. In what follows a pair of programs will be presented. The first, given a matrix A, stores the LU factors in place of the A array, using the maximal elements in a column as the pivot. The second, given the LU factors and a matrix of right-hand sides, B, produces solution vectors in place of the B array.

Up to this point the first of the elementary row transformations, permutation of rows, has not been used in any reduction algorithm. Strategies involving the search for pivotal elements introduce row interchanges which can be expressed as permutation matrices. For example, algorithms employing a column search for pivot elements can be implemented by simply interchanging the row containing the maximal element with what would be the pivot row in the absence of any search. Thus, if factoring is the goal, the reduction leads to triangular matrices whose product is a row permutation, P, of the given matrix A; that is,

$$LU = PA.$$

To use L and U in subsequent computations, it is necessary to preserve the matrix P in some form.

A different point of view, which amounts to the same thing, is not to interchange rows but to factor the given matrix A into two nontriangular but simply solved matrices $\hat{L}$ and $\hat{U}$. The rows of each of these can be reordered by P; that is, $P\hat{L} = L$ and $P\hat{U} = U$. Thus

$$P\hat{L} \cdot PL\hat{U} = PA \qquad \text{or} \qquad \hat{L} \cdot P\hat{U} = A.$$

Therefore, two direct solutions, with the second followed by a permutation, solve the problem. Specifically,

$$\hat{L}y = b,$$
$$P\hat{U}x = y.$$

In the algorithm to be presented the final permutation is accomplished by using a temporary vector in which the solutions are developed in the natural, indicial order.

The systems $\hat{L}\bar{y} = \bar{b}$ and $\hat{U}\bar{y} = \bar{x}$ are readily solved since both contain an equation with only one variable, an equation with two variables, etc. However, in order to solve such systems, it is necessary to know what rows have only one element, what rows two, etc. The storage of such structural information is tantamount to storing the P matrix, as was required when the actual rows were interchanged. This latter approach—where the order of pivot-row choice is preserved in a vector P—is the one used in the program.

Example Write a FORTRAN IV function sub-program FACTOR(N,A,P) which takes as input a square matrix A of dimension N, computes the factors $\hat{L}$, $\hat{U}$ of A, stores the factors in place of A, and records the order of pivot row selection in the N-dimensional vector P. In addition, the determinant of A should be produced as a direct result.

If the pivot rows are selected in their natural order, the algorithm is readily described in the indented notation used earlier.

$$\Delta = 1.0$$
$$k = 1, 2, \ldots, n - 1$$
$$\Delta = \Delta \times a_{kk}$$
$$i = k, k + 1, \ldots, n$$
$$a_{ik} = a_{ik}/a_{kk}$$
$$j = k + 1, k + 2, \ldots, n$$
$$a_{ij} = a_{ij} - a_{ik} \cdot a_{kj}$$

There are several complications when the pivot selection order is varied. There is a search section added and the actual row used in the kth pivot step is recorded in P_k which initially contains k.

$$i = 1, 2, \ldots, n$$
$$P_i = i$$
$$\Delta = 1.0$$
$$k = 1, 2, \ldots, n - 1$$
$$M = |a_{P_k,k}|$$
$$m = P_k$$
$$i = k, k + 1, \ldots, n$$

If $|a_{P_i,k}| > M$, $\begin{cases} M = |a_{P_i,k}| \\ m = i \\ \Delta = -\Delta \end{cases}$

If $M = 0$, singular

$$t = P_k$$
$$P_k = P_m$$
$$P_m = t$$
$$i = k + 1, k + 2, \ldots, n$$
$$a_{P_i,k} = a_{P_i,k}/a_{P_k,k}$$
$$j = k + 1, k + 2, \ldots, n$$
$$a_{P_i,j} = a_{P_i,j} - a_{P_i,k} \times a_{P_k,j}$$

$$\Delta = \Delta \times a_{P_k,k}$$
$$\Delta = \Delta \times a_{P_n,n}$$

This statement of the procedure is so close to a program that the transition to statements will be made directly without a flow chart. In the interest of efficiency there are instances in the program where the subscript value is first substituted into another integer variable before it is used as a subscript.

```
          DIMENSION A(3,3),P(3),B(3,4)
          INTEGER P
          DATA A/1.,2.,-2.,2.,-4.,4.,8.,-6.,2./
          DET=FACTOR(3,A,P)
          WRITE(5,100) DET,(P(I),(A(I,J),J=1,3),I=1,3)
100       FORMAT(F10.5/3(I10,3F10.5/))
          DATA B/-3.,0.,4.,1.,0.,0.,0.,1.,0.,0.,0.,1./
          CALL LU(A,P,3,B,4)
          WRITE(5,101) ((B(I,J),J=1,4),I=1,3)
101       FORMAT(4F10.5/)
          END

          REAL FUNCTION FACTOR(N,A,P)
          DIMENSION A(3,3),P(3)
          REAL MAX
          INTEGER P,PK,PI
          DO 10 I=1,N
10        P(I)=I
          N1=N-1
          DELTA=1.0
          DO 11 K=1,N1
          PK=P(K)
          MAX =ABS(A(PK,K))
          DO 12 I=K,N
          PI=P(I)
          IF(ABS(A(PI,K)).LT.MAX) GO TO12
          MAX=ABS(A(PI,K))
          PK=I
          DELTA=-DELTA
12        CONTINUE
          IF(MAX.GT.0) GO TO13
          FACTOR=0
          RETURN
13        PI=P(K)
          P(K)=P(PK)
          P(PK)=PI
          K1=K+1
          DO 14 I=K1,N
          PI=P(I)
          A(PI,K)=A(PI,K)/A(PK,K)
          DO 14 J=K1,N
14        A(PI,J)=A(PI,J)-A(PI,K)*A(PK,J)
11        DELTA=DELTA*A(PK,K)
          FACTOR=DELTA*A(P(N),N)
          RETURN
          END

          SUBROUTINE LU(A,P,N,B,M)
          DIMENSION A(3,3),P(3),B(3,4),T(3)
          INTEGER P,H
          DO 11 K=1,M
          T(1)=B(P(1),K)
          DO 10 I=2,N
          H=P(I)
          T(I)=B(H,K)
          ILIM=I-1
          DO 10 J=1,ILIM
10        T(I)=T(I)-A(H,J)*T(J)
          T(N)=T(N)/A(P(N),N)
          B(N,K)=T(N)
          DO 11 I=2,N
          IC=N+1-I
          H=P(IC)
          IP1=IC+1
          DO 12  J=IP1,N
12        T(IC)=T(IC)-A(H,J)*T(J)
          T(IC)=T(IC)/A(H,IC)
11        B(IC,K)=T(IC)
          END

              OUTPUT

       -32.00000
              3  -0.50000    4.00000    9.00000
              1  -1.00000    0.0       -4.00000
              2  -2.00000    4.00000    2.00000
        1.00000    0.50000    0.87500    0.62500
        2.00000    0.25000    0.56250    0.68750
       -1.00000    0.0       -0.25000   -0.25000
```

The second program is essentially two applications of the back solution algorithm that was discussed earlier. Again there is the complication that the permutation matrix P, developed in the factoring procedure and encoded as a vector, must be used in the two solutions. To see the basic process without this complication the two procedures $LY = B$ and $UX = Y$ are shown below in indented form. Matrices X, Y, and B are indicated, instead of the earlier use of vectors, because these algorithms will be designed to solve m systems of equations at a time. Thus the m right-hand-side vectors can be expressed as an $n \times m$ matrix B. This notation was used in the earlier numerical example where B was a specific solution and the identity matrix. The elements of both L and U will be indicated by a_{ij} and the elements of B, X, and Y by b_{ij}.

$$LY = B$$
$$k = 1, 2, \ldots, m$$
$$i = 2, 3, \ldots, n$$
$$j = 1, 2, \ldots, i - 1$$
$$b_{ik} = b_{ik} - a_{ij} \cdot b_{jk}$$

$$UX = Y$$
$$k = 1, 2, \ldots, m$$
$$b_{nk} = b_{nk}/a_{nn}$$
$$i = n - 1, n - 2, \ldots, 1$$
$$j = i, i + 1, \ldots, n$$
$$b_{ik} = b_{ik} - a_{ij} \cdot b_{jk}$$
$$b_{ik} = b_{ik}/a_{ii}$$

In combining these two "triangular solutions," entire columns may be completely solved, or, in other words, the iteration on k may have both solutions in its scope.

The effect of the addition of the permutation vector is similar for "half" solutions. Quite simply, the ith-row index is not i but is selected from P_i. The $LY = B$ algorithm with this addition is shown below.

$$k = 1, 2, \ldots, m$$
$$i = 2, 3, \ldots, n$$
$$h = P_i$$
$$j = 1, 2, \ldots, i - 1$$
$$b_{hk} = b_{hk} - a_{hj} \cdot b_{P_jk}$$

The transformation of these algorithms with a **FORTRAN** subroutine $LU(A, P, N, B, M)$ is straightforward. Here A is the matrix of dimension N containing the two factors, P the permutation vector of dimension N, and B the matrix of right-hand sides with N rows and M columns.

Notice in the actual program the substitution is not made immediately back into the B array. The element $b_{h,k} = b_{P_i,k}$ is first substituted into a temporary array T. That is, the substitution $T_i = b_{P_i,k}$ is made and in subsequent accesses the values are taken from the T array. Finally, after a solution column has been obtained the substitution $b_{i,k} = T_i$ is made, which effectively accomplishes the necessary reordering of elements, that is, the application of the permutation matrix P.

When the FACTOR procedure is applied to the example matrix, permuted factors are obtained. The permutation vector is $P = (3, 1, 2)$, representing the matrix

$$\begin{bmatrix} 0 & 0 & 1 \\ 1 & 0 & 0 \\ 0 & 1 & 0 \end{bmatrix}.$$

The resultant matrix is

$$\begin{bmatrix} -\frac{1}{2} & 4 & 9 \\ -1 & 0 & -4 \\ -2 & 4 & 2 \end{bmatrix},$$

from which

$$\hat{L} = \begin{bmatrix} -\frac{1}{2} & 1 & 0 \\ -1 & 0 & 1 \\ 1 & 0 & 0 \end{bmatrix},$$

$$\hat{U} = \begin{bmatrix} 0 & 4 & 9 \\ 0 & 0 & -4 \\ -2 & 4 & 2 \end{bmatrix}.$$

Solving $\hat{L}\bar{y} = \bar{b}$ when

$$\bar{b} = \begin{bmatrix} -3 \\ 0 \\ 4 \end{bmatrix}$$

yields

$$\bar{y} = \begin{bmatrix} -1 \\ 4 \\ 4 \end{bmatrix}$$

and $\hat{U}\bar{x} = \bar{y}$ gives

$$\bar{x} = \begin{bmatrix} 2 \\ -1 \\ 1 \end{bmatrix}.$$

The latter vector, when multiplied by P, results in the correct solution vector

$$\begin{bmatrix} 1 \\ 2 \\ -1 \end{bmatrix}.$$

Notice, in passing, that

$$\hat{L} \cdot P\hat{U} = \begin{bmatrix} -\frac{1}{2} & 1 & 0 \\ -1 & 0 & 1 \\ 1 & 0 & 0 \end{bmatrix} \begin{bmatrix} -2 & 4 & 2 \\ 0 & 4 & 9 \\ 0 & 0 & -4 \end{bmatrix} = \begin{bmatrix} 1 & 2 & 8 \\ 2 & -4 & -6 \\ -2 & 4 & 2 \end{bmatrix} = A.$$

Special treatment for tri-diagonal coefficient matrices

In many physical systems, the coefficient matrix is tri-diagonal; that is, all terms are zeros except for those along the main diagonal and on either side of it. For example, consider the following set of equations in matrix form:

$$\begin{bmatrix} a_{11} & a_{12} & 0 & 0 & \cdots & 0 \\ a_{21} & a_{22} & a_{23} & 0 & \cdots & 0 \\ 0 & a_{32} & a_{33} & a_{34} & \cdots & 0 \\ \vdots & & & & & \vdots \\ 0 & 0 & 0 & \cdots & a_{n,n-1} & a_{n,n} \end{bmatrix} \begin{bmatrix} x_1 \\ x_2 \\ x_3 \\ \vdots \\ x_n \end{bmatrix} = \begin{bmatrix} b_1 \\ b_2 \\ b_3 \\ \vdots \\ b_n \end{bmatrix}$$

It is evident that from the first equation x_1 can be expressed as a function of x_2, and this may be substituted into the second. If this is repeated, the last equation can be solved directly for x_n, and by back-substitution, the other x's can be found directly in a set of n calculations. For such systems it is often economically advisable to take advantage of this saving in computing time.

To develop the algorithm, let us first normalize the set of equations with respect to the diagonal element so that the system can be expressed as

$$\begin{aligned} x_1 - A_1 x_2 &= C_1 \\ -B_2 x_1 + x_2 - A_2 x_3 &= C_2 \\ -B_3 x_2 + x_3 - A_3 x_4 &= C_3 \\ \vdots \qquad\qquad & \vdots \\ -B_n x_{n-1} + x_n \qquad &= C_n \end{aligned}$$

or generally, each has the form expressed for the ith equation as,

$$-B_i x_{i-1} + x_i - A_i x_{i+1} = C_i.$$

The first can be solved for x_1 as

$$x_1 = C_1 + A_1 x_2$$

and the ith for x_i as,

$$x_i = C_i + A_i x_{i+1} + B_i x_{i-1}.$$

We seek a recursion relation, built upon successive substitution, which provides a solution for the ith unknown in terms of the $(i + 1)$th. That is

$$x_i = A'_i x_{i+1} + B'_i.$$

It is now necessary to obtain expressions for A'_i and B'_i which will make this equation true. The expression for x_1 is already in this form. From it we conclude that

$$A'_i = A_1 \quad \text{and} \quad B'_1 = C_1.$$

The equation for x_2 is

$$x_2 = C_2 + A_2 x_3 + B_2 x_1$$

or with the relation for x_1 introduced

$$x_2 = \frac{A_2 x_3}{1 - B_2 A'_i} x_3 + \frac{C_2 + B_2 B'_1}{1 - A_2 A'_1}.$$

Comparing this to the general recursion formula, it is evident that

$$A'_2 = \frac{A_2}{1 - B_2 A'_1} \quad \text{and} \quad B'_2 = \frac{B_2 B'_1 + C_2}{1 - B_2 A'_1}.$$

Repeating the process for x_3 leads to a similar form of expression, from which the general relations can be induced for the ith unknown:

$$x_i = A'_i x_{i+1} + B'_i$$

with

$$A'_i = \frac{A_i}{1 - B_i A'_{i-1}} \quad \text{and} \quad B'_i = \frac{B_i B'_{i-1} + C_i}{1 - B_i A'_{i-1}}.$$

Beginning, then, at $i = 1$, all N of the values of A'_i and B'_i can be computed once and for all from the coefficient matrix. It is then a simple matter to compute the unknowns from the general expression. It is necessary, of course, to begin the calculation for x_n as the value of $A'_n = 0$, making

$$x_n = B'_n.$$

Then x_{n-1} can be solved for in terms of x_n and the remaining x's can be computed in their turn. Once the modified coefficients are found, this requires only N simple calculations. In cases where the matrix expression is solved for repeated applications, as for example, a stepwise solution of certain types of differential equations, the saving in time can be quite significant.

Iterative methods

Iterative methods similar to those illustrated for equations involving one unknown can be used when the functions involve many variables. The problems of convergence of the iteration are often compounded by the additional variables, but such methods do have a virtue that the elimination methods do not possess. Error is not propagated. Since the general procedure is one of taking an approximate solution to a set of equations and then improving it, computational errors which are introduced during the improvement simply mean that the next approximate solution is not quite so good as it might be. If the process is assumed to converge, the error could conceivably result in an additional iterative cycle, but no loss in accuracy of the final solution would occur. When the number of equations is very large, the propagated error in elimination methods becomes a serious problem, and iterative methods are useful to obtain the solution or improve the solution after an approximate solution has been determined by elimination.

Given n simultaneous equations,

$$f_1(x_1, x_2, \ldots, x_n) = 0$$
$$f_2(x_1, x_2, \ldots, x_n) = 0$$
$$\vdots$$
$$f_n(x_1, x_2, \ldots, x_n) = 0$$

which are not necessarily linear, one scheme for the improvement of an initial approximate solution $x_1^{(0)}, x_2^{(0)}, x_3^{(0)}, \ldots, x_n^{(0)}$ would be to apply Newton's method to each equation individually.

As the functions are in terms of several variables, the Taylor series expansion of each function in terms of these variables necessitates the use of partial derivatives. For a Taylor series of degree one, expanded about $x_i^{(k)}$, the set of functions is expressed as:

$$f_1(x_1, x_2, \ldots, x_n) = [f_1(x_1, x_2, \ldots, x_n)]^{(k)} + (x_1 - x_1^{(k)})\left(\frac{\partial f_1}{\partial x_1}\right)^{(k)}$$
$$+ (x_2 - x_2^{(k)})\left(\frac{\partial f_1}{\partial x_2}\right)^{(k)} + \cdots$$
$$+ (x_n - x_n^{(k)})\left(\frac{\partial f_1}{\partial x_n}\right)^{(k)} + R_1,$$

$$f_2(x_1, x_2, \ldots, x_n) = [f_2(x_1, x_2, \ldots, x_n)]^{(k)} + (x_1 - x_1^{(k)})\left(\frac{\partial f_2}{\partial x_1}\right)^{(k)}$$
$$+ (x_2 - x_2^{(k)})\left(\frac{\partial f_2}{\partial x_2}\right)^{(k)} + \cdots$$
$$+ (x_n - x_n^{(k)})\left(\frac{\partial f_2}{\partial x_n}\right)^{(k)} + R_2.$$

The index of the approximation is indicated by a superscript. The remainder term is R and will be subsequently dropped. There will be similar equations for all of the functions through $f_n(x_1, x_2, \ldots, x_n)$. In matrix form the entire set can be written as:

$$f_j(\overline{x}) = f_j(\overline{x}^{(k)}) + \sum_{i=1}^{n} (x_i - x_i^{(k)}) \left[\frac{\partial f_j(\overline{x}^{(k)})}{\partial x_k} \right], \qquad j = 1, 2, \ldots, n$$

Formulas for the improvement of the components of $\overline{x}$ (in this case x_1) are constructed on the basis of the following question: What values of x_1 will cause f_1 to be zero for a set of $x_2, x_3, \ldots, x_n$? Under this restriction, $x_2 = x_2^{(k)}$, $x_3 = x_3^{(k)}$, etc. Consequently, f_1 is set to zero, and the Taylor series is simply

$$0 = [f_1(x_1, x_2, \ldots, x_n)]^{(k)} + (x_1^{(k+1)} - x_1^{(k)}) \left(\frac{\partial f_1}{\partial x_1} \right)^{(k)},$$

which leads directly to the improvement formula for x_1,

$$x_1^{(k+1)} = x_1^{(k)} - \frac{[f_1(x_1, x_2, \ldots, x_n)]^{(k)}}{(\partial f_1/\partial x_1)^{(k)}}.$$

Following the procedure for $f_2, f_3, \ldots, f_n$, the improvement formulas for the other variables are:

$$x_2^{(k+1)} = x_2^{(k)} - \frac{[f_2(x_1, x_2, \ldots, x_n)]^{(k)}}{(\partial f_2/\partial x_2)^{(k)}}$$

$$\vdots$$

$$x_n^{(k+1)} = x_n^{(k)} - \frac{[f_n(x_1, x_2, \ldots, x_n)]^{(k)}}{(\partial f_n/\partial x_n)^{(k)}}$$

These then are the iterative formulas of Newton's method for several variables. The solution procedure would be to use equation 1 to improve x_1, then equation 2 to improve x_2, etc., and to repeat this improvement until all the equations are simultaneously satisfied. Again, as a practical matter, the values of the functions will probably not be zero, but the sum of the absolute values of the differences from zero (the residuals) should be acceptably small.

In the particular case of interest, i.e. for linear equations, the root improvement formula is of a simple form. For the ith equation,

$$f_i(x_1, \ldots, x_n) = a_{i1}x_1 + a_{i2}x_2 + \cdots + a_{ii}x_i + \cdots + a_{in}x_n - a_{i,n+1} = 0,$$

the improvement formula is

$$x_i^{(k+1)} = x_i^{(k)} - \frac{\sum_{j=1}^{n} a_{ij}x_j - a_{i,n+1}}{a_{ii}} = x_i^{(k)} - \frac{r_i^{(k)}}{a_{ii}}.$$

In this case this same result can be obtained by solving f_1 for x_1; f_2 for x_2; and so forth. Employing the generalities of Newton's method will allow us to extend this analysis to other numerical techniques.

The numerator of the rightmost term will be zero if the equation is satisfied, and is called the residual, $r_i^{(k)}$. When Newton's iteration formula is applied to a straight line, the improved approximation is the root, but, of course, the function actually depends on n variables and hence the repeated trials are necessary.

Reverting to substitution forms and neglecting the superscripts, one proceeds by computing

$$\sum_{j=1}^{n} a_{ij}x_j - a_{i,n+1} = r_i,$$

$$x_i \leftarrow x_i - \frac{r_i}{a_{ii}},$$

for $i = 1, 2, \ldots, n$. At the same time, $\sum_{i=1}^{n} |r_i|$ is computed, and when this sum is sufficiently close to zero the solution is complete. Otherwise, the improvement cycle is repeated until there is convergence, or until some maximum number of iterations have been completed. This method is called the Gauss-Seidel method of solving linear equations.

In the Gauss-Seidel method, when a new value of x_i is formed, it immediately replaces the old value. This is sometimes referred to as a *single-step* method. In *total-step* methods, an improved approximation is made only after all of the improvement formulas have been evaluated. The total-step method for solving simultaneous equations is called Jacobi's method. As is apparent in the example which follows, Jacobi's method converges more slowly than the Gauss-Seidel method. Because of this, and because it is easier to update a variable as soon as a new value is found on a digital computer, only the Gauss-Seidel method should be given serious consideration.

For the set of equations

$$\begin{aligned}
2x_1 + x_2 \qquad\qquad &= 2, \\
x_1 + 4x_2 + 2x_3 &= 0, \\
2x_2 + 4x_3 &= 0,
\end{aligned}$$

compute the values of the unknowns by iteration.

The improvement formulas are

$$x_1 = 1 - \tfrac{1}{2}x_2, \qquad x_2 = -\tfrac{1}{4}x_1 - \tfrac{1}{2}x_3, \qquad x_3 = -\tfrac{1}{2}x_2.$$

With the total-step method (Jacobi's method) and initial approximations of 0, 0, 0, the values of the unknowns for four successive steps are:

k	x_1	x_2	x_3
0	0	0	0
1	1	0	0
2	1	$-.250$	0
3	1.125	$-.250$	.125
4	1.125	$-.344$	.125

Using the Gauss-Seidel method, the values after four steps are closer to the correct roots 1.20, $-.40$, .20.

k	x_1	x_2	x_3
0	0	0	0
1	1	$-.25$	.125
2	1.125	$-.344$	.172
3	1.172	$-.379$	.189
4	1.189	$-.395$	.198

Example Write a subroutine which attempts to solve a system of linear equations by the Gauss-Seidel method. If the solution is not obtained in 100 trials, the computation is terminated and the subroutine returns an integer, INDEX = 1. This causes the main program to print the comment 'AFTER 100 ITERATIONS NO SOLUTION FOUND'. For a successful solution, INDEX = 2 is returned, and the values of the functions are printed. The necessary input information is the number of equations, n, the name of the augmented matrix, a, the criterion for convergence, ϵ, and the initial approximation to the solution vector $x_1, x_2, \ldots, x_n$. (The flow chart, the subroutine, and a calling program are given in Fig. 6.6.) Convergence has been satisfied only when the sum of the absolute values of the residuals is less than the test value EPS. Three sets of input were supplied to the example program. Notice that convergence was not achieved for the third set.

The Gauss-Seidel method is a *single-step* method in that the solution is improved after the evaluation of each equation. Other *total-step* methods compute an improved approximation only after all equations have been evaluated. Among the latter is the method of *steepest descent*, which is similar to Newton's method, but the assumption that each equation may be treated as a function of a single variable is not made. If one pictures the approximation as being a point on a surface, the path of steepest descent from the point is projected (analogous to the tangent in Newton's method) to determine the next approximation.

Relaxation

The method of relaxation is still another iterative approach to the solution of linear equations. It is particularly suitable for hand calculation. Successive iteration is performed in an effort to cause all of the residuals to vanish simultaneously, or to put it more practically, to cause the sum of the absolute values of the residuals to be small. One advantage which this method offers for hand calculation is that it allows the person doing the computation to intervene and make judgments which improve convergence. Furthermore, the computations become simple bookkeeping operations using simple arithmetic steps.

To demonstrate the method, we return to the set of equations

$$\sum_{j=1}^{n} a_{ij}x_j = b_i, \qquad i = 1, 2, \ldots, n.$$

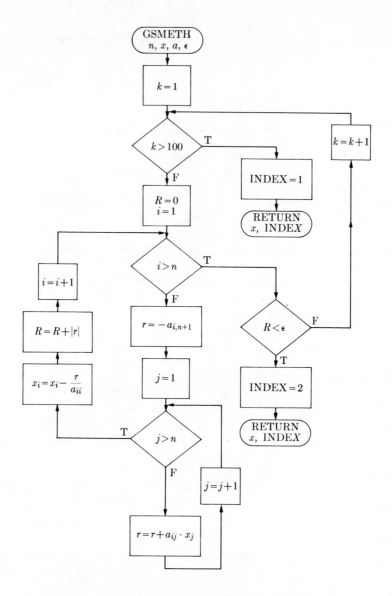

Fig. 6.6   Gauss-Seidel method subroutine.   (a) Flow chart.

The residual has been defined as

$$r_i = \sum_{j=1}^{n} a_{ij}x_j - b_i, \qquad i = 1, 2, \ldots, n.$$

It is evident that when the correct x_j are found, all r_i will go to zero.

```
C        MASTER GAUS
C        MAINLINE PROGRAM FOR GAUSS-SEIDEL METHOD
         DIMENSION X(20),A(20,21)
C        N IS THE NUMBER OF UNKNOWNS
C        EPS IS THE ACCURACY OF THE ITERATION
5        FORMAT(I9,F9.0)
10       FORMAT(8F9.0)
15       FORMAT(5X,38HAFTER 100 ITERATIONS NO SOLUTION FOUND///)
20       FORMAT(5X,2HX(,I2,3H) =,F15.6//)
4        READ(5,5) N,EPS
         NP1=N+1
         DO 11 I=1,N
11       READ(5,10) (A(I,J),J=1,NP1)
         READ(5,10) (X(I),I=1,N)
C        X IS THE TRIAL SOLUTION
         CALL GSMETH(N,NP1,A,X,EPS,INDEX)
21       FORMAT(4F9.5)
         GO TO (12,13),INDEX
12       WRITE(6,15)
         GO TO 14
13       DO 9 I=1,N
9        WRITE(6,20) I,X(I)
         GO TO 4
14       STOP
         END

         SUBROUTINE GSMETH(N,M,A,X,EPS,INDEX)
         DIMENSION X(20),A(20,21)
         DO 1 K=1,100
         CAPR=0.
         DO 2 I=1,N
         R=-A(I,M)
         DO 3 J=1,N
3        R=R+A(I,J)*X(J)
         X(I)=X(I)-R/A(I,I)
2        CAPR=CAPR+ABS(R)
         IF(CAPR.LT.EPS) GO TO 4
1        CONTINUE
         INDEX=1
         GO TO 5
4        INDEX=2
5        RETURN
         END
```

                    INPUT

```
  3,.001,
  4.,2.,-6.,0.,
  0.,4.,3.,7.,
  -1.,0.,6.,5.,
   1.,1.,1.,
   3,.0001,
  -3.,8.,5.,6.,
  2.,-7.,4.,9.,
  1.,9.,6.,1.,
   1.,1.,1.,
   3,.0001,
  2.,-7.,4.,9.,
  1.,9.,-6.,1.,
  -3.,8.,5.,6.,
   1.,1.,1.,
```

                    OUTPUT

```
  X( 1) =        1.000000
  X( 2) =        1.000000
  X( 3) =        1.000000
  X( 1) =       -1.450847
  X( 2) =       -0.789833
  X( 3) =        1.593224
  AFTER 100 ITERATIONS NO SOLUTION FOUND
```

Fig. 6.6 (b) Program.

An initial set of x_i, called $x_i^{(0)}$, are chosen, and the residuals, $r_i^{(0)}$, are calculated. The x_i are altered in an effort to reduce the residuals. To facilitate this, a residual table such as the following is constructed, by tabulating the amount each residual will change when each of the x's is changed by unity. What the table shows is that when x_1 (and only x_1) is changed, r_1 is changed by a_{11}; r_2 by a_{21}; r_3 by a_{31}, and so on.

Δx_1	Δx_2	Δx_3	Δx_n	Δr_1	Δr_2	Δr_3	Δr_n
1	0	0	...0	a_{11}	a_{21}	a_{31}	... a_{n1}
0	1	0	...0	a_{12}	a_{22}	a_{32}	... a_{n2}
0	0	1	...0	a_{13}	a_{23}	a_{33}	... a_{n3}
$\vdots$			$\vdots$	$\vdots$			$\vdots$
0	0	0	...1	a_{1n}	a_{2n}	a_{3n}	... a_{nn}

It is apparent that the elements in the residual table can be represented as elements of the transpose of the coefficient matrix.

When all of the residuals have been calculated for the initial x_i, the relaxation process proceeds by selecting a large residual, say r_k, and changing the corresponding x_k by an amount sufficient to cause that residual to be zero. The change in x_k required can be found quickly by inspection of the residual table. One soon learns that the residuals should be "over relaxed," that is, altered enough to change the sign of the residual, or more than enough to set it to zero. A change in x_k alters the other residuals in accordance with the coefficients in row k in the residual table.

The next highest residual is then "relaxed" and the process continued until all residuals are sufficiently small to provide the required accuracy. In each step an entry is made to show the change in x_i, and the current value of each residual. The changes in x are added to the initial value to give the current value. At that time the residuals appearing in the table can be checked.

The process can best be described by an example. Returning to the sample problem

$$2x_1 + x_2 \qquad = 2$$
$$x_1 + 4x_2 + 2x_3 = 0$$
$$2x_2 + 4x_3 = 0,$$

we find that the residuals are:

$$r_1 = 2x_1 + x_2 - 2,$$
$$r_2 = x_1 + 4x_2 + 2x_3 - 0,$$
$$r_3 = 2x_2 + 4x_3 - 0;$$

and the residual table is:

Δx_1	Δx_2	Δx_3	Δr_1	Δr_2	Δr_3
1	0	0	2	1	0
0	1	0	1	4	2
0	0	1	0	2	4

Now let us carry out a few steps in the calculation, beginning with a set of trial values $x_1 = x_2 = x_3 = 0$. Entries will be made in tabular form.

Step	x_1	x_2	x_3	r_1'	r_2	r_3

Trial entries are entered and the residuals calculated.

0.	0	0	0	-2	0	0

To facilitate computation we multiply all values by 100:

1.	0	0	0	-200	0	0

r_1 is reduced to zero by altering x_1 by $+100$:

2.	100			0	100	0

Notice that the value of the change in x_1 is entered along with the current value of the residual. Calculation of the change in residuals was done by referring to the residual table. Now r_2 can be reduced. This time it will be made less than zero by changing x_2 by -30:

3.		-30		-30	-20	-60

As r_3 is now largest, change x_3 by $+16$:

4.			16	-30	12	4

Continue the process:

5.	16			$+2$	28	4
6.		-8		-6	-4	-12
7.			4	-6	4	4
8.	4			2	8	4
9.		-2		0	0	0

At this point, the current value of the solution vector can be found by applying the sum of the changes to the trial values, and the residuals can be checked by the residual equations.

10.	120	-40	20	0	0	0

Correcting for the factor of 100 introduced in step 2, the solution is

$$x_1 = 1.20; \quad x_2 = -0.40; \quad \text{and} \quad x_3 = 0.20.$$

This is exact, and the residuals are exactly zero. In the usual case the residuals are small, but not zero. Had this been the case, and had more precision been required, a second factor of 100 could have been introduced and the process continued.

Simultaneous nonlinear equations

When there are nonlinear terms in the simultaneous equations, the problem is considerably more complicated. For one thing, there may be several values of an unknown which satisfy the set of equations. One means of solving the nonlinear set of equations is to apply the method of successive approximations, described in Chapter 4.

Starting with the n equations in n unknowns,

$$f_1(x_1, x_2, \ldots, x_n) = 0$$
$$f_2(x_1, x_2, \ldots, x_n) = 0$$
$$\vdots \qquad\qquad \vdots$$
$$f_n(x_1, x_2, \ldots, x_n) = 0$$

each is solved for x_i, by rewriting the equations as:

$$x_1 = F_1(x_1, x_2, \ldots, x_n)$$
$$x_2 = F_2(x_1, x_2, \ldots, x_n)$$
$$\vdots \qquad\qquad \vdots$$
$$x_n = F_n(x_1, x_2, \ldots, x_n)$$

A trial set of x_i are then selected in the vicinity of the solution, and the iteration proceeds, as in the one-dimensional case, with the improvement formulas:

$$x_1^{(k+1)} = F_1(x_1^{(k)}, x_2^{(k)}, \ldots, x_n^{(k)})$$
$$x_2^{(k+1)} = F_2(x_1^{(k)}, x_2^{(k)}, \ldots, x_n^{(k)})$$
$$\vdots \qquad\qquad \vdots$$
$$x_n^{(k+1)} = F_n(x_1^{(k)}, x_2^{(k)}, \ldots, x_n^{(k)})$$

As in the one-dimensional case, the success of this method depends upon convergence. We shall examine the conditions for convergence for the case of two unknowns. First, the functions F_1 and F_2 are expanded in a linear Taylor series about $x_1^{(0)}$ and $x_2^{(0)}$. The series is evaluated at the first approximation, $x_1^{(1)}, x_2^{(1)}$:

$$F_1(x_1^{(1)}, x_2^{(1)}) = F_1(x_1^{(0)}, x_2^{(0)}) + (x_1^{(1)} - x_1^{(0)})\frac{\partial \overline{F}_1}{\partial x_1} + (x_2^{(1)} - x_2^{(0)})\frac{\partial \overline{F}_1}{\partial x_2},$$

$$F_2(x_1^{(1)}, x_2^{(1)}) = F_2(x_1^{(0)}, x_2^{(0)}) + (x_1^{(1)} - x_1^{(0)})\frac{\partial \overline{F}_2}{\partial x_1} + (x_2^{(1)} - x_2^{(0)})\frac{\partial \overline{F}_2}{\partial x_2}.$$

A bar over the $\overline{F}$ has been used to indicate that the function has been evaluated at some points ζ_1 and ζ_2 in the interval such that

$$x_1^{(0)} < \zeta_1 < x_1^{(1)} \qquad \text{and} \qquad x_2^{(0)} < \zeta < x_2^{(1)}.$$

From the improvement formulas

$$x_1^{(1)} = F_1(x_1^{(0)}, x_2^{(0)}), \qquad x_2^{(1)} = F_2(x_1^{(0)}, x_2^{(0)}),$$

and

$$x_1^{(2)} = F_1(x_1^{(1)}, x_2^{(1)}), \qquad x_2^{(2)} = F_2(x_1^{(1)}, x_2^{(1)}).$$

Introducing these expressions into the Taylor series results in the equations

$$x_1^{(2)} - x_1^{(1)} = (x_1^{(1)} - x_1^{(0)})\frac{\partial \overline{F}_1}{\partial x_1} + (x_2^{(1)} - x_2^{(0)})\frac{\partial \overline{F}_1}{\partial x_2},$$

$$x_2^{(2)} - x_2^{(1)} = (x_1^{(1)} - x_1^{(0)})\frac{\partial \overline{F}_2}{\partial x_1} + (x_2^{(1)} - x_2^{(0)})\frac{\partial \overline{F}_2}{\partial x_2}.$$

Taking absolute values of these expressions, and noting that the sum of the absolute values of two terms is greater than or equal to the absolute values of their sum, we may write:

$$|x_1^{(2)} - x_1^{(1)}| \leq |x_1^{(1)} - x_1^{(0)}|\left|\frac{\partial \overline{F}_1}{\partial x_1}\right| + |x_2^{(1)} - x_2^{(0)}|\left|\frac{\partial \overline{F}_1}{\partial x_2}\right|,$$

$$|x_2^{(2)} - x_2^{(1)}| \leq |x_1^{(1)} - x_1^{(0)}|\left|\frac{\partial \overline{F}_2}{\partial x_1}\right| + |x_2^{(1)} - x_2^{(0)}|\left|\frac{\partial \overline{F}_2}{\partial x_2}\right|.$$

Summing the two expressions gives

$$|x_1^{(2)} - x_1^{(1)}| + |x_2^{(2)} - x_2^{(1)}| \leq |x_1^{(1)} - x_1^{(0)}|\left[\left|\frac{\partial \overline{F}_1}{\partial x_1}\right| + \left|\frac{\partial \overline{F}_2}{\partial x_1}\right|\right]$$

$$+ |x_2^{(1)} - x_2^{(0)}|\left[\left|\frac{\partial \overline{F}_1}{\partial x_2}\right| + \left|\frac{\partial \overline{F}_2}{\partial x_2}\right|\right].$$

Assuming that the partial derivatives exist and are continuous in the neighborhood of the root, there must be some value, K, such that:

$$\left[\left|\frac{\partial \overline{F}_1}{\partial x_1}\right| + \left|\frac{\partial \overline{F}_2}{\partial x_1}\right|\right] \leq K \quad \text{and} \quad \left[\left|\frac{\partial \overline{F}_1}{\partial x_2}\right| + \left|\frac{\partial \overline{F}_2}{\partial x_2}\right|\right] \leq K, \quad \text{with } K > 0.$$

Carrying the iteration through several steps,

$$|x_1^{(2)} - x_1^{(1)}| + |x_2^{(2)} - x_2^{(1)}| \leq K[|x_1^{(1)} - x_1^{(0)}| + |x_2^{(1)} - x_2^{(0)}|]$$

$$|x_1^{(3)} - x_1^{(2)}| + |x_2^{(3)} - x_2^{(2)}| \leq K[|x_1^{(2)} - x_1^{(0)}| + |x_2^{(1)} - x_2^{(0)}|]$$

$$\vdots$$

$$|x_1^{(n)} - x_1^{(n-1)}| + |x_2^{(n)} - x_2^{(n-1)}| \leq K[|x_1^{(n-1)} - x_1^{(n-2)}| + |x_2^{(n-1)} - x_2^{(n-2)}|].$$

Solving backwards to express the last term on the left side in terms of the first term on the right, we obtain:

$$|x_1^{(n)} - x_1^{(n-1)}| + |x_2^{(n)} - x_2^{(n-1)}| \leq K^n[|x_1^{(1)} - x_1^{(0)}| + |x_2^{(1)} - x_2^{(0)}|].$$

For convergence to the root, the left side must approach zero. This will happen only if $K < 1$. Thus the iteration will converge to the root if the initial choice of the variables is in a region where the derivatives exist and are continuous, and for which

$$\left[\left|\frac{\partial F_1}{\partial x_1}\right| + \left|\frac{\partial F_2}{\partial x_1}\right|\right] < 1, \quad \left[\left|\frac{\partial F_1}{\partial x_1}\right| + \left|\frac{\partial F_2}{\partial x_1}\right|\right] < 1$$

everywhere in the region.

While it has been shown that $K < 1$ is a sufficient condition for convergence, it has not been proved that the solution will always diverge if $K > 1$.

To demonstrate the application of the method, consider the following set of equations:

$$x_1^2 + 2x_1x_2 + 10x_1 = 15,$$
$$x_1 + 5x_2 - \tfrac{1}{2}x_2^2 = 9.$$

There is a root at $x_1 = 1$ and $x_2 = 2$. Writing the equations in the form for iteration,

$$x_1 = 1.5 - 0.1x_1^2 - 0.2x_1x_2 = F_1,$$
$$x_2 = 1.8 - 0.2x_1 + 0.1x_2^2 = F_2,$$

$$\frac{\partial F_1}{\partial x_1} = -0.2x_1 - 0.2x_2, \qquad \frac{\partial F_1}{\partial x_2} = -0.2x_1,$$

$$\frac{\partial F_2}{\partial x_1} = -0.2, \qquad \frac{\partial F_2}{\partial x_2} = 0.2x_2.$$

Convergence is guaranteed for $x_1 < 1.5$ and $x_2 < 2.5$. Starting with $x_1 = 0.5$ and $x_2 = 1.0$, calculations were carried out for four steps. The results are reported in the table below, with values on the left obtained using total steps, while the method of single steps was used to produce the results in the right-hand columns:

Step	Method of total steps		Method of single steps	
	x_1	x_2	x_1	x_2
0	0.5000	1.0000	0.5000	1.0000
1	1.3750	1.8000	1.3750	1.6250
2	0.8159	1.8490	.8641	1.8598
3	1.1317	1.9787	1.104	1.9251
4	.9241	1.9652	.9531	1.9800

Although convergence is faster for the single-step method, convergence with the method of successive approximation is slow and restricted. Applying Newton's method directly to the system of equations offers an alternative. Once again the formulation begins with the expansion of the functions in a linear Taylor series. For the kth iteration the matrix form of the expressions is:

$$f_j(\bar{x}) = f_j(\bar{x}^{(k)}) + \sum_{i=1}^{n} (x_i - x_i^{(k)})\left[\frac{\partial f_j(\bar{x}^{(k)})}{\partial x_i}\right] + R_j, \quad j = 1, 2, \ldots, n.$$

For the iteration to converge to the root $\bar{r}$ requires:

$$\lim_{k \to \infty} (\bar{r} - \bar{x}^{(k)}) \to 0.$$

Dropping the remainder term and evaluating the Taylor series at the root,

$$f_j(\bar{r}) = f_j(\bar{x}^{(k)}) + \sum_{i=1}^{n} (r_i - x_i^{(k)}) \left[\frac{\partial f_j}{\partial x_i} (\bar{x}^{(k)}) \right] = 0,$$

or

$$-f_j(\bar{x}^{(k)}) = \sum_{i=1}^{n} (r_i - x_i^{(k)}) \left[\frac{\partial f_j}{\partial x_i} (\bar{x}^{(k)}) \right], \quad j = 1, 2, \ldots, n.$$

Defining error terms after k iterations as

$$e_1^{(k)} = r_1 - x_1^{(k)}$$
$$e_2^{(k)} = r_2 - x_2^{(k)}$$
$$\vdots$$
$$e_n^{(k)} = r_n - x_n^{(k)}$$

the iteration formula becomes

$$-f_j(\bar{x}^{(k)}) = \sum_{i=1}^{n} e_i^{(k)} \left[\frac{\partial f_j(\bar{x}^{(k)})}{\partial x_i} \right], \quad j = 1, 2, \ldots, n,$$

or, in expanded form,

$$\frac{\partial f_1}{\partial x_1} (\bar{x}^{(k)}) e_1^{(k)} + \frac{\partial f_1}{\partial x_2} (\bar{x}^{(k)}) e_2^{(k)} \cdots + \frac{\partial f_1}{\partial x_n} (\bar{x}^{(k)}) e_n^{(k)} = -f_1(\bar{x}^{(k)})$$

$$\frac{\partial f_2}{\partial x_1} (\bar{x}^{(k)}) e_1^{(k)} + \frac{\partial f_2}{\partial x_2} (\bar{x}^{(k)}) e_2^{(k)} \cdots + \frac{\partial f_2}{\partial x_n} (\bar{x}^{(k)}) e_n^{(k)} = -f_2(\bar{x}^{(k)})$$

$$\vdots$$

$$\frac{\partial f_n}{\partial x_1} (\bar{x}^{(k)}) e_1^{(k)} + \frac{\partial f_n}{\partial x_2} (\bar{x}^{(k)}) e_2^{(k)} \cdots + \frac{\partial f_n}{\partial x_n} (\bar{x}^{(k)}) e_n^{(k)} = -f_n(\bar{x}^{(k)}).$$

Introduction of the error term has resulted in transforming the nonlinear equations into a set of simultaneous equations which are linear in $e_1, e_2, \ldots, e_n$. It is possible to solve the set of equations for the error terms providing the determinant of the coefficient matrix, say C, is nonsingular:

$$\det C = \begin{vmatrix} \dfrac{\partial f_1}{\partial x_1} & \dfrac{\partial f_1}{\partial x_2} & \cdots & \dfrac{\partial f_1}{\partial x_n} \\[2mm] \dfrac{\partial f_2}{\partial x_1} & \dfrac{\partial f_2}{\partial x_2} & \cdots & \dfrac{\partial f_2}{\partial x_n} \\[2mm] \vdots & & & \vdots \\[2mm] \dfrac{\partial f_n}{\partial x_1} & \dfrac{\partial f_n}{\partial x_2} & \cdots & \dfrac{\partial f_n}{\partial x_n} \end{vmatrix} \neq 0.$$

To begin the calculation process, initial values of $\bar{x}^{(0)}$ are assumed. Functions and the appropriate partial derivatives are evaluated to form the set of linear equations in the error,

$$C\bar{e}^{(k)} = -\bar{f}(\bar{x}^{(k)}).$$

Employing a Gauss-Jordan reduction, or the Gauss-Seidel method, the errors $\bar{e}^{(k)}$ are found. Using the definition of the error terms, and assuming that the new value of $\bar{x}$ is approaching a root, improved values of $\bar{x}$ are found by the relation

$$x_i^{(k+1)} = x_i^{(k)} + e_i^{(k)}.$$

With the improved x_i found, new values of the functions and the derivatives can be computed, and a new set of errors determined. The process is repeated until convergence is satisfactory.

A numerical example should be helpful in demonstrating the method. Consider again the set of nonlinear equations

$$f_1(\bar{x}) = x_1^2 + 2x_1x_2 + 10x_1 - 15 = 0,$$
$$f_2(\bar{x}) = x_1 + 5x_2 - 0.5x_2^2 - 9 = 0;$$

$$\frac{\partial f_1}{\partial x_1} = 2x_1 + 2x_2 + 10, \qquad \frac{\partial f_1}{\partial x_2} = 2x_1$$

$$\frac{\partial f_2}{\partial x_1} = 1, \qquad \frac{\partial f_2}{\partial x_2} = 5 - x_2.$$

With trial values of $x_1^{(0)} = 0.5$ and $x_2^{(0)} = 1.0$, the equations in $\bar{e}$ are

$$13.e_1^{(0)} + e_2^{(0)} = 8.75,$$
$$e_1^{(0)} + 4e_2^{(0)} = 4.00,$$

from which

$$e_1^{(0)} = 0.6078, \qquad e_2^{(0)} = 0.8481,$$

and

$$x_1^{(1)} = x_1^{(0)} + e_1^{(0)} = 1.1078, \qquad x_2^{(1)} = x_2^{(0)} + e_2^{(0)} = 1.8481.$$

Results for several iterations are given in Table 6.1. It is evident that convergence is much more rapid than it was for the method of successive approximation. The third iteration has produced results within .01% of the true root (1, 2).

For convergence with Newton's method (sometimes called the Newton-Raphson method) it is necessary that the first and second derivatives of the functions be continuous and finite, and that the determinant of the matrix C not

Table 6.1

Calculation of Roots of Nonlinear Equations by the Newton-Raphson Method

i	x_1	x_2	$\dfrac{\partial f_1}{\partial x_1}$	$\dfrac{\partial f_1}{\partial x_2}$	$\dfrac{\partial f_2}{\partial x_1}$	$\dfrac{\partial f_2}{\partial x_2}$	$-f_1$	$-f_2$	e_1	e_2
0	.5	1.	13.	1	1	4	8.75	4	.6078	.8481
1	1.1078	1.8481	15.9118	2.2156	1	3.1519	−1.3999	.3644	.1643	−.1535
2	.9543	2.0124	15.9334	1.9086	1	2.9876	.7054	.0085	.0456	−.0125
3	.9999	1.9999								

vanish. Consequently, it is necessary, as it was in the case of a single unknown, to choose the first approximation close to the root.

Yet another form of solution for nonlinear systems of equations can be generated by applying Newton's method to improve the unknowns directly. In this application, the improvement formulas are identical to those used for iterative solutions to linear equations derived earlier in the chapter.

$$x_1^{(k+1)} = x_1^{(k)} - \frac{f_1(\overline{x}^{(k)})}{\partial f_1(\overline{x}^{(k)})/\partial x_1}$$
$$\vdots \qquad \qquad \vdots$$
$$x_n^{(k+1)} = x_n^{(k)} - \frac{f_n(\overline{x}^{(k)})}{\partial f_n(\overline{x}^{(k)})/\partial x_n}$$

Following an assumption of initial values of $\overline{x}$, the iteration proceeds using a single-step method to hasten convergence.

We have treated three methods for the solution of systems of nonlinear algebraic equations: successive approximation, Newton's method with iteration of an error term, and Newton's method with iteration of the unknown. Each is iterative, and each requires that the initial approximations be near enough the root to insure convergence to the root of interest. Often it is easier to resort to the brute force of trial-and-error methods to obtain the roots, rather than search out values near the roots and follow one of the more elegant techniques described above.

Exercises

1. Rewrite SLEQ so that the computation of new elements within a row proceeds from right to left and reduces all the elements, even those resulting in zeros. This version requires three statements less than SLEQ does.

2. Write a computer program which inverts a matrix by evaluating each coefficient of the inverse according to the equation

$$\alpha_{ij} = \frac{\Delta_{ji}}{\Delta}.$$

The determinant Δ will be obtained by evaluating co-factors. The numerator of the expression, Δ_{ji}, is the minor of element a_{ji} of the matrix to be inverted.

3. Consider the following set of linear equations:

$$5x_1 - 4x_2 = 2$$
$$-4x_1 + 10x_2 - 5x_3 = 0$$
$$-5x_2 + 6x_3 = -1$$

a) Obtain a solution of the equations for the x_i by the Gauss-Seidel method.
b) Obtain a solution using Gaussian elimination.
c) Solve by matrix inversion.
d) Solve by the method of relaxation.

4. The "back solution" is the solution of a specialized system of equations such as

$$x_1 + a_{12}x_2 + a_{13}x_3 = a_{14} \tag{1}$$

$$x_2 + a_{23}x_3 = a_{24} \tag{2}$$

$$x_3 = a_{34} \tag{3}$$

Write the algorithm, using the "indented" form described in this chapter, by substituting x_3 in equation (2) to determine x_2 and x_3 in equation (1) to determine x_1. Extend the algorithm to n equations instead of three.

5. The basic operation of the forward solution, with division, is

$$a'_{ij} = a_{ij} - a_{ik}a_{kj},$$

with appropriate ranges for i, j, and k. There is a variant of this basic elimination scheme, called the Crout method, which carries out these basic steps in a different order. The algorithm for a single column of right-hand sides is:

$$k = 1, 2, \ldots, n - 1$$
$$j = k + 1, \ldots, n + 1$$
$$a_{kj} = a_{kj}/a_{kk}$$
$$i = k + 1$$
$$j = k + 1, \ldots, n + 1$$
$$1 = 1, 2, \ldots, k$$
$$a_{ij} = a_{ij} = a_{il}a_{lj}$$

Write a program which carries out the forward solution by means of this algorithm.

6. Show that interchanging two rows (or columns) of a square matrix changes the sign of the determinant.

7. Write a program which uses the Gauss-Jordan reduction scheme to solve n simultaneous linear equations where the pivots in the reduction are taken in order along the main diagonal of the coefficient matrix. Obtain a print-out of the values of the elements of the unknowns; the inverse of the coefficient matrix; and the value of the determinant of the coefficient matrix. Test each pivot; if it is less than a value ϵ print this out and go to the next case.

Using your program solve the following two cases.

a) *Case 1: n* = 3; ϵ, the test value for the magnitude of the pivot, is $1.0 \cdot E - 4$. The augmented matrix is:

$5.00 \cdot E - 3$	$2.00 \cdot E - 4$	$3.00 \cdot E - 4$	$1.00 \cdot E - 3$
$5.00 \cdot E - 5$	$1.00 \cdot E - 10$	$5.00 \cdot E - 5$	$4.30 \cdot E - 6$
$2.00 \cdot E - 3$	$1.00 \cdot E - 4$	$3.00 \cdot E - 1$	$2.70 \cdot E - 2$

b) *Case 2: n* = 4; $\epsilon = 1.0 \cdot E - 4$. The augmented matrix is:

5.	2.	3.	1.	4.
2.	5.	3.	0.	0.
4.	7.	11.	0.	16.
3.	1.	1.	5.	2.

8. Solve the following simultaneous equations using the Gauss-Jordan method with the diagonals as pivot elements:

$$x_1 + 50x_2 = 3,$$
$$10x_1 + x_2 = 4.$$

9. Solve Problem 8 with a simple row interchange and compare the accuracy of the two solutions.

10. Solve the following set of linear equations by a hand calculation using a Gauss-Jordan procedure with a simple row interchange:

$$x_1 + 5x_2 - 10x_3 = 2,$$
$$5x_1 + x_2 + 6x_3 = 3,$$
$$-10x_1 + 6x_2 + x_3 = -4.$$

Rows will be interchanged by searching the current pivot column for the maximal element. Record the order of pivot-row choice in an N-dimensional vector P, as was done in the example program.

11. Write a program which solves a system of linear equations by the Gauss-Jordan method but which uses as a pivot element the largest (in magnitude) of the reduced elements in the *pivot column*. Interchange rows to put the selected row in the pivot position. Does this approach involving a limited maximal element solve the zero-divisor problem? Will this method require the rearrangement of the solution vector at the end of the procedure? Use the program to solve the system of equations in Case 1 of Problem 8.

12. Consider the set of equations in Problem 10.
a) Following the factoring method, determine the matrices L and U. Show that $LU = A$, and obtain the matrix X composed of the solution vector and the inverse of A.
b) Following the interchange algorithm, find $\hat{L}$, P and $\hat{U}$. Show that $\hat{L}P\hat{U} = A$ and obtain the solution vector $\bar{x}$.

13. Modify the Gauss-Seidel iterative program to make a total-step algorithm; that is, the current trial solution vector is used for the evaluation of the entire set of equations before the values are improved.

14. Apply the Newton-Raphson method to the following set of nonlinear equations:

$$x^2 + y^2 = 4,$$
$$xy = 1.$$

Start the iteration with $x^{(0)} = 2$ and $y^{(0)} = 0$. How many other roots would you anticipate?

15. Solve the following set of nonlinear equations by applying the method of successive approximation in the vicinity of $0 < x_1 < 1$:

$$f_1(x_1, x_2) = x_1 - .4e^{x_2} - 0.09 = 0$$
$$f_2(x_1, x_2) = \cos x_1 - x_2 - 0.3825 = 0$$

16. Recall the form of Newton's method in which the improvement is accomplished by iteration of the unknowns. Apply it to the set of equations given in Problem 15, searching for a solution in the region

$$-1 < x_1 < 0.$$

17. Write a computer program to solve the algorithm formulated for a tri-diagonal coefficient matrix:

$$x_i = A'_i x_{i+1} + B'_i$$

where

$$A'_i = \frac{A_i}{1 - B_i A'_{i-1}}$$

$$B'_i = \frac{B_i B'_{i-1} + C_i}{1 - B_i A'_{i-1}}$$

The A_i and B_i relate to the general equation for the system

$$-B_i x_{i-1} + x_i - A_i x_{i+1} = C.$$

Use the program to obtain a solution for the following linear system:

$$\begin{bmatrix} 4 & -1 & 0 & 0 & 0 & 0 \\ -1 & 4 & -1 & 0 & 0 & 0 \\ 0 & -1 & 4 & -1 & 0 & 0 \\ 0 & 0 & -1 & 4 & -1 & 0 \\ 0 & 0 & 0 & -1 & 4 & -1 \\ 0 & 0 & 0 & 0 & -1 & 4 \end{bmatrix} \begin{bmatrix} x_1 \\ x_2 \\ x_3 \\ x_4 \\ x_5 \\ x_6 \end{bmatrix} = \begin{bmatrix} 0 \\ 1 \\ 1 \\ 1 \\ 1 \\ 0 \end{bmatrix}.$$

18. Write a computer program for solving the two simultaneous nonlinear equations of Problem 14 by Newton's method in which iteration is performed on the unknowns. Continue until the error in x is no more than 10^{-10}. Start with trial values of $x = 3$ and $y = 1$.

approximation

In the preceding chapters the criterion used in approximating a function by a polynomial was that the polynomial agree exactly with the function at a prescribed number of points. As mentioned before, when this criterion is used, the problem is one of *interpolation*, and the polynomial is called an interpolating polynomial. When other criteria are used to find polynomials (or other forms) to represent functions, the problem is one of *approximation*. A fairly obvious need for other criteria arises when the given points are empirical, such as observations from an experiment. In such cases it does not seem reasonable to ask that the approximating function pass through the given points since these values are known to contain some measurement error. A numerical process analogous to the draftsman's technique of drawing a smooth curve through the "center" of a collection of plotted points is needed. Ideally, this process should take into account the reliability of the observations, if such information can be quantified, so that the more reliable points will have a greater effect or *weight* in the production of an approximating function. In fact, this idea of weighting has applications beyond the considerations of reliability; for instance, it may be used to give greater emphasis to points within a particular interval. However, in the initial introduction to least-square approximations it is assumed that the given values are of equal importance.

Least-squares polynomial approximation

The draftsman's procedure of visually fitting a curve to a set of points may be described as follows. (1) By observation of the plotted points, a curve (i.e., a function) is selected, and (2) the curve is then drawn so that there are about as many points above the line as below; however, in this division, the distance of the points from the line is taken into account. In Fig. 7.1, these distances, or deviations, are labeled $d_0, d_1, \ldots, d_N$.

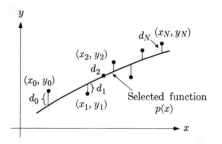

Fig. 7.1

These deviations are the differences between the selected function, $p(x)$, evaluated at a certain abscissa, and the given ordinate corresponding to the same abscissa:

$$d_0 = p(x_0) - y_0$$
$$d_1 = p(x_1) - y_1$$
$$\vdots$$
$$d_k = p(x_k) - y_k$$
$$\vdots$$
$$d_N = p(x_N) - y_N.$$

For the points above $p(x)$ the deviations are negative, and for those below $p(x)$ they are positive. It is tempting to say that the curve is "centered" if the negative deviations are equal to the positive deviations or, more simply, if

$$\sum_{k=0}^{N} d_k = \sum_{k=0}^{N} \left(p(x_k) - y_k \right) = 0;$$

but a simple two-point case (Fig. 7.2) illustrates that this condition can be satisfied without yielding the desired curve. Here

$$d_0 = -d_1$$

so that

$$\sum_{k=0}^{1} d_k = 0,$$

but the straight line $p(x)$ is obviously not as good an approximation as the dashed line shown. The alternative criterion that $\sum_{k=0}^{1} |d_k|$ be as small as possible would have given the better result. In general, of course, $\sum_{k=0}^{N} |d_k|$ will not be zero as it was in this example, but the requirement that it be a minimum avoids the difficulties encountered when the sum of deviations is set to zero. However, the absolute values give rise to a problem. Differentiation is useful in the search for minimum values, but the absolute-value function is not differentiable at its minimum. Another function which is dependent only on the magnitude of a number is the square, $f(x) = x^2$, and it is differentiable. Moreover, the square of a variable (or even the sum of the squares of several variables) does not have a maximum—only a minimum. In other words, the square increases without bound as the magnitude of the argument increases. These, then, are reasons for using the least-

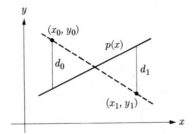

Fig. 7.2

squares criterion to determine an approximating function $p(x)$;* i.e.,

$$\sum_{k=0}^{N} d_k^2 = \sum_{k=0}^{N} \left(p(x_k) - y_k\right)^2 = \text{minimum}.$$

For the purposes at hand, the postulated function $p(x)$ will be limited to polynomials. All that needs to be determined then by looking at the given points is an estimate of the degree, n, of the polynomial. The method to be developed is just as applicable when $p(x)$ is a linear combination of functions other than polynomials.

In summary, if the given points are

x_0	y_0
x_1	y_1
$\vdots$	$\vdots$
x_N	y_N

---

* It is possible to reach the same conclusion using the principles of probability. If we assume that the measurements of the data were all made with equal care, and that for any value of x the data have a normal distribution, the probability that the errors d_k will fall within a small interval δy is given for x_0 as

$$\text{Pr}\,(x_0) = \frac{1}{\sigma\sqrt{2\pi}} \, [\exp - d_k^2/2\sigma^2]\,\delta y.$$

The standard deviation σ is a measure of the precision of the measurements and is a constant for the data. Similar expressions can be written for the probabilities Pr (x_0), Pr (x_1), ..., Pr (x_N).

As the separate measurements are all independent events, the probability for all of the x's is the product of their separate probabilities, or

$$P = \text{Pr}\,(x_0) \cdot \text{Pr}\,(x_1) \cdots \text{Pr}\,(x_N),$$

$$P = \left(\frac{1}{\sigma\sqrt{2\pi}}\right)^{N+1} \left[\exp\left(-\frac{1}{2\sigma^2}\sum_{k=0}^{N} d_k^2\right)\right](\delta y)^{N+1}.$$

A measure of the goodness of fit of the selected curve, $P(x_k)$, is for P to be a maximum. This will occur when

$$\sum_{k=0}^{N} d_k^2 = \text{minimum}.$$

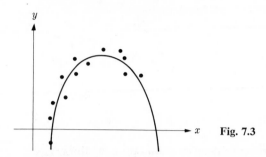

Fig. 7.3

and a plot of these points looks like the graph in Fig. 7.3, it would seem reasonable that the inserted curve could be represented by the parabola (i.e., the second-degree polynomial) $p(x) = a_0 + a_1 x + a_2 x^2$. The problem is to find the best values for a_0, a_1, a_2, and the least-squares criterion for "best" is to be used:

$$F = \sum_{k=0}^{N} d_k^2 = \sum_{k=0}^{N} \left(p(x_k) - y_k \right)^2$$

$$= \sum_{k=0}^{N} (a_0 + a_1 x_k + a_2 x_k^2 - y_k)^2 = \text{minimum}.$$

It is necessary to find the values of a_0, a_1, a_2 for which $F(a_0, a_1, a_2)$ is minimum. If F were a function of only one variable instead of three, the procedure would be the familiar one of finding the value of the variable for which the derivative is equal to zero. With three variables it is necessary to find the values for which the three partial derivatives are simultaneously zero:

$$\frac{\partial F}{\partial a_0} = 0, \qquad \frac{\partial F}{\partial a_1} = 0, \qquad \frac{\partial F}{\partial a_2} = 0.$$

Since F is a sum of terms, the derivative of F will be the sum of the derivatives of the terms.

$$\frac{\partial F}{\partial a_0} = \sum_{k=0}^{N} 2(a_0 + a_1 x_k + a_2 x_k^2 - y_k) = 0$$

$$\frac{\partial F}{\partial a_1} = \sum_{k=0}^{N} 2x_k(a_0 + a_1 x_k + a_2 x_k^2 - y_k) = 0$$

$$\frac{\partial F}{\partial a_2} = \sum_{k=0}^{N} 2x_k^2(a_0 + a_1 x_k + a_2 x_k^2 - y_k) = 0$$

This is a set of simultaneous linear equations in the unknowns a_0, a_1, a_2. Since one may multiply a linear equation by a constant without affecting the equality, all the equations above may be multiplied by $\frac{1}{2}$, with the result that the 2 will be eliminated. If the summation operator is applied to each term within the parentheses separately and the term involving y_k moved to the right-hand side, the equations are in a more conventional form.

$$\left(\sum_{k=0}^{N} 1\right) a_0 + \left(\sum_{k=0}^{N} x_k\right) a_1 + \left(\sum_{k=0}^{N} x_k^2\right) a_2 = \sum_{k=0}^{N} y_k$$

$$\left(\sum x_k\right) a_0 + \left(\sum x_k^2\right) a_1 + \left(\sum x_k^3\right) a_2 = \sum x_k y_k$$

$$\left(\sum x_k^2\right) a_0 + \left(\sum x_k^3\right) a_1 + \left(\sum x_k^4\right) a_2 = \sum x_k^2 y_k$$

The first term in the first equation may look strange. When the expression being summed does not depend on the index of summation as in $\sum_{k=0}^{N} a_0$, it may be written $a_0 \sum_{k=0}^{N} 1$, which is simply $a_0(N + 1)$.

Thus the coefficients of the second-degree polynomial are obtained by simply computing the coefficients of the linear equations above (called *normal equations*) and solving the system to determine a_0, a_1, a_2. Since the solution of a linear system has already been discussed, only the algorithm for computing the elements of the augmented matrix need be developed.

The first equation can be written

$$\sum_{j=0}^{2} \left(\sum_{k=0}^{N} x^j\right) a_j = \sum_{k=0}^{N} y_k,$$

the second

$$\sum_{j=0}^{2} \left(\sum_{k=0}^{N} x^{j+1}\right) a_j = \sum_{k=0}^{N} x_k y_k,$$

and the third

$$\sum_{j=0}^{2} \left(\sum_{k=0}^{N} x^{j+2}\right) a_j = \sum_{k=0}^{N} x_k^2 y_k.$$

The exponents increased by one for each equation, as did the exponent of x_k on the right-hand side. Expressing these changes in terms of i, an equation index (row index), one can write the entire system as

$$\sum_{j=0}^{n} \left(\sum_{k=0}^{N} x_k^{i+j}\right) a_j = \sum_{k=0}^{N} x_k^i y_k, \qquad i = 0, 1, 2, \ldots, n.$$

The degree of the polynomial, n, would be two for the example considered, but the formula above is general. The usual case is $n < N$, but when $n = N$ one obtains the coefficients of an interpolating polynomial, since a polynomial of degree n can pass through $n + 1$ points with $\sum_{k=0}^{N} d_k^2 = 0$. For $n > N$, an independent set of equations will not be obtained.

The parenthesized sums in the last form of the system are the coefficients. A minor complication is introduced by the fact that matrix-element subscripts conventionally start with 1; that is, the upper left-hand element in a matrix, b, is $b_{1,1}$, not $b_{0,0}$. This discrepancy in the notation can be adjusted by writing

$$\sum_{j=1}^{n+1} \left(\sum_{k=0}^{N} x_k^{i+j-2}\right) a_{j-1} = \sum_{k=0}^{N} x_k^{i-1} y_k, \qquad i = 1, 2, \ldots, n, n + 1.$$

Now the computation of the augmented matrix can be described by the following steps:

1. Compute

$$b_{ij} = \sum_{k=0}^{N} x_k^{i+j-2} = \sum_{k=0}^{N} x_k^{i-1} x_k^{j-1}$$

for

$$i = 1, 2, \ldots, n + 1,$$

and for each i,

$$j = 1, 2, \ldots, n + 1.$$

2. Compute the right-hand sides:

$$b_{i,n+2} = \sum_{k=0}^{N} x_k^{i-1} y_k, \qquad \text{for} \quad i = 1, 2, \ldots, n + 1.$$

Strictly speaking, for this algorithm to be valid for any x_k, one must assume that the result of computing 0^0 is unity. If it is further observed that the coefficient matrix is symmetric, that is, $a_{ij} = a_{ji}$, then some saving in computation time can be effected.

Before proceeding to a programming example, let us illustrate this method by a simple numerical problem.

Approximate the points,

x	y
0	1.00
1	3.85
2	6.50
3	9.35
4	12.05

by a first-degree polynomial ($n = 1$). Since five points are given, $N = 4$. The augmented matrix to be computed is:

$$\begin{bmatrix} \sum\limits_{k=0}^{4} 1 & \sum\limits_{k=0}^{4} x_k & \sum\limits_{k=0}^{4} y_k \\ \sum\limits_{k=1}^{4} x_k & \sum\limits_{k=0}^{4} x_k^2 & \sum\limits_{k=0}^{1} x_k y_k \end{bmatrix} = \begin{bmatrix} 5 & 10 & 32.75 \\ 10 & 30 & 93.10 \end{bmatrix}$$

Using the Gauss-Jordan method, one obtains the augmented matrix in diagonal form:

$$\begin{bmatrix} 250 & 0 & 257.5 \\ 0 & 50 & 138.0 \end{bmatrix},$$

from which

$$a_0 = \frac{257.5}{250} = 1.03 \quad \text{and} \quad a_1 = \frac{138}{50} = 2.76.$$

The straight-line approximation is

$$y = 1.03 + 2.76x.$$

Example Write a program which applies the least-squares criterion to determine the coefficients of an nth-degree polynomial which approximates the given points $(x_1, y_1), (x_2, y_2), \ldots, (x_N, y_N)$, where $n \leq N$. Assume that the linear equations can be solved by use of the external function SLEQ1 described earlier.

The principal task is to compute the coefficients of the augmented matrix from the given points. After this has been accomplished, the linear equations must be solved, and finally the solutions, which are the desired coefficients, must be printed. There is a variety of computational sequences which would accomplish this calculation. The one selected is not the shortest in expression, but it is more efficient in terms of computing time than algorithms which can be described in fewer statements. The flow chart and the program for this example are given in Fig. 7.4. Since all the elements of the matrix b are sums, before the process of accumulation, these elements must be initially set to zero. The somewhat cryptic block containing $b = 0$ indicates this step of setting the matrix to zero. The step could, of course, be written as an iteration, but it is convenient to write the common operation of setting array elements to zero more succinctly. The variables p_1 and p_2 are used to develop partial products; p_1 has the value x_k^{i-1} and p_2 the value x_k^{j-1}. Thus the box in the innermost repetition might be written

$$b_{ij} = b_{ij} + x_k^{i-1} x_k^{j-1},$$

and if b_{ij} is initially zero, the repetition of this computation for $k = 0, 1, \ldots, N$ produces the proper value for the (i, j)-element. The box immediately above the one just described computes the values of the right-hand sides, and the indicated computation is carried out once for each row for each value of k. Only the elements of the lower triangle, including the main diagonal, are computed, since the column index is $j = 1, 2, \ldots, i$. Therefore the section starting with ① transfers the lower triangle elements to the upper triangle. This step is permissible since the matrix is symmetric about the main diagonal. The final step after the normal equations are solved is to transfer the augmented column, which now represents the desired solutions, to the linear array $a_0, \ldots, a_n$ and print these results. The fourth-degree polynomial obtained from the data shown in the program is

$$p(x) = 0.7053 - 1.2539x - 1.0167x^2 + 0.7153x^3 + 0.0243x^4.$$

The approximation or curve-fitting procedure described above is the basis for the statistical technique called *regression analysis*, although the use of this name generally implies, in addition, the evaluation of certain quantities which are measures of how well the postulated function represents the given points.

Although the discussion has been confined to curve fitting with polynomials, the method of least squares is by no means limited to these forms. For example,

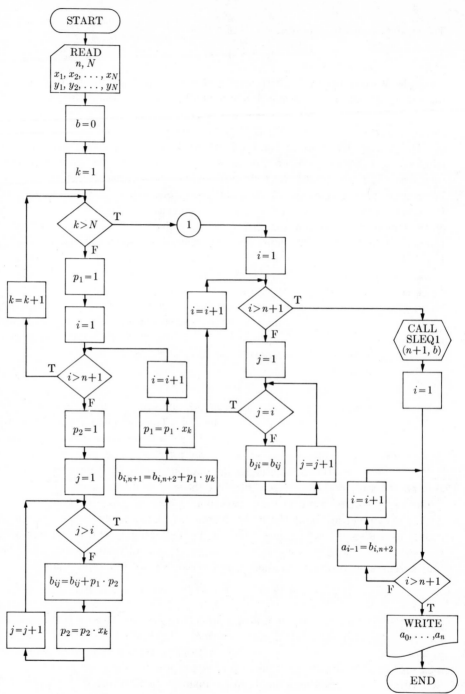

Fig. 7.4   Least-squares curve fitting.   (a) Flow chart.

```
C         MASTER LEAST SQUARE
C         A LEAST SQUARES APPROXIMATION FOR A POLYNOMIAL OF
C         DEGREE N WITH CAPN POINTS
          DIMENSION X(100),Y(100),B(20,21),A(20)
          INTEGER DET,CAPN
10        FORMAT(2I9)
20        FORMAT(14F6.2)
30        FORMAT(5X,5HA(0)=,F10.4)
40        FORMAT(5X,2HA(,I2,2H)=,F10.4)
          READ(5,10) N,CAPN
          READ(5,20) (X(K),Y(K),K=1,CAPN)
          DET=2
          NP1=N+1
          DO 2 I=1,NP1
          B(I,N+2)=0.
          DO 2 J=1,I
2         B(I,J)=0.
          DO 4 K=1,CAPN
          P1=1.0
          DO 4 I=1,NP1
          P2=1.0
          DO 3 J=1,I
          B(I,J)=B(I,J)+P1*P2
3         P2=P2*X(K)
          B(I,N+2)=B(I,N+2)+P1*Y(K)
4         P1=P1*X(K)
          DO 5 I=1,NP1
          DO 5 J=1,I
5         B(J,I)=B(I,J)
          CALL SLEQ1(N+1,DET,B,DELTA)
          DO 6 I=1,NP1
          I1=I-1
6         WRITE(6,40) I1,B(I,N+2)
          STOP
          END
```

```
          INPUT

4,7,
 -.45,1.0,-.15,.866,.15,.5,.45,0.,.75,-.5,.9,-.707,1.35,-1.0,
```

```
          OUTPUT

     A( 0) =      0.7053
     A( 1) =     -1.2539
     A( 2) =     -1.0167
     A( 3) =      0.7153
     A( 4) =      0.0243
```

Fig. 7.4   (b) Program.

suppose the coefficients a and b were to be selected to fit data with a curve of the form

$$y = ae^{bx},$$

subject to the criterion of the method of least squares. If the normal equations are formulated directly by requiring

$$\sum_{k=0}^{n} (ae^{bx_k} - y_k)^2 = \text{minimum},$$

they will be nonlinear in the coefficients a and b. If the equation of the curve is first written as

$$\log_e y = \log_e a + bx,$$

then the normal equation formed in terms of the variables $\log_e y$ and x will be linear in $\log_e a$ and b.

Near-singular sets of equations

Now that the least-squares method of producing polynomials seems to be settled, it is necessary to point out that it frequently gives rise to a certain problem. As the number of normal equations increases, the determinant of the matrix of coefficients often becomes very small. We noted in the last chapter that this leads to a near-singular system of equations which, in other terms, means that one or more of the linear equations can *almost* be expressed as a linear combination of the other equations. For two equations, this situation is represented geometrically by straight lines which are almost, but not quite, parallel. The reason that this state of affairs tends to occur can be seen by examining a general coefficient from the normal equations:

$$b_{ij} = \sum_{k=0}^{N} x^{i+j-2}$$

Assuming that $x_0, x_1, x_2, \ldots, x_N$ tend to be evenly distributed in some interval, say $(0, 1)$, one can represent the sum above by the shaded area in the bar graph of Fig. 7.5; that is, the sum is composed of rectangles with base 1 and altitude x_k^{i+j-2}. The exponent $(i + j - 2) \geq 0$ is a constant, and $0 \leq x_k \leq 1$. Still subject to the assumption that the x_k's are evenly distributed in $(0, 1)$, this sum can be closely approximated by an integral,

$$b_{ij} = \sum_{k=0}^{N} x_k^{i+j-2} \approx (N + 1) \int_0^1 x^{i+j-2} \, dx,$$

which is simply the area under the continuous curve x^{i+j-2}, and the $(N + 1)$-factor is necessary since the "base" of the integration is length 1 rather than $N + 1$. The integral is readily evaluated:

$$\int_0^1 x^{i+j-2} \, dx = \left[\frac{x^{i+j-1}}{i + j - 1} \right]_0^1 = \frac{1}{i + j - 1}$$

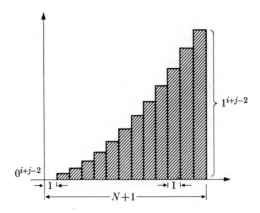

Fig. 7.5

and

$$b_{ij} \approx \frac{N+1}{i+j-1}.$$

The determinant of the system of normal equations is approximated by

$$\Delta = \begin{vmatrix} \dfrac{N+1}{1} & \dfrac{N+1}{2} & \cdots & \dfrac{N+1}{n+1} \\[2mm] \dfrac{N+1}{2} & & & \\[1mm] \vdots & & & \\[1mm] \dfrac{N+1}{n+1} & & \cdots & \dfrac{N+1}{2n+1} \end{vmatrix}$$

$$= (N+1)^{n+1} \begin{vmatrix} 1 & \dfrac{1}{2} & \dfrac{1}{3} & \dfrac{1}{4} & \cdots & \dfrac{1}{n+1} \\[2mm] \dfrac{1}{2} & \dfrac{1}{3} & \dfrac{1}{4} & & & \\[2mm] \dfrac{1}{3} & \dfrac{1}{4} & \dfrac{1}{5} & & & \\[2mm] \dfrac{1}{n+1} & & & & \cdots & \dfrac{1}{2n+1} \end{vmatrix}.$$

The matrix whose determinant is the factor on the right is a well-known one, called the *Hilbert matrix*. The determinant of this matrix of order n can be indicated by Δ_n^H. Using this notation, one has

$$\Delta = (N+1)^{n+1} \Delta_{n+1}^H.$$

The Hilbert matrix is so well known because the system of equations which it represents rapidly becomes near-singular as the number of equations increases. This can be seen by evaluating the determinant for various numbers of equations:

$$\Delta_1^H = 1$$
$$\Delta_2^H = 1 \cdot \tfrac{1}{3} - \tfrac{1}{2} \cdot \tfrac{1}{2} = \tfrac{1}{12} = 8.3 \times 10^{-2}$$
$$\Delta_3^H = 1(\tfrac{1}{3} \cdot \tfrac{1}{5} - \tfrac{1}{4} \cdot \tfrac{1}{4}) - \tfrac{1}{2}(\tfrac{1}{2} \cdot \tfrac{1}{5} - \tfrac{1}{3} \cdot \tfrac{1}{4}) + \tfrac{1}{3}(\tfrac{1}{2} \cdot \tfrac{1}{4} - \tfrac{1}{3} \cdot \tfrac{1}{3})$$
$$\qquad = 4.6 \times 10^{-4}$$
$$\Delta_4^H = 1.7 \times 10^{-7}$$
$$\Delta_5^H = 3.7 \times 10^{-12}$$
$$\Delta_6^H = 5.4 \times 10^{-18}$$
$$\vdots$$
$$\Delta_9^H = 9.7 \times 10^{-43}$$

Suppose, for example, that $N + 1 = 20$ and $n + 1 = 9$; then

$$\Delta = (20)^9 \Delta_9^H = 5.12 \times 10^{11} \times 0.97 \times 10^{-42} = 5 \times 10^{-31}.$$

A determinant of such a small magnitude would not permit solution by the elimination methods that have been described.

This difficulty can be resolved by producing a system of equations which are independent, such as a system represented by a diagonal coefficient matrix. To achieve this result it is necessary to express the desired polynomial in terms of coordinate functions other than 1, x, x^2, ..., x^n; that is, the desired function is

$$p(x) = a_0 P_0(x) + a_1 P_1(x) + a_2 P_2(x) + \cdots,$$

where $P_0(x)$, $P_1(x)$, $P_2(x)$ are the first three members of a family of orthogonal polynomials.

Chebyshev economization

Thus far the discussion has concentrated on the use of the algebraic polynomials as the approximating functions. It turns out that the use of other criteria for the determination of approximating functions leads quite naturally to the use of other coordinate functions. There is a general approach to this subject which is indicated in a later section. At this point, a specific problem is considered which leads to the employment of Chebyshev polynomials as coordinate functions.

It would seem desirable to approximate a function uniformly well over a particular interval of interest, say (a, b); that is, the error in approximating the function $f(x)$ should ideally be about the same anywhere in (a, b) and not large at some points and very small at others. After all, since the approximation is to be evaluated at points anywhere in (a, b) (otherwise this would not be the interval of interest), it will be the maximum error in the interval that will limit the usefulness of the approximation. As before, it simplifies matters if the interval considered is $(-1, 1)$, but this is not an essential restriction since, if the original z was in the range $a \leq z \leq b$, then a new variable,

$$x = \frac{2z - b - a}{b - a}$$

is in the range $-1 \leq x \leq 1$.

To illustrate that the coordinate functions 1, x, x^2, x^3, ... do not have this uniform error property, suppose that

$$p(x) = a_0 + a_1 x + a_2 x^2 + a_3 x^3 + a_4 x^4$$

is a good approximation to $f(x)$ for $-1 \leq x \leq 1$. The graphs of the five coordinate functions involved are shown in Fig. 7.6. The approximation $p(x)$ may be regarded as a weighted sum of these functions. Note that in this interval the maximum magnitudes of these functions occur at -1 and 1. Given that $p(x)$ is a good approximation, dropping the term $a_4 x^4$ will apparently cause considerable error near the endpoints but almost no additional error near $x = 0$, where x^4 has small values. It is apparent that the error will always be concentrated at the endpoints when these functions are used to produce approximations.

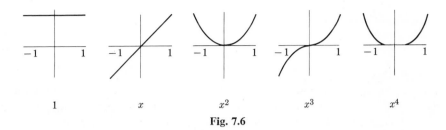

Fig. 7.6

The criterion can be altered at this point to remove the emphasis from the uniform distribution of error and simply state that the maximum error in the interval is to be minimized. It is obvious, though, that this smallest error will not be achieved if the error is concentrated and not spread through the interval.

The question now arises, what sort of coordinate functions would be useful to minimize the maximum error? In view of the preceding discussion, two useful properties would be: (1) The maxima and minima should not always occur at the same points in the interval, and (2) the magnitudes of the extreme values of a given function should be equal—otherwise the sums of such functions might lead to concentration of the error.

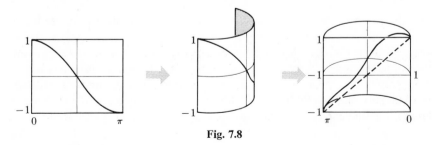

Fig. 7.7

The trigonometric functions, sine and cosine, have these properties in the interval $(0, \pi)$. In particular, the set 1, $\cos \theta$, $\cos 2\theta$, $\cos 3\theta$, $\cos 4\theta$, ... exhibits these properties, as shown in Fig. 7.7. These functions could be used as coordinate functions, but as transcendental functions their computation requires approximation, and it is more convenient to transform $\cos n\theta$ to polynomials in x. This transformation can be viewed as a projection of these curves from a cylinder to the xy-plane, which contains the cylindrical axis. The projection of $\cos \theta$ in this manner is indicated in Fig. 7.8. The dashed line is the projection on the plane.

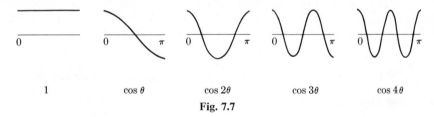

Fig. 7.8

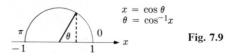

$$x = \cos \theta$$
$$\theta = \cos^{-1} x$$

Fig. 7.9

A look at the top of diagram (c) clearly shows the transformations between x and θ (Fig. 7.9). As a result, $y = \cos \theta$ is transformed in the xy-plane to $y = \cos(\cos^{-1} x) = x$. This is the first Chebyshev polynomial; it is usually written

$$T_1(x) = x.$$

The second Chebyshev polynomial is obtained by transforming $y = \cos 2\theta = 2\cos^2 \theta - 1$, so that

$$T_2(x) = 2\cos^2(\cos^{-1} x) - 1 = 2x^2 - 1.$$

Similarly,

$$\cos 3\theta = 4\cos^3 \theta - 3\cos \theta,$$
$$T_3(x) = 4x^3 - 3x.$$

These trigonometric identities become more complicated, and it is useful to devise a recursion formula, i.e., a formula which expresses the polynomial desired in terms of the lower-order polynomials. If the formulas for the sum and difference of two angles are used for $\cos(n\theta + \theta)$ and $\cos(n\theta - \theta)$, the results are

$$\cos(n\theta + \theta) = \cos((n+1)\theta) = \cos(n\theta)\cos(\theta) - \sin(n\theta)\sin(\theta),$$
$$\cos(n\theta - \theta) = \cos((n-1)\theta) = \cos(n\theta)\cos(\theta) + \sin(n\theta)\sin(\theta).$$

Adding these two expressions gives

$$\cos(n+1)\theta = 2\cos n\theta \cos \theta - \cos(n-1)\theta$$

which, in terms of the T's, is

$$T_{n+1}(x) = 2T_n(x) \cdot x - T_{n-1}(x).$$

Now $T_4(x)$ can be determined from the polynomials already known:

$$T_4(x) = 2T_3(x) \cdot x - T_2(x)$$
$$= 2(4x^3 - 3x)x - (2x^2 - 1)$$
$$= 8x^4 - 8x^2 + 1.$$

The Chebyshev polynomials produced thus far, with one more added, are listed in Table 7.1. Because the relations will be used shortly, $1, x, x^2, x^3, x^4, x^5$ are also expressed in terms of the Chebyshev polynomials. These relations were obtained by the straightforward algebraic manipulation of the left-hand column.

The graph of these Chebyshev polynomials would be very similar to the cosine curves with maximum and minimum values of ± 1. The curves would be somewhat compressed at the ends of the interval $(-1, 1)$. If these polynomials are divided

Table 7.1

$T_0 = 1$	$1 = T_0$
$T_1 = x$	$x = T_1$
$T_2 = 2x^2 - 1$	$x^2 = (\frac{1}{2})(T_0 + T_2)$
$T_3 = 4x^3 - 3x$	$x^3 = (\frac{1}{4})(3T_1 + T_3)$
$T_4 = 8x^4 - 8x^2 + 1$	$x^4 = (\frac{1}{8})(3T_0 + 4T_2 + T_4)$
$T_5 = 16x^5 - 20x^3 + 5x$	$x^5 = (\frac{1}{16})(10T_1 + 5T_3 + T_5)$

by the high-order coefficient, which is 2^{n-1} for T_n, the greatest magnitude assumed by the resulting polynomial with high-order coefficient unity is $1/2^{n-1}$.

It still is not obvious that a function which is a combination of Chebyshev polynomials will have the smallest maximum error in $(-1, 1)$. Indeed, such a statement cannot be made because any function in $1, x, x^2, \ldots$ can be written as a combination of $1, T_1, T_2, T_3, \ldots$ by merely using the right-hand column of Table 7.1. Such an algebraic change would not alter the error properties of the function. However, it is possible to show rather simply that there is no polynomial with leading coefficient unity which will have a smaller maximum magnitude in $(-1, 1)$ than a Chebyshev polynomial of the same order (also with leading coefficient unity). This property is shown by contradiction. Assume that there is such an nth-degree polynomial $P_n(x)$ which has a smaller maximum magnitude. The contradiction arises when the difference $D = P_n(x) - T_n(x)/2^{n-1}$ is formed. The polynomial $T_n(x)/2^{n-1}$ takes its maximum magnitude $n + 1$ times in $(-1, 1)$. (Remember its $\cos n\theta$ origin.) If $P_n(x)$ is between the maxima and minima of $T_n(x)/2^{n-1}$, as it must be if it has a smaller maximum magnitude, then the difference D must change signs n times; that is, D must be negative at the maxima of $T_n(x)/2^{n-1}$ and positive at the minima. But both $T_n(x)/2^{n-1}$ and $P_n(x)$ have high-order terms of x^n, and hence the difference will be a polynomial of degree $n - 1$. However, $(n - 1)$-degree polynomials cannot have n zeros (or roots) as this one must have. Therefore there cannot be such a polynomial as $P_n(x)$. The logic of this proof may emerge more clearly if a graph representing a specific case is drawn. If $T_4/8 = x^4 - x^2 + \frac{1}{8}$ and the postulated

$$P_4(x) = x^4 + a_3x^3 + a_2x^2 + a_1x + a_0$$

with extreme values of smaller magnitude are plotted on the same graph, the difference, $D = P_4 - T_4/8$ (represented by a dashed line), can be drawn as shown in Fig. 7.10. The difference has four roots in the interval, but this is a contradiction since a third-degree polynomial,

$$D = P_4 - \frac{T_4}{8} = a_3x^3 + (a_2 + 1)x^2 + a_1x + (a_0 - \frac{1}{8})$$

can have only three roots.

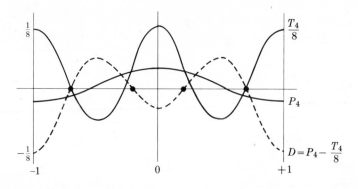

Fig. 7.10

This property of smallest extreme values is used in the economization procedure and is best introduced by an example. Suppose that it is desired to approximate e^x in $(-1, 1)$ with a maximum error of 0.01. An examination of Taylor's series shows that terms through the fifth degree are necessary:

$$e^x \approx 1 + x + \frac{x^2}{2} + \frac{x^3}{6} + \frac{x^4}{24} + \frac{x^5}{120},$$

$$|R_6(x)| = \left| \frac{e^\zeta x^6}{720} \right| \leq \frac{e}{720} = 0.0038.$$

Dropping the fifth-degree term would permit an error of $e/120 \approx 0.0226$, and hence this term must be included. But an economization of terms can be accomplished if the polynomial is written in terms of Chebyshev polynomials and then one or more high-order Chebyshev terms are dropped. The writing in this form is accomplished by substituting for x, x^2, x^3, x^4, x^5 the equivalent in terms of T_1, T_2, T_3, T_4, T_5 and then collecting terms:

$$e^x \approx \tfrac{81}{64}T_0 + \tfrac{217}{192}T_1 + \tfrac{13}{48}T_2 + \tfrac{17}{384}T_3 + \tfrac{1}{192}T_4 + \tfrac{1}{1920}T_5.$$

Since the magnitude of $T_n(x)$ does not exceed unity, dropping the rightmost two terms introduces at most the additional error of $\tfrac{1}{192} + \tfrac{1}{1920} = \tfrac{11}{1920} \approx 0.0057$. Making these deletions, one obtains

$$e^x \approx \tfrac{81}{64}T_0 + \tfrac{217}{192}T_1 + \tfrac{13}{48}T_2 + \tfrac{17}{384}T_3,$$

and the total error is at most $0.0038 + 0.0057 = 0.0095$. This result can be converted to x's again by using the left-hand column in Table 7.1:

$$e^x \approx \tfrac{1}{384}(382 + 383x + 208x^2 + 68x^3),$$

which produces the desired result without the original fourth- and fifth-degree terms.

This technique of Chebyshev economization has been the basis for obtaining many useful approximations. Another example (from Hastings) is

$$\sin \frac{\pi}{2} x = 1.5706268x - 0.6432292x^3 + 0.0727102x^5$$

where $-1 \leq x \leq 1$ and the maximum error $= 0.0001$.

A computer program designed to carry out this economization procedure is an interesting algorithm. The obvious procedure is to incorporate the two transformation columns of Table 7.1 in the program and then to accumulate the Chebyshev coefficients by transforming each of the polynomial coefficients by means of the table. After economization, the reverse transformation back to polynomial coefficients is done by using the table of Chebyshev polynomials. To illustrate this approach with a sample, consider the polynomial

$$p(x) = 1 + 2x + 3x^2 + 4x^3.$$

In tabular form the transformation is:

	$\times T_0$	$\times T_1$	$\times T_2$	$\times T_3$
$1 \cdot 1 =$	1			
$2 \cdot x =$		2		
$3 \cdot x^2 =$	$\frac{3}{2}$		$\frac{3}{2}$	
$4 \cdot x^3 =$		3		1
Total	$\frac{5}{2}$	5	$\frac{3}{2}$	1

Thus $p(x) = \frac{5}{2}T_0 + 5T_1 + \frac{3}{2}T_2 + T_3$, and the change back to polynomial form would be a similar tabular process. Actually, tables need not be used since the transformation can be carried out by means of the recursion relations. In a slightly rewritten form, these are

$$xT_n = \tfrac{1}{2}T_{n+1} + \tfrac{1}{2}T_{n-1} \quad \text{for} \quad n \geq 1 \quad \text{and} \quad xT_0 = T_1.$$

The use of these formulas to effect the transformation from the ordinary polynomial form to the Chebyshev form is more apparent if a "mixed" term is considered; that is, a term which is the product of a coefficient, a power of x, and a Chebyshev polynomial, e.g., $cx^n T_k(x)$. The primitive transformations $T_0(x) = 1$ and $T_1(x) = x$ result in $cx^n = cx^n T_0 = cx^{n-1}T_1(x)$ while the recursion formula results in

$$cx^n T_k(x) = cx^{n-1}(xT_k(x))$$

$$= cx^{n-1}\left(\frac{T_{k+1}(x)}{2} + \frac{T_{k-1}(x)}{2}\right)$$

$$= \frac{c}{2}x^{n-1}T_{k+1}(x) + \frac{c}{2}x^{n-1}T_{k-1}(x).$$

Notice that the exponents of x are reduced and the degree of the Chebyshev polynomials are increased by this process. Repeated application will eliminate the x terms entirely, leaving the original polynomial expressed in terms of Chebyshev polynomials only. The following example illustrates the transformation of

$$p(x) = 1 + 2x + 3x^2 + 4x^3,$$

and shows that the new coefficients can be developed in the same array as the original coefficients:

1	$2x$	$3x^2$	$4x^3$
1	$2xT_0$	$3xT_1$	$4x^2T_1$
1	$2xT_0$	$3xT_1$	$4x\left(\dfrac{T_2}{2} + \dfrac{T_0}{2}\right)$
1	$4xT_0$	$3xT_1$	$2xT_2$
1	$4T_1$	$3xT_1$	$2\left(\dfrac{T_3}{2} + \dfrac{T_1}{2}\right)$
1	$5T_1$	$3xT_1$	T_3
T_0	$5T_1$	$3\left(\dfrac{T_2}{2} + \dfrac{T_0}{2}\right)$	T_3
$\frac{5}{2}T_0$	$5T_1$	$\frac{3}{2}T_2$	T_3

If the original polynomial coefficients are in the array $c_0, \ldots, c_n$, then the algorithm can be described by the double iteration:

Let

$$i = n, n-1, \ldots, 2,$$

and for each i, $j = n, n-1, \ldots, i$,

$$c_j \leftarrow \frac{c_j}{2},$$

$$c_{j-2} \leftarrow c_{j-2} + c_j.$$

The process of retransforming from the Chebyshev basis back to conventional polynomial form is very similar. The recursive formula on a mixed term is employed in this case to decrease the degree of the Chebyshev polynomials and increase the powers of x; that is,

$$cx^n T_k(x) = c\left(2x^{n+1}T_{k-1}(x) - x^n T_{k-2}(x)\right)$$
$$= 2cx^{n+1}T_{k-1}(x) - cx^n T_{k-2}(x).$$

The process is illustrated by retransforming the result of the previous example:

$\frac{5}{2}T_0$	$5T_1$	$\frac{3}{2}T_2$	T_3
$\frac{5}{2}T_0$	$5T_1$	$\frac{3}{2}(2xT_1 - T_0)$	T_3
T_0	$5T_1$	$3xT_1$	T_3
T_0	$5T_1$	$3xT_1$	$2xT_2 - T_1$
T_0	$4T_1$	$3xT_1$	$2xT_2$
T_0	$4xT_0$	$3xT_1$	$4x^2T_1 - 2xT_0$
1	$2x$	$3x^2$	$4x^3$

Again, a double iteration describes the procedure:

Let

$$i = 3, 4, \ldots, n,$$

and for each $i, j = i, i + 1, \ldots, n$,

$$c_{j-2} \leftarrow c_{j-2} - c_j,$$
$$c_j \leftarrow 2c_j.$$

Example Write a program which reads $(n + 1)$ coefficients $c_0, c_1, \ldots, c_n$ and a value ϵ_1 which is the maximum additional error that is allowed. The printed output should be the economized coefficients $c_0, c_1, \ldots, c_m$, where $m \leq n$ and $\epsilon_2 \leq \epsilon_1$. The value of ϵ_2 is the amount of additional error introduced by the Chebyshev economization. The flow chart and the program are presented in Fig. 7.11. The section from **START** to ① converts the given polynomial coefficients to Chebyshev coefficients. The section from ① to ② determines how many high-order terms can be dropped before the additional error exceeds ϵ_1. From ② to **START** the array elements $c_0, \ldots, c_m$ are retransformed to coefficients of $1, x, x^2, \ldots, x^m$. Since **FORTRAN** permits only positive increments for iteration variables, the iterations that "count down" are modified so that the iteration variable meets this constraint. This change is simply made; for example, if the iteration is originally

$$i = n, n - 1, \ldots, k,$$

then the expression $(n - i)$ will take the desired values if

$$i = 0, 1, \ldots, n - k.$$

The program tests for $n < 2$ and $m < 2$ for in both of these cases, the transformation need not be done since the two forms are identical in the first two terms.

A unifying approach

The two approximation techniques just described seem radically different, and yet they can be understood as special cases of general procedures. The criterion for selecting an approximating function $p(x)$ was that the sum of the squared deviations be a minimum:

$$\sum_{k=0}^{N} d_k^2 = \sum_{k=0}^{N} (p(x_k) - y_k)^2 = \text{minimum}.$$

Although the technique was not applied in the earlier discussion, it was mentioned that it might be useful to weight some of the points (x_k, y_k); that is, attach a multiplier, w_k, which will increase or decrease the contribution of a specific point in the process of obtaining the minimum. With this addition,

$$\sum_{k=0}^{N} w_k[p(x_k) - y_k]^2 = \text{minimum},$$

where it is assumed that w_k is always nonnegative. If, in addition, the y-function is a known function rather than a set of points, the summation above can be

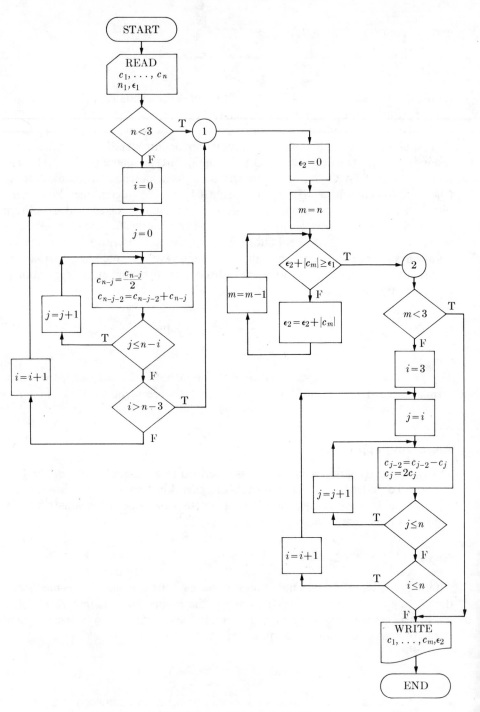

Fig. 7.11   Economizing coefficients.   (a) Flow chart.

```
C   CHEBYSHEV ECONOMIZATION PROGRAM
    DIMENSION C(10)
10  READ(5,14) N,EPS1,(C(I),I=1,N)
14  FORMAT(I2,8F10.5)
    IF (N.LT.3) GO TO 1
    ILIM=N-3
    DO 11 I=0,ILIM
    DO 11 J=0,I
    C(N-J)=C(N-J)/2.0
11  C(N-J-2)=C(N-J-2)+C(N-J)
1   EPS2=0.
    M=N
    WRITE(6,15) (I,C(I),I=1,N)
12  IF(EPS2+ABS(C(M)).GE.EPS1) GO TO 2
    EPS2=EPS2+ABS(C(M))
    M=M-1
    GO TO 12
2   IF(M.LT.3) GO TO 3
    DO 13 I=3,M
    DO 13 J=I,M
    C(J-2)=C(J-2)-C(J)
13  C(J)=2.0*C(J)
3   WRITE(6,15) (I,C(I),I=1,M)
15  FORMAT(6(I2,F10.6))
    GO TO 10
    END

    INPUT

4,.00001,1.,2.,3.,4.,
6,.0062,1.,1.,.5,.166667,.041667,.008333,

    OUTPUT

1  2.500000 2  5.000000 3  1.500000 4  1.000000
1  1.000000 2  2.000000 3  3.000000 4  4.000000
1  1.268229 2  1.130729 3  0.268229 4  0.043750 5  0.005208 6  0.000521
1  0.999999 2  0.999478 3  0.536459 4  0.175000
```

Fig. 7.11 (b) Program.

replaced by integration. Again, the interval is transformed to $(-1, 1)$ for convenience:

$$\int_{-1}^{1} w(x)[p(x) - y(x)]^2 \, dx = \text{minimum}.$$

With these changes, the augmented matrix (i.e., the coefficients and right-hand sides of the normal equations) would be:

$$\int_{-1}^{1} w(x)x^0 x^0 \, dx \qquad \int_{-1}^{1} w(x)x^0 x^1 \, dx \qquad \int_{-1}^{1} w(x)x^0 x^2 \, dx \qquad \int_{-1}^{1} w(x)x^0 y(x) \, dx$$

$$\int_{-1}^{1} w(x)x^1 x^0 \, dx \qquad \int_{-1}^{1} w(x)x^1 x^1 \, dx \qquad \int_{-1}^{1} w(x)x^1 x^2 \, dx \qquad \int_{-1}^{1} w(x)x^1 y(x) \, dx$$

$$\int_{-1}^{1} w(x)x^2 x^0 \, dx \qquad \int_{-1}^{1} w(x)x^2 x^1 \, dx \qquad \int_{-1}^{1} w(x)x^2 x^2 \, dx \qquad \int_{-1}^{1} w(x)x^2 y(x) \, dx$$

This form comes from the partial differentiation process, with $p(x)$ assumed to be the polynomial

$$p(x) = a_0 + a_1 x + a_2 x^2.$$

The next step is to reduce the coefficient matrix to diagonal form so that the linear equations represented by the matrix are independent and directly solvable.

This matrix reduction would not be necessary if the off-diagonal coefficients,

$$\int_{-1}^{1} w(x)x^i x^j\, dx, \qquad i \neq j,$$

were zero at the outset. If two functions $p(x)$ and $q(x)$ have this property,

$$\int_{-1}^{1} w(x)p(x)q(x)\, dx = 0,$$

then they are said to be *orthogonal*. As observed, this property would certainly simplify the task of finding the coefficients of the least-squares polynomial.

The functions 1, x, x^2, x^3, ... are coordinate functions which could be named as follows:

$$P_0 = 1$$
$$P_1 = x$$
$$P_2 = x^2$$
$$P_3 = x^3$$

With this notation, an off-diagonal element is

$$\int_{-1}^{1} w(x)P_i P_j\, dx, \qquad i \neq j.$$

It has been verified that these functions are not orthogonal, but the expression does raise the possibility that one could use another set which is orthogonal.

In the development of Chebyshev polynomials, emphasis was placed on the error introduced at the endpoints of $(-1, 1)$ when $P_0, P_1, P_2, \ldots$ were used. This error can be reduced by giving it relatively more weight than the error elsewhere. In other words, one selects a weighting function which has its largest values near ± 1. The function

$$w(x) = \frac{1}{\sqrt{1 - x^2}}$$

has such a property. Hopefully, then, coordinate functions which are polynomials can be found which permit this weighting function and are also orthogonal. With this weighting function, the coefficients are of the form

$$\int_{-1}^{1} \frac{p_r(x)q_{r-1}(x)}{\sqrt{1 - x^2}}\, dx,$$

where $p_r(x)$ and $q_{r-1}(x)$ are polynomials of degree r and $r - 1$ or less, respectively. The form of $w(x)$ suggests that the problem will be simplified if the trigonometric substitution $x = \cos \theta$ is made:

$$-\int_{\pi}^{0} \frac{p_r(\cos \theta)q_{r-1}(\cos \theta) \sin \theta\, d\theta}{\sqrt{1 - \cos^2 \theta}} = \int_{0}^{\pi} p_r(\cos \theta)q_{r-1}(\cos \theta)\, d\theta.$$

If the rth-degree polynomial is orthogonal to *any* polynomial of lesser degree, q_{r-1}, then $p_r, p_{r-1}, \ldots, p_0$ is a family of orthogonal polynomials. The first step in determining the form of p_r is to rewrite

$$q_{r-1} (\cos \theta) = a_0 + a_1 \cos \theta + a_2 \cos^2 \theta + \cdots + a_{r-1} \cos^{r-1} \theta$$

as

$$q_{r-1} (\cos \theta) = b_0 + b_1 \cos \theta + b_2 \cos 2\theta + \cdots + b_{r-1} \cos (r - 1)\theta.$$

The second column of Table 7.1, which shows $1, x, x^2, x^3, \ldots$ expressed in terms of Chebyshev polynomials, becomes the following set of trigonometric identities when $\cos \theta$ is written for x:

$$\cos^0 \theta = 1$$
$$\cos^1 \theta = \cos \theta$$
$$\cos^2 \theta = \tfrac{1}{2}(1 + \cos 2\theta)$$
$$\cos^3 \theta = \tfrac{1}{4}(3 \cos \theta + \cos 3\theta)$$
$$\vdots$$

If we replace the powers of $\cos \theta$ in the first form by these identities and collect terms, we obtain the second form. Using the second form, we find that $p_r (\cos \theta)$ is a member of an orthogonal family if

$$\int_0^\pi p_r (\cos \theta)[b_0 + b_1 \cos \theta + b_2 \cos 2\theta + \cdots + b_{r-1} \cos (r - 1)\theta]\, d\theta = 0.$$

However, for this to be true, each term must be zero or

$$\int_0^\pi p_r (\cos \theta)b_k \cos k\theta\, d\theta = 0, \qquad k = 0, 1, \ldots, r - 1.$$

By consulting a table of definite integrals one finds

$$\int_0^\pi \cos r\theta \cos k\theta\, d\theta = 0, \qquad k \neq r,$$

so that $p_r (\cos \theta) = C_r \cos r\theta$. The constant C_r is chosen to be unity for the ordinary Chebyshev polynomial. Resubstituting $\cos \theta = x$, one obtains

$$p_r(x) = T_r(x) = \cos (r \cos^{-1} x).$$

This was the result obtained in the section on economization, but the procedure here has been more general. The selection of other weighting functions gives rise to other families of orthogonal polynomials which are, however, not so readily obtained as those due to Chebyshev.

If the Chebyshev polynomials are used as coordinate functions in which the postulated least-squares polynomial is expressed (for example, $p(x) = a_0 T_0 + a_1 T_1 + a_2 T_2$), then the orthogonality condition guarantees that the off-diagonal coefficients of the normal equations are zero:

$$\int_{-1}^{1} \frac{T_i T_j}{\sqrt{1 - x^2}}\, dx = 0 \qquad \text{when} \quad i \neq j.$$

The resulting system of independent equations requires no elimination procedure:

$$\left[\int_{-1}^{1} \frac{T_0 T_0}{\sqrt{1 - x^2}}\, dx\right] a_0 \qquad\qquad = \int_{-1}^{1} \frac{T_0 y(x)}{\sqrt{1 - x^2}}\, dx$$

$$\left[\int_{-1}^{1} \frac{T_1 T_1}{\sqrt{1 - x^2}}\, dx\right] a_1 \qquad\qquad = \int_{-1}^{1} \frac{T_1 y(x)}{\sqrt{1 - x^2}}\, dx$$

$$\left[\int_{-1}^{1} \frac{T_2 T_2}{\sqrt{1 - x^2}}\, dx\right] a_2 = \int_{-1}^{1} \frac{T_2 y(x)}{\sqrt{1 - x^2}}\, dx$$

Since

$$\int_{-1}^{1} \frac{T_0^2}{\sqrt{1 - x^2}}\, dx = \int_{0}^{\pi} (\cos^0 \theta)^2\, d\theta = \pi$$

$$\int_{-1}^{1} \frac{T_r^2}{\sqrt{1 - x^2}}\, dx = \int_{0}^{\pi} (\cos r\theta)^2\, d\theta = \frac{\pi}{2}, \qquad r = 1, 2, 3, \ldots,$$

the desired coefficients are simply

$$a_0 = \frac{1}{\pi} \int_{-1}^{1} \frac{y(x)}{\sqrt{1 - x^2}}\, dx,$$

$$a_1 = \frac{2}{\pi} \int_{-1}^{1} \frac{x y(x)}{\sqrt{1 - x^2}}\, dx,$$

$$a_2 = \frac{2}{\pi} \int_{-1}^{1} \frac{(2x^2 - 1) y(x)}{\sqrt{1 - x^2}}\, dx.$$

The integrals have complicated matters slightly; only sums were involved in the normal equations as they were first derived. The integrations can be carried out numerically, of course, and there is a computational advantage to the orthogonal expansion in addition to the elimination of the "near-singular" problem. The advantage is that since the equations are independent, one can compute additional terms in the postulated function without having to recompute those already obtained.

The burden of numerical integration can be removed if the selected coordinate functions are orthogonal under summation; that is,

$$\sum_{k=0}^{N} p_r(x_k) p_s(x_k) = 0 \qquad \text{for} \qquad r \neq s.$$

The best-known functions of this type are not polynomials but the familiar trigonometric functions, the sine and cosine. If the families, 1, $\cos \theta$, $\cos 2\theta$, $\cos 3\theta$, ..., $\cos N\theta$ and $\sin \theta$, $\sin 2\theta$, $\sin 3\theta$, ..., $\sin (N-1)\theta$, are evaluated only at the values

$$\theta_0 = 0$$
$$\theta_1 = \frac{\pi}{N}$$
$$\theta_2 = \frac{2\pi}{N}$$
$$\theta_3 = \frac{3\pi}{N}$$
$$\vdots$$
$$\theta_{2N-1} = \frac{(2N-1)\pi}{N},$$

then

$$\sum_{k=0}^{2N-1} \sin r\theta_k \sin s\theta_k = \begin{cases} 0 & r \neq s \\ N & r = s \neq 0 \end{cases}$$

$$\sum_{k=0}^{2N-1} \sin r\theta_k \cos s\theta_k = 0$$

$$\sum_{k=0}^{2N-1} \cos r\theta_k \cos s\theta_k = \begin{cases} 0 & r \neq s \\ N & r = s \neq 0 \\ 2N & r = s = 0 \end{cases}$$

The arguments are equally spaced in increments of π/N, and if values of the function to be approximated can be obtained at θ_0, θ_1, θ_2, ..., θ_{2N-1}, then the coefficients of these coordinate functions can be obtained by summation. This is the basis of discrete Fourier analysis.

Since $T_r = \cos r (\cos^{-1} x)$, the Chebyshev polynomials can also be orthogonal under summation if

$$x_0 = \cos \theta_0$$
$$x_1 = \cos \theta_1$$
$$x_2 = \cos \theta_2$$
$$\vdots$$
$$x_{2N-1} = \cos \theta_{2N-1}$$

But in this case the x's are not equally spaced, and it is not often convenient to determine values of an empirical function corresponding to these abscissas. It would be useful to have a family which does not require integration, consists of polynomials, and yet has equally spaced abscissas. The Gram (or Gram-Schmidt) polynomials constitute such an orthogonal family and, although they are not developed here, the first three members are given so that their general form may be seen. The total interval is of length $2Mh$, where h is the size of the equal intervals shown in Fig. 7.12:

$$p_0(t, 2M) = 1$$

$$p_1(t, 2M) = \frac{t}{M}$$

$$p_2(t, 2M) = \frac{3t^2 - M(M+1)}{M(2M-1)}$$

$$\vdots$$

$$p_M(t, 2M) = \ldots$$

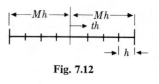

Fig. 7.12

The summation index is t and the orthogonality relation is

$$\sum_{t=-M}^{M} p_r(t, 2M)p_s(t, 2M) = 0, \qquad \text{where} \qquad r \neq s.$$

Discrete Fourier series

It is often useful to be able to approximate a function with a series of trigonometric functions called Fourier series; they are particularly useful when the function to be represented is periodic. In its most general form, the Fourier-series representation of a function known throughout the interval 0 to π is

$$f(x) = \frac{a_0}{2} + \sum_{k=1}^{\infty} (a_k \cos kx + b_k \sin kx),$$

where the coefficients are given by the expressions

$$a_k = \frac{2}{\pi} \int_0^{\pi} f(x) \cos kx \, dx, \quad k = 0, 1, 2, \ldots, \infty$$

$$b_k = \frac{2}{\pi} \int_0^{\pi} f(x) \sin kx \, dx, \quad k = 1, 2, \ldots, \infty.$$

Obviously, in a practical case k will be finite. These expressions result from the orthogonality of trigonometric functions in the interval 0 to π after the fashion discussed in the previous section. Specifically, the orthogonality relations are:

$$\int_0^{\pi} \cos kx \cos mx \, dx = \begin{cases} 0 & k \neq m \\ \dfrac{\pi}{2} & k = m \neq 0 \\ \pi & k = m = 0, \end{cases}$$

$$\int_0^{\pi} \sin kx \sin mx \, dx = \begin{cases} 0 & k \neq m \\ \dfrac{\pi}{2} & k = m \neq 0, \end{cases}$$

$$\int_0^{\pi} \sin kx \cos mx \, dx = 0.$$

Often series representations employ only the sine functions or only the cosine functions.

For a function to be represented by a Fourier series it must meet several conditions within the interval:

1. it must have onl one value for each x except at discontinuities;
2. it must be finite
3. it must be conti uous or have only a finite number of discontinuities; and
4. it must not have an infinite number of maxima and minima.

The development and properties of Fourier Series are extensively treated in many texts on intermediate mathematics.

Our concern here is with the development of Fourier series when the function is known at a discrete number of equally spaced points. Values might be expressed as data or a curve over an interval. By a change in argument, any interval can be transformed to the range 0 to 2π. The interval is then divided into $2N$ equally spaced parts.

In terms of the transformed variable θ, values of the function y_k are given at the abscissa points having spacing

$$\Delta\theta = \frac{2\pi}{2N} = \frac{\pi}{N}.$$

There are $2N$ known values of the function through which the series will be fitted. In terms of the ordinate. y, these are:

$$y_0 = f(0)$$
$$y_1 = f(\Delta\theta)$$
$$y_2 = f(2\,\Delta\theta)$$
$$\vdots \qquad \vdots$$
$$y_k = f(k\,\Delta\theta)$$
$$\vdots \qquad \vdots$$
$$y_{2N-1} = f\big((2N-1)\,\Delta\theta\big)$$

The discrete Fourier series is then written in terms of the continuous variable θ. Written to include both sines and cosines, the approximation to the function is

$$f(\theta) = \frac{a_0}{2} + a_1\cos\theta + a_2\cos 2\theta + \cdots + a_m\cos m\theta + b_1\sin\theta$$

$$+ b_2\sin 2\theta + \cdots + b_m\sin m\theta$$

$$= \frac{a_0}{2} + \sum_{j=1}^{m}(a_j\cos j\theta + b_j\sin j\theta).$$

With the function known at $2N$ values of θ, it is possible to fit the curve through all of the $2N - 1$ points. However, we may wish to use only enough coordinate functions to fit a smooth curve to the data, determining the coefficients according to the least-squares criterion. Consequently, we require that $m < N$. That is, we wish to determine the coefficients so as to cause

$$\sum_{k=0}^{2N-1} (f(\theta) - y_k)^2$$

to be a minimum.

Following the procedure of the last section, the Fourier series is used to represent the function $f(\theta)$. The sum-of-the-squares expression is then expanded and the derivatives, taken with respect to the coefficients, are set equal to zero. Because of the orthogonality of the functions, only the diagonal terms remain on the left-hand sides of the normal equations, all other terms being zero. Hence, the coefficients can be found directly. They are:

$$a_j = \frac{1}{N} \sum_{k=0}^{2N-1} f(\theta_k) \cos j\theta_k, \qquad j = 0, 1, 2, \ldots, m,$$

$$b_j = \frac{1}{N} \sum_{k=0}^{2N-1} f(\theta_k) \sin j\theta_k, \qquad j = 1, 2, \ldots, m,$$

where

$$\theta_k = k\,\Delta\theta = k\left(\frac{\pi}{N}\right).$$

As an example, we shall fit a discrete Fourier series to the following data:

k	0	1	2	3	4	5	6	7
θ_k	0	$\pi/4$	$\pi/2$	$3\pi/4$	π	$5\pi/4$	$3\pi/2$	$7\pi/4$
$f(\theta_k)$	0	2.707	5.000	2.707	0.	1.293	3.000	1.293

The interval is 2π, with $N = 4$. Appropriate sums are then made and the coefficients for a series of $m = 2$ are

$$a_0 = \frac{1}{N} \sum_{k=0}^{2N-1} f(\theta_k) = \frac{16.000}{4.} = 4.$$

$$a_1 = \frac{1}{N} \sum_{k=1}^{2N-1} f(\theta_k) \cos \theta_k = \frac{0.}{4.} = 0.$$

$$a_2 = -2.; \quad b_1 = 1.; \quad b_2 = 0.$$

The resulting Fourier-series approximation is

$$f(\theta) = 2 + \sin \theta - 2 \cos 2\theta.$$

This is not surprising, as the data were generated from this expression.

Using a development like that presented for the case of an even number of points, formulas can be developed for an odd number of points, $2N + 1$. The Fourier series is:

$$f(\theta) = \frac{a_0}{2} + \sum_{j=1}^{m} (a_j \cos j\theta + b_j \sin j\theta), \qquad m < N,$$

with the coefficients given by the relations

$$a_j = \frac{2}{2N + 1} \sum_{k=0}^{2N} f(\theta_k) \cos j\theta_k, \qquad j = 0, 1, 2, \ldots, m,$$

$$b_j = \frac{2}{2N + 1} \sum_{k=0}^{2N} f(\theta_k) \sin j\theta_k, \qquad j = 1, 2, \ldots, m,$$

where

$$\theta_k = k \left(\frac{2\pi}{2N + 1} \right).$$

In the formulas given for the coefficients, it is necessary to compute a series of products, each distinct, and then compute the appropriate summations. As the interval spacing of the abscissa is constant, it is possible to reduce the number of computations by taking advantage of the recursive character of trigonometric functions. Forms for 12, 24, or 48 points were developed when it was necessary to do harmonic analysis with hand calculation. Using recursion relations permits computer calculations to be done by simple looping procedures.

To demonstrate the recursive character of these computations, consider again the expression for the coefficients of the sine series when the number of points is even:

$$Nb_j = \sum_{k=0}^{2N-1} f(\theta_k) \sin j\theta_k, \qquad j = 1, 2, \ldots, m.$$

As the intervals are uniform, the summation can be written in terms of the first θ, θ_1.

$$Nb_j = f(\theta_0) \sin j(0) + f(\theta_1) \sin j\theta_1 + f(\theta_2) \sin 2j\theta_1 + \cdots$$
$$+ f(\theta_{2N-1}) \sin j(2N - 1)\theta_1$$
$$= \sum_{k=0}^{2N-1} f(\theta_k) \sin jk\theta_1$$

A recurrence relation can be developed for the sines by using the familiar relation

$$\sin (A + B) = 2 \sin A \cos B - \sin (A - B).$$

Introducing, $A = (m - 1)j\theta_1$ and $B = j\theta_1$, this relation becomes

$$\sin mj\theta_1 = 2 \sin (m - 1)j\theta_1 \cos j\theta_1 - \sin (m - 2)j\theta_1.$$

Then with $m = 2$,

$$\sin 2j\theta_1 = 2 \sin j\theta_1 \cos j\theta_1;$$

with $m = 3$,

$$\sin 3j\theta_1 = 2 \sin 2j\theta_1 \cos j\theta_1 - \sin j\theta_1$$
$$= 4 \sin j\theta_1 \cos^2 j\theta_1 - \sin j\theta_1;$$

with $m = 4$,

$$\sin 4j\theta_1 = 8 \sin j\theta_1 \cos^3 j\theta_1 - 4 \sin j\theta_1 \cos j\theta_1;$$

and so on. Introducing these relations into the equation for the sine coefficients b_j for the case of $N = 2$ gives

$$2b_j = \sin j\theta_1[f(\theta_1) + f(\theta_2)(2 \cos j\theta_1) + f(\theta_3)(4 \cos^2 j\theta_1 - 1)].$$

We wish to replace the right-hand summation with a function which can be computed by a simple looping calculation, using a recursion formula. Calling this function $U_m(\theta)$, we require that it satisfy the relation,

$$U_{2N-1} (\sin j\theta_1) = Nb_j = \sum_{k=0}^{2N-1} f(\theta_k) \sin jk\theta_1.$$

By substituting the recursion relation for the sine series in the summation on the right-hand side, it is possible to express U_{2N-1} in terms of trigonometric functions of $j\theta_1$. For example, in the case of $N = 2$,

$$U_{2N-1} = U_3 = f(\theta_1) + f(\theta_2)2 \cos j\theta_1 + f(\theta_3) 4 \cos^2 j\theta_1 - f(\theta_3).$$

By observing that a recursion formula for U_m must follow the form of the recursion relation for the sine, and by repeated evaluation of U_{2N-1}, a recursion relation for U_m can be induced as

$$U_m = 2 \cos j\theta_1 U_{m-1} - U_{m-2} + f(\theta_{2N-m}), \qquad m = 2, 3, \ldots, 2N - 1,$$

with the first two values determined as

$$U_0 = 0, \qquad U_1 = f(\theta_{2N-1}).$$

Beginning with U_0 and U_1, the recursion relation is repeatedly applied until the desired U_{2N-1} is found. A computer program with a simple looping calculation would accomplish this. Validity of the recursion formula can be demonstrated by computing U_3 with the recursion relation and comparing it to the appropriate expression for the summation given above.

A similar form can be developed for the coefficients in the cosine series. Multiplying both sides of the expression for the cosine coefficients by $\sin j\theta_1$ gives:

$$Na_j \sin j\theta_1 = \sum_{k=0}^{2N-1} f(\theta_k) \cos kj\theta_1 \sin j\theta_1$$
$$= f(\theta_0) \sin j\theta_1 + \sum_{k=1}^{2N-1} f(\theta_k) \cos kj\theta_1 \sin j\theta_1.$$

Employing the trigonometric identity

$$\sin (A - B) = \sin A \cos B - \cos A \sin B,$$

with $A = mj\theta_1$ and $B = j\theta_1$, we can construct a recurrence relation which is useful here. Its form is

$$\cos mj\theta_1 \sin j\theta_1 = \sin mj\theta_1 \cos j\theta_1 - \sin (m - 1)j\theta_1.$$

Making the suitable change in argument and substituting for the summation in the expression for the cosine coefficients yields

$$Na_j \sin j\theta_1 = f(\theta_0) \sin j\theta_1 + \sum_{k=1}^{2N-1} f(\theta_k) [\cos j\theta_1 \sin kj\theta_1 - \sin (k - 1)\theta_1]$$

or

$$Na_j = f(\theta_0) + \left(\frac{\cos j\theta_1}{\sin j\theta_1}\right) \sum_{k=1}^{2N-1} f(\theta_k) \sin kj\theta_1 - \frac{1}{\sin j\theta_1} \sum_{k=1}^{2N-1} f(\theta_k) \sin (k - 1)\theta_1.$$

With the definition of U_m introduced into the above, the cosine coefficients can be expressed as

$$Na_j = f(\theta_0) + \cos j\theta_1 U_{2N-1} - U_{2N-2}.$$

Using the recursion relation to compute U_{2N-1} and U_{2N-2} we can compute the coefficients in the cosine series with a simple looping process.

Example Write a program which computes the coefficients of a Fourier series for data supplied at an even number of points. To indicate the form of the series an index, L, is read in. For a sine series $L = 1$; for a cosine series, $L = 2$; and for a sine-cosine series $L = 3$. Values of the function are supplied at equally spaced values of the variable $\theta_1 = \pi/n$. A total of $2n$ values are supplied, $f(\theta_1), f(\theta_2),$ $\dots, f(\theta_{2n})$. An integer MM is read in to indicate the highest term in the series. Coefficients are computed for the series,

$$f(\theta) = a_0/2 + a_1 \cos \theta + a_2 \cos 2\theta + \cdots + a_{mm} \cos (mm)\theta + b_1 \sin \theta$$
$$+ b_2 \sin 2\theta + \cdots + b_{mm} \sin (mm)\theta.$$

Computation of the coefficients will be computed by repeated application of the recursion relation for the function U_m, described in the preceding section.

Because it is not possible to use zero subscripts in a FORTRAN program, subscripts on θ are from 1 to $2n$, rather than from 0 to $2n - 1$. This has been carried out in the flow chart as well. In addition it was necessary to introduce the index k in the computation of coefficients in the cosine series to accommodate the coefficient a_0.

A flow chart and program for this example are shown in Fig. 7.13. The program was used to fit data generated by the expression

$$f(\theta) = 2 + \sin \theta - 2 \cos 2\theta,$$

given in the numerical example. Results for a sine-cosine series ($MM = 2, L = 3$)

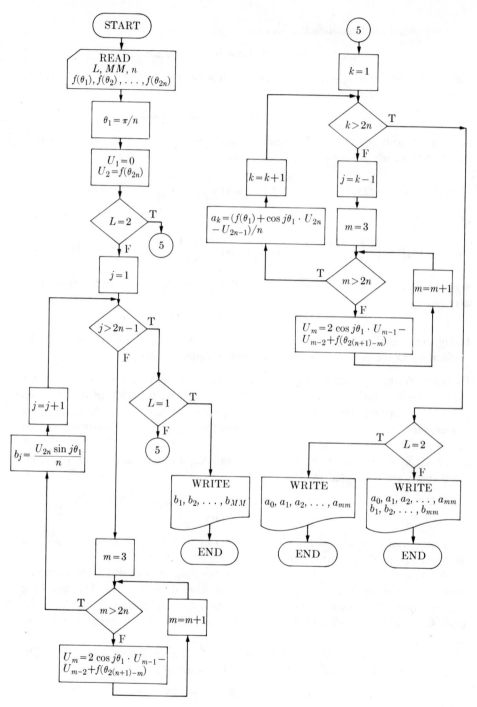

Fig. 7.13 Determination of coefficients for Fourier series. (a) Flow chart.

```
C       MASTER FOURIER
C       CURVE FITTING WITH FOURIER SERIES
        DIMENSION FUN(10),A(10),B(10),U(10)
        READ(5,17) L,MM,N
        NN=2*N
        READ(5,18) (FUN(K),K=1,NN)
        AN=N
        PI=3.141593
        THET1=PI/AN
        U(1)=0.0
        U(2)=FUN(NN)
        IF(L.EQ.2) GO TO 5
C
C       FOURIER SINE SERIES
C
        NNM1=NN-1
1       DO 3 J=1,NNM1
        DO 2 M=3,NN
        AJ=J
2       U(M)=(2.0*COS(AJ*THET1))*U(M-1)-U(M-2)+FUN(NN-M+2)
3       B(J)=(U(NN)*SIN(AJ*THET1))/AN
        IF(L.EQ.1) GO TO 8
C
9       FORMAT('          ...FOURIER SERIES COEFFICIENTS...')
C
5       DO 7 K=1,NN
        AJ=K-1
        DO 6 M=3,NN
6       U(M)=(2.0*COS(AJ*THET1))*U(M-1)-U(M-2)+FUN(NN-M+2)
7       A(K)=(FUN(1)+U(NN)*COS(AJ*THET1)-U(NN-1))/AN
C
8       WRITE(6,9)
        IF(L-2) 10,11,12
10      DO 101 J=1,NN
101     A(J)=0.0
        GO TO 12
11      DO 111 J=1,NNM1
111     B(J)=0.0
12      WRITE(6,13)A(1)
13      FORMAT(8H A(0) IS,F8.4)
14      DO 16 J=1,MM
16      WRITE(6,15)J,A(J+1),J,B(J)
15      FORMAT(3H A(,I1,4H) IS,F8.4,5X,2HB(,I1,4H) IS,F8.4)
17      FORMAT(3I9)
18      FORMAT(8F9.0)
        STOP
        END

        INPUT

3,2,4,
0.,2.707106,5.,2.707106,0.,1.292844,3.,1.292844,

        OUTPUT

          ...FOURIER SERIES COEFFICIENTS...
A(0) IS   4.0000
A(1) IS   0.0000      B(1) IS   1.0000
A(2) IS  -2.0000      B(2) IS   0.0000
```

Fig. 7.13 (b) Program.

with data given for eight equally spaced arguments ($N = 4$) yielded the coefficients

$$A(0) = 4.0000 \qquad B(1) = 1.0000$$
$$A(1) = 0.0 \qquad B(2) = 0.0$$
$$A(2) = -2.0000$$

Exercises

1. If a "least-absolute-value" criterion were used, would you expect the resulting approximation to be the same as that produced by the least-squares criterion? Explain your answer in terms of the relative weights of the deviations.

2. Devise a procedure to determine the coefficients a and b, for fitting a curve by the method of least squares to an expression having the form

$$y = ae^{bx}.$$

3.   An experiment on the cooling of a body yields the following results, t being the time in seconds from the beginning of the experiment and θ being the difference in temperature in degrees Fahrenheit between the body and its surrounding medium:

t	0	4	8	14	20	28	36	44	56
θ	40.0	36.4	32.4	27.3	23.0	18.1	14.4	11.4	8.0

Use the procedure developed in Problem 2 to fit a curve of the form

$$\theta = A + Be^{ct}$$

to these data. [*Note:* The coefficient A must be determined from an alternate procedure. Use the information that the temperature after a long time reached a *steady-state* value of 6°F.]

4.   The computation of the coefficients of the normal equations can be described in five statements. Although the program is not so efficient as the one illustrated, the statements do provide a concise description of the algorithm. Write these five executable statements. Assume, if necessary, that $0^0 \equiv 1$.

5.   Use Chebyshev polynomials to find an economized series which will approximate

$$\sin\left(\frac{\pi}{2}\right) x \qquad \text{for} \qquad -1 \le x \le 1,$$

with a maximum error of 0.001, and compare this with the error in the Taylor-series polynomial of equal degree.

6.   Verify the transformation of the example

$$e^x \approx 1 + x + \frac{x^2}{2} + \frac{x^3}{6} + \frac{x^4}{24} + \frac{x^5}{120}$$

to Chebyshev form by writing the Taylor-series coefficients in array form and then using the recursion formula to produce the corresponding array of Chebyshev coefficients.

7.   It is desired to approximate a function F in an interval (A, B) by using the first three Chebyshev coordinate functions. Assuming that a numerical integration subroutine, INTEG(A,B,N), is available, write a program to compute:

$$a_0 = \frac{1}{\pi} \int_{-1}^{1} \frac{g(x)}{\sqrt{1 - x^2}}\, dx,$$

$$a_1 = \frac{2}{\pi} \int_{-1}^{1} \frac{xg(x)}{\sqrt{1 - x^2}}\, dx,$$

$$a_2 = \frac{2}{\pi} \int_{-1}^{1} \frac{(2x^2 - 1)g(x)}{\sqrt{1 - x^2}}\, dx.$$

Here

$$F(Z) = f(z) = f\left(\frac{(b - a)x + b + a}{2}\right) = g(x),$$

and N is the number of applications of the integrating procedure.

8. To extend the previous problem to compute $a_0, a_1, \ldots, a_n$, one must systematically produce the Chebyshev polynomials. From the definition of these polynomials, suggest a method of systematically evaluating the higher-order polynomials without having to write each one explicitly.

9. The amplitude of a vibrating mass was measured experimentally. From the graph of a single period of the oscillations, the following displacements, a, were given at the times, t. (Both are expressed in dimensionless form.)

t	0	1	2	3	4	5	6	7	8	9	10	11	12
a	0.	.61	.98	.95	.15	.28	.01	−.25	−.16	−.92	−.97	−.60	0.

Determine the first three coefficients of a Fourier sine series fitted to the data. Sketch each of the curves.

10. If the function to be approximated is tabulated at $0, \pi/N, 2\pi/N, \ldots, (2N-1)\pi/N$, one can determine the coefficients of the approximation

$$f(\theta) = a_0 + a_1 \cos\theta + a_2 \cos 2\theta + \cdots + b_1 \sin\theta + b_2 \sin 2\theta + \cdots,$$

without resorting to numerical integration. Flow-chart a procedure to compute a specified number of these coefficients directly from the expressions given for the Fourier-series coefficients a_j and b_j.

11. Using the data given in the numerical example for discrete Fourier series, determine the coefficients for a sine-cosine series with $m = 2$, computing the appropriate values of U_m by hand.

12. Data taken to determine the pressure drop along a gas-pipe line is given in the table.

Distance, x, feet	100	500	1000	2000	5000	10000	20000
Pressure drop from $x = 0$, Δp, psi	0.4	1.8	3.6	7.3	14.9	34.2	68.0

Assume that the result can be approximated by a straight line as

$$\Delta p = a_0 + a_1 x$$

and that

$$a_0 = 0.$$

It is felt that the readings at small x are more reliable than those at large x. As a result, the pressure drop will be weighted more heavily near $x = 0$. A weighting factor, w, is assumed to be linear in x. At $x = 0$, $w = 1$ and at $x = 20,000$, w is assigned 0.5. Thus

$$w(x) = 1 - (0.000025)x.$$

Incorporate these weighting factors into the least-squares approximation to find a_1. Repeat with equal weighting for each pressure drop.

chapter eight

linear eigenvalue problems

Analysis of some physical systems leads to sets of linear algebraic equations which can only be solved uniquely when the value of a parameter within the equations is known. This parameter is called an *eigenvalue* (or *characteristic value*), and the solution associated with each eigenvalue is its *eigenvector*. These eigenvectors describe critical configurations, or *modes*, of the system. Eigenvalue problems occur in the analysis of vibrating masses, buckling of structures, and AC circuits, to identify a few applications. A simple problem in free vibration will be treated here to demonstrate the development and character of eigenvalue problems.

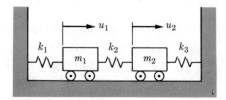

Fig. 8.1

Example problem Two masses, m_1 and m_2, are constrained at each end by linear springs having stiffnesses k_1 and k_3, respectively. The masses are connected to each other by a third spring of stiffness k_2, as shown in Fig. 8.1. Newton's second law can be applied to each mass. The forces applied to each mass through the springs can be determined by imagining that each mass is disturbed from its equilibrium state by extending or compressing the springs. It is assumed that the system is frictionless. Denoting the displacement from the equilibrium state as u_1 and u_2, respectively, the equations of motion are:

$$m_1 \frac{d^2 u_1}{dt^2} = -k_1 u_1 + k_2(u_2 - u_1),$$

$$m_2 \frac{d^2 u_2}{dt^2} = -k_2(u_2 - u_1) - k_3 u_2,$$

where t is time. For natural (or *free*) vibration, the system will vibrate at a single frequency ω_n, and the oscillation will be sinusoidal, having an amplitude x and an associated phase angle γ. Expressing these mathematically, the displacements are

$$u_1 = x_1 \sin (\omega_n t - \gamma),$$
$$u_2 = x_2 \sin (\omega_n t - \gamma).$$

These expressions can now be substituted into the equations of motion with the result

$$-m_1 \omega_n^2 x_1 + k_1 x_1 + k_2 x_1 - k_2 x_2 = 0,$$
$$-m_2 \omega_n^2 x_2 + k_2 x_2 + k_3 x_2 - k_2 x_1 = 0.$$

To simplify this problem, assume the following:

$$m_1 = m_2 = m \quad \text{and} \quad k_1 = k_2 = k_3 = k.$$

We can define a dimensionless frequency as

$$\lambda = \frac{m \omega_n^2}{k}.$$

Introducing these into the equations of motion, we obtain

$$(2 - \lambda)x_1 - x_2 = 0,$$
$$-x_1 + (2 - \lambda)x_2 = 0.$$

At this point, there are two equations in the unknowns x_1, x_2, and λ. It is evident that the best one could do is to establish a value of λ from these equations, and obtain a relation between x_1 and x_2. These equations are linear in $\bar{x}$, but notice that the right-hand sides are all zero. If the set is solved using Cramer's rule, it is apparent that the determinant of the coefficients of the left-hand side must be zero. Otherwise the amplitude vector $\bar{x}$ would be zero, a trivial result. Therefore,

$$\begin{vmatrix} 2 - \lambda & -1 \\ -1 & 2 - \lambda \end{vmatrix} = 0.$$

Now the determinant can be evaluated, yielding a polynomial equation in λ. There will be two roots in this case, λ_1 and λ_2. In general there will be as many roots as there are degrees of freedom in the physical system. These are the eigenvalues for the problem. Associated with each eigenvalue will be a solution vector, $\bar{x}$, an eigenvector. There will be two eigenvectors in this problem. Once an eigenvalue has been found, the set of n equations has been reduced to a set of $n - 1$ equations in n unknowns. (It is expected that a nonvanishing determinant of order $n - 1$ will exist.) It is then possible to determine $n - 1$ components of $\bar{x}$, expressing them in terms of the nth component. In this form, the eigenvector is said to be normalized. Each eigenvector will describe a configuration of the system under the critical condition imposed by the eigenvalue.

We now return to the simple example being considered, to determine the physical significance of these remarks. Expanding the determinant results in a quadratic equation in λ,

$$\lambda^2 - 4\lambda + 3 = 0.$$

This is called the *characteristic equation*. Its roots are the eigenvalues,

$$\lambda_1 = 1 \quad \text{and} \quad \lambda_2 = 3.$$

In physical terms, these are the two natural frequencies of the system in dimensionless form. From the definition of λ the frequencies can be found. The lower frequency is

$$\omega_1 = \sqrt{k/m},$$

and the higher frequency is

$$\omega_2 = \sqrt{3k/m}.$$

Introducing the eigenvalues into either of the equations of motion allows us to determine the eigenvectors. As the order of the nonvanishing determinant is now 1, we are only able to solve for x_1 in terms of x_2. For $\lambda_1 = 1$,

$$x_1 = x_2,$$

and for $\lambda_2 = 3$,

$$x_1 = -x_2.$$

In normalized form the first eigenvector is

$$\bar{x}^{(1)} = [1, 1],$$

and the second is

$$\bar{x}^{(2)} = [1, -1].$$

Each describes the displacement of the masses when the system is vibrating in one of its two *natural modes*. In the first mode the two masses are displaced by the same amount in the same direction. In the second mode the masses are displaced by equal amounts but in opposite directions. The frequency of the second mode is higher than that of the first. Actual values for the displacements will depend upon the initial conditions required for the problem. The complete solution will involve both modes and both frequencies, being expressed as the sum of all the solutions, or

$$u_1 = x_1^{(1)} \sin(\omega_1 t - \gamma^{(1)}) + x_1^{(2)} \sin(\omega_2 t - \gamma^{(2)}),$$
$$u_2 = x_2^{(1)} \sin(\omega_1 t - \gamma^{(1)}) + x_2^{(2)} \sin(\omega_2 t - \gamma^{(2)}).$$

There are six unknowns $x_1^{(1)}$, $x_1^{(2)}$, $x_2^{(1)}$, $x_2^{(2)}$, $\gamma^{(1)}$, and $\gamma^{(2)}$, where the superscript relates to the first or second eigenvalues. Using the relations between x_1 and x_2 obtained from the equations of motion, the number of unknowns can be reduced to four, with the second displacement equation becoming

$$u_2 = x_1^{(1)} \sin(\omega_1 t + \gamma^{(1)}) - x_1^{(2)} \sin(\omega_2 t - \gamma^{(2)}).$$

To determine the motion completely we require four initial conditions; for example, the position and velocity of each mass at zero time. The behavior of this system with both modes present is interesting but beyond the scope of the present discussion.

The general eigenvalue problem

Problems which require computer solutions will be like the example problem, but with more degrees of freedom, and with more eigenvalues to be found. The general form of the linear eigenvalue problem in matrix notation is

$$A\bar{x} = \lambda B\bar{x}.$$

Here, A and B are square matrices of order n. It is necessary to determine some or all of the n scalars, λ, which satisfy this relation, and the eigenvectors, $\bar{x}$, associated with each of them. We shall confine ourselves to systems in which the matrices are symmetric and nonsingular. It will be helpful to premultiply both sides of the equation by B^{-1}; then

$$B^{-1}A\bar{x} = \lambda B^{-1}B\bar{x},$$

or

$$H\bar{x} = \lambda\bar{x},$$

where

$$H = B^{-1}A.$$

In the formal solution of the problem the matrix equation is written

$$(A - \lambda B)\bar{x} = 0.$$

In order for $\bar{x}$ to exist, the determinant of $(A - \lambda B)$ must vanish. Expanding the determinant yields the characteristic equation, an nth-degree polynomial in λ. We can obtain each λ_j as a root of the characteristic equation, and then substitute it back into the matrix equation to find the corresponding eigenvector, $\bar{x}_j$.

Solving by iteration

It is evident that the classical method soon becomes impractical as the size of the matrices becomes very large. One frequently used approach is to solve by iteration. Operations are applied to the problem in the form

$$H\bar{x} = \lambda\bar{x}.$$

A trial vector, $\bar{x}^{(0)}$, is assumed. The matrix multiplication on the left-hand side is executed, producing a vector $\lambda\bar{x}^{(1)}$. This process is repeated until the product, $\lambda\bar{x}$ is proportional to $\bar{x}$. This happens when the vector $\bar{x}$ reproduces itself in the multiplication. The scalar multiplier in this proportion is an eigenvalue. To demonstrate this, consider the simple eigenvalue problem

$$2x_1 + 2x_2 + 2x_3 = \lambda x_1,$$
$$2x_1 + 5x_2 + 5x_3 = \lambda 3x_2,$$
$$2x_1 + 5x_2 + 11x_3 = \lambda 2x_3.$$

Both A and B are symmetric, and B is also diagonal. In terms of the H matrix,

$$H\bar{x} = \begin{bmatrix} 2 & 2 & 2 \\ \frac{2}{3} & \frac{5}{3} & \frac{5}{3} \\ 1 & \frac{5}{2} & \frac{11}{2} \end{bmatrix} \begin{bmatrix} x_1 \\ x_2 \\ x_3 \end{bmatrix} = \lambda \begin{bmatrix} x_1 \\ x_2 \\ x_3 \end{bmatrix}.$$

Starting with a trial vector $\bar{x}^{(0)} = [1, 0, 0]$, the matrix multiplication is carried out. In each case a normalized vector will be used. Here, it has been normalized with respect to x_1. Carrying out the arithmetic,

$$\begin{bmatrix} 2 & 2 & 2 \\ \frac{2}{3} & \frac{5}{3} & \frac{5}{3} \\ 1 & \frac{5}{2} & \frac{11}{2} \end{bmatrix} \begin{bmatrix} 1 \\ 0 \\ 0 \end{bmatrix} = \begin{bmatrix} 2 \\ \frac{2}{3} \\ 1 \end{bmatrix} = 2.00 \begin{bmatrix} 1.00 \\ 0.333 \\ 0.500 \end{bmatrix}.$$

Notice that after the multiplication the vector is again normalized. In the process the scalar 2 was extracted. This corresponds to the first approximation to the eigenvalue, while the first approximation to the eigenvector is [1.0, 0.333, 0.500], which is quite different from the initial guess. Using the new vector, the products are again taken, and normalized. Each time the total change of the components in the eigenvector is compared to some test value, and the process is repeated until the change is sufficiently small. Several iterations are carried out below,

$$\begin{bmatrix} 1.000 \\ .333 \\ .555 \end{bmatrix} \rightarrow \begin{bmatrix} 3.667 \\ 2.056 \\ 4.583 \end{bmatrix} \rightarrow 3.667 \begin{bmatrix} 1.000 \\ .561 \\ 1.250 \end{bmatrix} \rightarrow 5.6212 \begin{bmatrix} 1.000 \\ .655 \\ 1.650 \end{bmatrix}.$$

After 12 more steps the calculation proceeds as,

$$\begin{bmatrix} 1.00000 \\ 0.69058 \\ 1.81183 \end{bmatrix} \rightarrow \begin{bmatrix} 7.00480 \\ 4.83735 \\ 12.69148 \end{bmatrix} \rightarrow 7.00480 \begin{bmatrix} 1.00000 \\ 0.69058 \\ 1.81183 \end{bmatrix}.$$

The vector is now reproducing itself to the scalar multiple 7.00480. An eigenvalue has been found, $\lambda = 7.00480$, and its corresponding eigenvector is [1.00000, 0.69058, 1.81183].

Some other trial vector might have been used (e.g., [0, 0, 1]), but it would converge to the same eigenvalue and produce the same eigenvector. It remains now to determine which of the eigenvectors has been found, and how the remaining eigenvalues can be determined.

Convergence

Each step of the iteration of vector $V^{(k)}$ produces a new vector $V^{(k+1)}$, according to the expression

$$V^{(k+1)} = HV^{(k)}.$$

Here the scalar multiplier has not been extracted. Let the trial vector $V^{(0)}$ be expanded as a weighted sum of all of the eigenvectors,

$$V^{(0)} = \sum_{j=1}^{n} C_j \bar{x}_j.$$

Carrying out several steps of iteration, we proceed:

$$V^{(1)} = HV^{(0)} = \sum_{j=1}^{n} C_j \lambda_j \bar{x}_j$$

$$V^{(2)} = HV^{(1)} = \sum_{j=1}^{n} C_j (\lambda_j)^2 \bar{x}_j$$

$$\vdots$$

$$V^{(k)} = HV^{(k-1)} = \sum_{j=1}^{n} C_j (\lambda_j)^k \bar{x}_j.$$

If λ_n is the eigenvalue having the largest absolute value, and if $C_n \neq 0$, the summation can be expressed as

$$V^{(k)} = C_n \lambda_n^{(k)} \left[\bar{x}_n + \sum_{j=1}^{n-1} \left(\frac{C_j}{C_n} \right) \left(\frac{\lambda_j}{\lambda_n} \right)^k \bar{x}_j \right].$$

As $|\lambda_j/\lambda_n| < 1$, $j \neq n$, the expression following the summation will approach zero for large k, leaving

$$V^{(k)} \approx C_n \lambda_n^{(k)} \bar{x}_n.$$

That is, the iteration will converge toward λ_n which is the eigenvalue with the largest absolute value. Furthermore, when k is large, the scalar ratios between vector elements will correspond to λ_n, or

$$\frac{V^{(k+1)}}{V^{(k)}} \rightarrow \lambda_n.$$

Several other points of interest can be observed in this demonstration. The number of iterations required to achieve a certain level of accuracy will decrease if:

1. $|C_j/C_n|$ is small. This can be accomplished by choosing as the initial trial vector a vector near $\bar{x}_n$—one which is relatively free of elements of the other vectors.

2. $|\lambda_j/\lambda_n|$ is small. This is a characteristic of the problem and cannot be affected by the computer.

Notice also that, if all elements of the largest eigenvector $\bar{x}_n$ were purged from the problem, convergence would be to the next highest eigenvector.

Example Write a computer program which, using the iterative procedure, computes the largest eigenvalue and its eigenvector for a system governed by the equation

$$H\bar{x} = \lambda \bar{x}.$$

The matrix H is of order n. Program and flow chart are presented in Fig. 8.2.

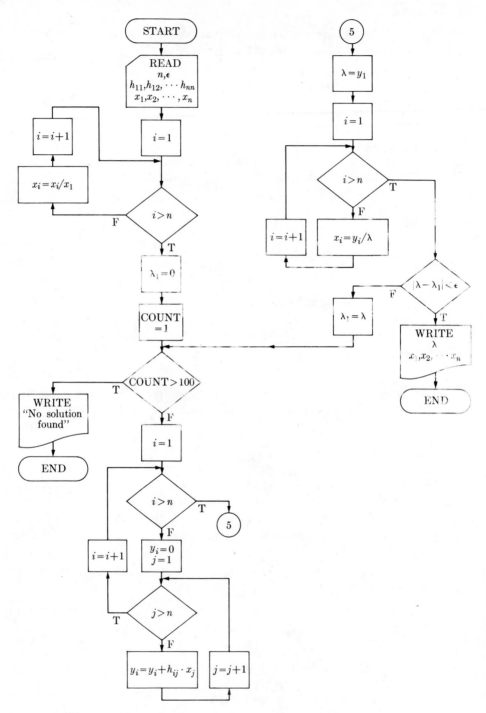

Fig. 8.2   Eigenvalue program.   (a) Flow chart.

```
C       MASTER EIGENVALUE
C       A PROGRAM TO COMPUTE AN EIGENVALUE BY ITERATION.
        DIMENSION X(20),Y(20),H(20,20)
        INTEGER COUNT
        READ(5,10)N,EPSI
C       N IS THE NUMBER OF EQUATIONS,EPSI IS THE
C       TEST FOR CONVERGENCE
C       READ THE H MATRIX
        DO 1 I=1,N
  1     READ(5,15)(H(I,J),J=1,N)
C       READ THE TRIAL EIGENVECTOR
        READ(5,15)(X(I),I=1,N)
C       NORMALIZE THE TRIAL EIGENVECTOR
        X1=X(1)
        DO 2 I=1,N
  2     X(I)=X(I)/X1
C       INITIALIZE THE EIGENVALUE
        EIGEI=0.
        DO 5 COUNT=1,100
        DO 3 I=1,N
        Y(I)=0.
        DO 3 J=1,N
  3     Y(I)=Y(I)+H(I,J)*X(J)
        EIGEN=Y(1)
        DO 4 I=1,N
  4     X(I)=Y(I)/EIGEN
        IF(ABS(EIGEN-EIGEI).LE.EPSI)GO TO 6
  5     EIGEI=EIGEN
        WRITE(6,30)
        GO TO 7
  6     WRITE(6,20)EIGEN
        WRITE(6,25)(X(I),I=1,N)
 10     FORMAT(I9,F9.0)
 15     FORMAT(6F9.0)
 20     FORMAT(5X,23HTHE LARGEST EIGENVALUE=,
       1F10.5,19H THE EIGENVECTOR IS,//)
 25     FORMAT(10X,8F10.5,//)
 30     FORMAT(///,5X,38HNO SOLUTION FOUND AFTER 100 ITERATIONS,//)
  7     STOP
        END

            INPUT

     3,.0001,
     2.,2.,2.,
     .6667,1.6667,1.6667,
      1.,2.5,5.5,
  5  1.,0.,0.,

            OUTPUT

    THE LARGEST EIGENVALUE=    7.00481 THE EIGENVECTOR IS
             1.00000    0.69059    1.81182
```

Fig. 8.2   (b) Program.

Supplied in addition to the elements of matrix H are a value, ϵ, used to determine when the process has converged satisfactorily, and a trial vector, $\bar{x}$. Although the vector $\bar{x}$ should be normalized with respect to x_1, a routine has been included to do this in the event this has not been done. The iteration proceeds by evaluating the product

$$H\bar{x} = \bar{y}.$$

The vector $\bar{y}$ is normalized, and the normalizing factor, y_1, is taken as the next approximation to λ. Convergence is based on comparing $|\lambda - \lambda_1|$ to ϵ. Here λ_1 is the previous approximation to the eigenvalue. When convergence is achieved, the eigenvalue and eigenvector are printed. If the process has not converged after 100 tries, a message "No solution found" is printed.

The program was used to compute the largest eigenvalue for the matrix

$$H = \begin{bmatrix} 2.0 & 2.0 & 2.0 \\ 0.6667 & 1.6667 & 1.6667 \\ 1.0 & 2.5 & 5.5 \end{bmatrix}.$$

We began with a trial vector $[1, 0, 0]$ and $\epsilon = 0.0001$, and the results were

$$\lambda = 7.0048 \qquad \text{and} \qquad \bar{x} = [1.00000, 0.69059, 1.81182].$$

Smallest eigenvalue

In some problems, such as the buckling of structures, the largest eigenvalue is most important, while in others (e.g., vibration problems) it is the smallest which is of most interest. We now turn to the problem of obtaining other eigenvalues.

It is possible to transform the problem so that the lowest eigenvalue will be found by direct iteration. This is accomplished by premultiplying both sides of the equation by the inverse of H and dividing through by λ to give

$$\frac{1}{\lambda} H^{-1} H \bar{x} = H^{-1} \bar{x}.$$

A new matrix expression has been formed

$$H^{-1} \bar{x} = \frac{1}{\lambda} \bar{x},$$

which can now be solved by iteration. Convergence will be to the largest value of $1/\lambda$, which will correspond to the smallest eigenvalue.

Intermediate eigenvalues

After determining the highest (or lowest) eigenvalue, it is possible to obtain intermediate eigenvalues by employing the orthogonality of the eigenvectors. For a proof of this we require that the matrices A and B be symmetric. Suppose we know an eigenvector $\bar{x}_j$, corresponding to the eigenvalue λ_j. We shall prove that a second eigenvector $\bar{x}_k$, corresponding to the eigenvalue λ_k, will satisfy the relation

$$\bar{x}_j^T B \bar{x}_k = 0.$$

In words, the eigenvectors are orthogonal with respect to the matrix B. The proof follows:

As $\bar{x}_j$ and $\bar{x}_k$ are solutions of the governing equations, we can write

$$A\bar{x}_j = \lambda_j B \bar{x}_j \qquad \text{and} \qquad A\bar{x}_k = \lambda_k B \bar{x}_k.$$

Premultiplying the first by $\bar{x}_k^T$ and the second by $\bar{x}_j^T$ and subtracting, we get

$$\bar{x}_k^T A \bar{x}_j - \bar{x}_j^T A \bar{x}_k = \lambda_j \bar{x}_k^T B \bar{x}_j - \lambda_k \bar{x}_j^T B \bar{x}_k.$$

Providing A and B are symmetric, the multiplication is commutative and we can write

$$(\lambda_j - \lambda_k)\bar{x}_k^T B \bar{x}_j = 0.$$

But $\lambda_j \neq \lambda_k$. Therefore

$$\bar{x}_k^T B \bar{x}_j = 0,$$

and the orthogonality with respect to B has been demonstrated.

Let us now return to the example problem. We have determined, to reasonable accuracy, the largest eigenvalue $\lambda_1 = 7.0048$, with its eigenvector $\bar{x}_1 = [1.000, 0.6906, 1.8118]$. Employing the orthogonality property, we obtain

$$[1.000 \quad 0.6906 \quad 1.8118] \begin{bmatrix} 1 & 0 & 0 \\ 0 & 3 & 0 \\ 0 & 0 & 2 \end{bmatrix} \begin{bmatrix} x_1 \\ x_2 \\ x_3 \end{bmatrix} = 0,$$

where the column vector is the second eigenvector, $\bar{x}_2$. Expansion of this expression leads to a scalar relation for the components of the second eigenvector $\bar{x}_2$,

$$x_1 + 2.0715x_2 + 3.6236x_3 = 0,$$

or

$$x_1 = -2.0715x_2 - 3.6236x_3.$$

This can be substituted in the matrix equation using the H matrix, as follows:

$$\begin{bmatrix} 2 & 2 & 2 \\ \frac{2}{3} & \frac{5}{3} & \frac{5}{3} \\ 1 & \frac{5}{2} & \frac{11}{2} \end{bmatrix} \begin{bmatrix} -2.0715x_2 - 3.6236x_3 \\ x_2 \\ x_3 \end{bmatrix} = \lambda \begin{bmatrix} x_1 \\ x_2 \\ x_3 \end{bmatrix}.$$

We carry out the arithmetic and produce the equations:

$$-6.1430x_2 - 9.2472x_3 = \lambda x_1,$$
$$0.2856x_2 - 0.7490x_3 = \lambda x_2,$$
$$0.4285x_2 + 1.8764x_3 = \lambda x_3.$$

The last two equations are an independent set in x_2 and x_3. They form a new system from which the first eigenvector has been removed, or *swept* out. Consequently the largest eigenvalue associated with this set is λ_2, and this can be obtained, along with its eigenvector $\bar{x}_2$, by again applying the iteration procedure (or, in this simple instance, by direct solution). At any rate, the original problem has been reduced to one of order $n - 1$ by utilizing the property that the eigenvectors are orthogonal with respect to the matrix B, thus eliminating the eigenvector $\bar{x}_1$

from the system. For the present problem, the two equations can be solved for the second eigenvalue

$$\lambda_2 = 1.63931$$

and its eigenvector

$$\bar{x}_2 = [4.4779, 1.000, -1.8074].$$

The components of x_2 and x_3 of the vector $\bar{x}_2$ come directly from the iteration. They were normalized in this case with respect to x_2. Using the equation resulting from the orthogonality condition, we found x_1 in terms of the remaining components; in this case, x_2 and x_3.

The third eigenvalue and eigenvector can be found by forming two orthogonal relations:

$$\bar{x}_1^T B \bar{x}_3 = 0 = x_1 + 2.0715x_2 + 3.6236x_3$$

and

$$\bar{x}_2^T B \bar{x}_3 = 0 = 4.4779x_1 + 3.0000x_2 - 3.8148x_3.$$

In the numerical example, the values of the components of the third eigenvector can be found directly in terms of the one remaining component, x_3. In the general case these two expressions can be substituted into the vector on the left-hand side of the equation

$$H\bar{x} = \lambda\bar{x}$$

to reduce the number of equations (and unknowns) to $n - 2$. The new set has now been nearly swept clean of eigenvectors $\bar{x}_1$ and $\bar{x}_2$, leaving a problem in which the largest eigenvalue is λ_3. Iteration will converge to λ_3. Eigenvectors $\bar{x}_1$ and $\bar{x}_2$ can be completely purged from the set only if they are known exactly. As they generally are not, some errors do accumulate, and these are compounded by repeated iteration. For this problem $\lambda_3 = 0.5225$ with the third eigenvector [2.9915, -3.1934, 1.0000]. These results agree well with those results obtained by matrix inversion and iteration to the smallest eigenvector, as described in the previous section.

As each mode is swept, any inaccuracy in the eigenvector will be passed on to the next eigenvector, with an increasing loss of accuracy. It is necessary that the first modes be found with extreme precision in order to obtain the desired accuracy in the other eigenvectors. Even then the method is of questionable use for determining more than three modes from the highest (or lowest) eigenvalue.

Solution with polynomial operators

It would be useful to be able to transform the matrix equation in such a way that the eigenvalue with the largest absolute value was not the largest eigenvalue for the original system, yet was an eigenvalue of the original system. Iteration then could proceed directly to the largest eigenvalue for the transformed system. In this way it would be possible to obtain an intermediate eigenvalue without the danger of accumulating errors through sweeping. Consider again the original eigenvalue problem

$$A\bar{x} = \lambda B\bar{x}$$

transformed to the iterative equation by premultiplying by B^{-1},

$$H\bar{x} = \lambda\bar{x}.$$

If both sides are premultiplied by H,

$$H^2\bar{x} = \lambda H\bar{x} = \lambda^2\bar{x},$$

or in general

$$H^n\bar{x} = \lambda^n\bar{x}, \qquad n = 0, 1, 2, \ldots$$

If $\bar{x}_j$ is an eigenvector of the original system, it is also an eigenvector of the transformed system. In fact, the expressions in various powers of H can be multiplied by coefficients and added together in various combinations, and $\bar{x}_j$ will be an eigenvector of the new system. For example, a matrix polynomial can be constructed of the form

$$(a_n H^n + \cdots + a_2 H^2 + a_1 H + a_0)\bar{x} = (a_n \lambda^n + \cdots + a_2 \lambda^2 + a_1 \lambda + a_0)\bar{x},$$

and $\bar{x}_j$ will be an eigenvector of the system.

It is possible that the system can now be altered in this way so that the eigenvalue with the largest absolute value is an intermediate eigenvalue of the original system.

Following a method proposed by Aitken, we introduce the linear polynomial operator

$$a_1 H + a_0 I.$$

If we set $a_1 = 1$, the matrix equation becomes

$$(H + a_0 I)\bar{x} = (\lambda + a_0)\bar{x}.$$

An iterative solution for this problem will converge to the largest absolute value of $\lambda + a_0$. As a_0 is arbitrary, it is possible to select it to cause the iteration to converge to a subordinate eigenvalue of the original system. To demonstrate, return to the example. An iteration on the original equation turned up the largest eigenvalue, 7.0048. If a_0 is arbitrarily chosen as -7.0, the scalar multiplier found by iteration of the modified matrix equation will be $\lambda - 7.0$. The choice of $a_0 = -7$ means that the eigenvalue which has been found, $\lambda_1 = 7.0048$, has little chance of producing the largest scalar multiplier in the modified equation. Introducing numerical values into the polynomial operator yields the matrix

$$\begin{bmatrix} 2.0 & 2.0 & 2.0 \\ .6667 & 1.6667 & 1.6667 \\ 1.0 & 2.5 & 5.5 \end{bmatrix} - 7\begin{bmatrix} 1.0 & 0. & 0. \\ 0. & 1.0 & 0. \\ 0. & 0. & 1.0 \end{bmatrix} = \begin{bmatrix} -5.0 & 2.0 & 2.0 \\ .6667 & -5.3333 & 1.6667 \\ 1.0 & 2.5 & -1.5 \end{bmatrix}$$

Using the modified matrix, the iterative solution can proceed from some trial vector, say [1.0, 0.0, 0.0]. Following forty-two iterative steps, the forty-third iteration is

$$\begin{bmatrix} -5.0 & 2.0 & 2.0 \\ 0.6667 & -5.3333 & 1.6667 \\ 1.0 & 2.5 & -1.5 \end{bmatrix} \begin{bmatrix} 1.00 \\ -1.079 \\ 0.341 \end{bmatrix} = \begin{bmatrix} -6.476 \\ 6.990 \\ -2.209 \end{bmatrix} = -6.476 \begin{bmatrix} 1.0 \\ -1.079 \\ 0.341 \end{bmatrix}.$$

As a result of the iteration, the third eigenvector to the problem has been produced with the scalar multiplier -6.476. Referring to the right-hand side of the modified equation, the third eigenvalue can be found. If $\lambda + a_0 = -6.476$ and $a_0 = -7.0$, then

$$\lambda_3 = 0.524.$$

This is the smallest eigenvalue in the system, and we have labeled it λ_3. Notice that the absolute value of the scalar multiplier in the modified equation, 6.476, is larger than the multiplier corresponding to λ_1 which would be .0048, corresponding to $\lambda_2(5.361)$.

It is possible to select a_0 so that a third eigenvalue is found using the linear polynomial operator. Alternatively, with two eigenvalues known, a second-degree polynomial operator can be formed. In an effort to eliminate λ_1 or λ_3, the polynomial coefficients can be found by expanding the factorial expression to form the polynomial.

$$p(\lambda) = (\lambda - \lambda_1)(\lambda - \lambda_3) = (\lambda - 7.0)(\lambda - .52),$$
$$p(\lambda) = \lambda^2 - 7.52\lambda + 3.64.$$

With the coefficients $a_1 = 1.$, $a_2 = -7.52$, and $a_3 = 3.64$, the polynomial operator becomes

$$\begin{bmatrix} 2.0 & 2.0 & 2.0 \\ 0.6667 & 1.6667 & 1.6667 \\ 1.0 & 2.5 & 5.5 \end{bmatrix}^2 - 7.52 \begin{bmatrix} 2.0 & 2.0 & 2.0 \\ 0.6667 & 1.6667 & 1.6667 \\ 1.0 & 2.5 & 5.5 \end{bmatrix} + 3.64 \begin{bmatrix} 1.0 & 0. & 0. \\ 0. & 1.0 & 0. \\ 0. & 0. & 1.0 \end{bmatrix}$$

$$= \begin{bmatrix} -4.0666 & -2.7066 & 3.2934 \\ -0.9023 & -0.6156 & 0.7445 \\ 1.6468 & 1.1168 & -1.3032 \end{bmatrix}.$$

Beginning the iteration with the trial vector [1.0, 0, 0] the result for the second iteration is

$$\begin{bmatrix} -4.0666 & -2.7066 & 3.2934 \\ -0.9023 & -0.6156 & 0.7445 \\ 1.6468 & 1.1168 & -1.3032 \end{bmatrix} \begin{bmatrix} 1.000 \\ 0.222 \\ -0.405 \end{bmatrix} = \begin{bmatrix} -6.001 \\ -1.338 \\ 0.242 \end{bmatrix} - 6.001 \begin{bmatrix} 1.000 \\ 0.223 \\ -0.404 \end{bmatrix}.$$

In two steps the iteration has produced the third eigenvector correctly. There was less than 0.5% alteration in the eigenvector in the third step for three-place accuracy. Convergence with the second-degree polynomial operator is obviously much more rapid than for the linear operator. While the scalar multiplier -6.001 corresponds to the eigenvalue with the largest absolute value in the problem which was solved, in terms of the eigenvalue for the original problem it is equal to $p(\lambda)$ or

$$\lambda^2 - 7.52\lambda + 3.64 = -6.001.$$

It is possible to solve the quadratic expression directly for λ, and if this is done a value of 1.64 results, which is λ_2. A more direct and accurate approach is to put the eigenvector just found into the original matrix expression, and determine the eigenvalue directly. In this instance this would be

$$\begin{bmatrix} 2.0 & 2.0 & 2.0 \\ 0.6667 & 1.6667 & 1.6667 \\ 1.0 & 2.5 & 5.5 \end{bmatrix} \begin{bmatrix} 1.000 \\ 0.223 \\ -0.404 \end{bmatrix} = \begin{bmatrix} 1.638 \\ 0.365 \\ 0.664 \end{bmatrix} = 1.638 \begin{bmatrix} 1.000 \\ 0.223 \\ -0.405 \end{bmatrix}.$$

The eigenvector has been reproduced, and the eigenvalue found directly. Both the value for λ_2 and the eigenvector are in agreement with our earlier computation.

These methods have the advantage of allowing a mode to be found for an eigenvalue of particular interest when the other eigenvalues are known to only rough approximations.

Trial-and-error solutions

In solving large eigenvalue problems one often resorts to a trial-and-error procedure for determining the eigenvalues. This has the advantage of working (with sufficient effort) for any linear problem with real matrices. First the problem is written in the form

$$(A - \lambda B)\bar{x} = 0.$$

If an eigenvalue were known, the determinant of the matrix $(A - \lambda B)$ would be zero.

Assuming trial values of λ, the determinant can be evaluated numerically. Then an interpolation procedure can be used to find a closer approximation to the eigenvalue. We might improve the approximation by using two trial values for λ, in a linear interpolation or a form of half-interval search. For example, in the sample problem we have been treating in this chapter, let the trial values of λ be 1 and 2. For $\lambda = 1$,

$$\det (A - \lambda B) = \begin{vmatrix} (2 - 1) & 2 & 2 \\ 2 & (5 - 3) & 5 \\ 2 & 5 & (11 - 2) \end{vmatrix} = -11,$$

and for $\lambda = 2$,

$$\det (A - \lambda B) = \begin{vmatrix} 0 & 2 & 2 \\ 2 & -1 & 5 \\ 2 & 5 & 7 \end{vmatrix} = +16.$$

There is a value of λ between 1 and 2 which will cause the determinant to be zero. Using a half-interval search technique, the determinant is evaluated for a trial λ of 1.5, and its sign determined. The result is -4.5, indicating that the correct λ lies between 1.5 and 2. The process is repeated until the interval is sufficiently small to produce the desired accuracy. Searching can then proceed for another eigenvalue with a new pair of trial values.

Three trial eigenvalues can be used with quadratic interpolation to find an improved value. Determinants would be evaluated for the three λ, and a second-degree polynomial fitted through the three points. A root would then be found to give an improved approximation. Determinants could be evaluated in the vicinity of the new value, and a second improvement made. This procedure would be repeated until the desired accuracy is obtained. To demonstrate, trial values of $\lambda = 1$, 1.5, and 2 will be used in the matrices of the sample problem. The determinants are -11, -4.5, and 16 respectively. These are shown in the graph of Fig. 8.3. A second-degree polynomial is fitted through these points. The polynomial is

$$D(\lambda) = a_0 + a_1\lambda + a_2\lambda^2,$$

where $D(\lambda)$ is the value of the determinant of $(A - B\lambda)$ evaluated for the trial λ.

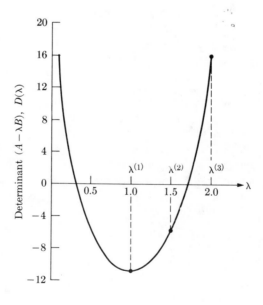

Fig. 8.3

For the present example:

$$D(\lambda) = 18 - 57\lambda + 18\lambda^2.$$

For a correct value of λ, $D(\lambda) = 0$. The roots of the polynomial will provide an improved approximation to λ. Roots of the above polynomial are at $\lambda = 0.391$ and $\lambda = 1.645$. It is the second which is in the region of interest. A second approximation can now be made in the neighborhood of $\lambda = 1.645$, and the process repeated. While convergence is more rapid with this method, it is not as simple to execute as a linear interpolation. Furthermore, the half-interval search provides a means of locating regions in which eigenvalues exist, by changing the approximation to the eigenvalue by a regular amount, and observing when the sign of the determinant changes.

When an eigenvalue has been found it is a simple matter to substitute this back into the eigenvalue problem

$$(A - \lambda B)\bar{x} = 0,$$

and solve the resulting set of linear equations for the eigenvector. It is necessary to normalize the eigenvector first by dividing through by a component of the eigenvector. This reduces the system to $n - 1$ simultaneous equations for the remaining components, which can be solved by a Gauss-Jordan reduction or another suitable method.

Once a value of λ has been found (say λ_a) with a trial-and-error search, the problem can be simplified by evaluating the expression

$$\frac{D(\lambda)}{\lambda - \lambda_a}.$$

That is, three trial values of λ are assumed and the coefficients of the second-degree polynomials are found by setting

$$\frac{D(\lambda_i)}{\lambda_i - \lambda_a} = a_2\lambda_i^2 + a_1\lambda_i + a_0, \qquad i = 1, 2, 3.$$

As λ_a is a root of the expanded determinant, then $\lambda - \lambda_a$ is a factor of the expression $D(\lambda)$. By dividing through by this factor, the order of the matrix is, in effect, reduced. When the second-degree polynomial has been found for the trial λ, interpolation, followed by successive improvement in the approximation, will produce a second eigenvalue. Suppose this second eigenvalue is λ_b; the matrix can again be *deflated* by dividing by the two factors $(\lambda - \lambda_a)$ and $(\lambda - \lambda_b)$. This has the effect of reducing the order of the original matrix by 2. The procedure again is to find, by quadratic interpolation, the value of λ which will cause

$$\frac{D(\lambda)}{(\lambda - \lambda_a)(\lambda - \lambda_b)}$$

to be equal to zero. It is not necessary to deflate the matrix each time, as the problem is simply to find the appropriate root of the polynomial using successive approximations which will cause the determinant of $(A - B\lambda)$ to be zero. Matrix deflation, however, does simplify the problem with each step by utilizing all of the information available. With sufficient perseverance, a trial-and-error search will produce all of the real eigenvalues from the matrix $(A - B\lambda)$. In treating physical systems one is usually interested in specific eigenvalues in regions which are known from the physical problem.

In addition to the procedures described, there are methods for obtaining all of the eigenvalues directly. For the case in which B is an identity matrix, that is,

$$A\bar{x} = \lambda\bar{x},$$

it is evident that if A can be transformed into a diagonal or triangular form without changing the eigenvalues, then the eigenvalues will be the diagonal elements of the transformed matrix. If the matrix D is a triangular transform of A, then the matrix equation is

$$\begin{bmatrix} d_{11} - \lambda & d_{12} & \cdots & d_{1n} \\ 0 & d_{22} - \lambda & \cdots & d_{2n} \\ \vdots & \vdots & & \vdots \\ 0 & 0 & \cdots & d_{nn} - \lambda \end{bmatrix} \begin{bmatrix} x_1 \\ x_2 \\ \vdots \\ x_n \end{bmatrix} = 0,$$

and the diagonal elements d_{ii} are roots of the characteristic equation found by setting the determinant of D to zero. When A is real and symmetric it can be transformed to diagonal form by Jacobi's method. This employs a series of iterations (successive rotations) designed to cause the off-diagonal elements to go to zero. A method due to Householder, also limited to a symmetric coefficient matrix, seeks, through a series of matrix manipulations, to transform A into a tridiagonal form. There are useful methods for handling eigenvalue problems with a nonsymmetric coefficient matrix. Called Rutishauser's LR transformation, and the QR transformation (due to J. G. F. Francis), they involve transformation of A into upper-triangular form by a succession of matrix operations. A discussion of these methods is beyond the scope of this book. [See, for example, Ralston's *A First Course in Numerical Analysis*.]

Exercises

1. Given the eigenvalue problem

$$3x_1 + 2x_3 = \lambda x_1,$$
$$5x_2 + x_3 = \lambda x_2,$$
$$2x_1 + x_2 + 3x_3 = \lambda x_3,$$

a) Obtain the three eigenvalues by direct evaluation.
b) Obtain the three eigenvectors directly.
c) Verify that the modes are orthogonal with respect to B. *Note: B = I.*

2. Using iterative techniques, find the mode (i.e., the eigenvector) which corresponds to the largest eigenvalue for the exercise

$$2x_1 + 2x_2 + 2x_3 = \lambda x_1,$$
$$2x_1 + 5x_2 + 5x_3 = \lambda 3x_2,$$
$$2x_1 + 5x_2 + 11x_3 = \lambda 2x_3.$$

Also compute the eigenvalue.

3. By matrix inversion obtain the eigenvector corresponding to the smallest eigenvalue of Exercise 2. Use the iteration method. Also, evaluate λ and test for orthogonality.

4. Continue Exercise 2 by finding the intermediate eigenvalue and its eigenvector using iterative techniques.

5. Use the matrix inversion program as a subroutine to the eigenvalue solution program of this chapter to determine the largest and smallest eigenvalues of a system, and their eigenvectors. Apply it to the system of Exercise 2.

6. Compute the second eigenvalue for the system in Exercise 2, using the largest eigenvalue in conjunction with a linear polynomial operator to modify the original system.

7. Use the first and second eigenvalues of Exercise 2 and solve for the third by introducing a matrix polynomial operator of degree 2.

8. Write a computer program to execute a trial-and-error solution to linear eigenvalue problems of order n. Utilize linear interpolation and a half-interval procedure. Read in the matrices, a value to test convergence, and the trial eigenvalues. Have the program determine the elements of the matrix $A - \lambda B$; then use the subroutine SLEQ1 (see Chapter 6) to evaluate the determinant and compute the eigenvector for each trial λ. After convergence, print out the eigenvalue and its eigenvector, and move along to the next approximation. Try the program using the eigenvalue of Exercise 2, and obtain all three eigenvalues.

9. By a hand calculation determine an eigenvalue for the problem of Exercise 2, using the trial-and-error method. Use a second-degree interpolating polynomial.

10. Apply the technique of *deflation* to obtain the second and third eigenvalues for Exercise 9.

chapter nine

the numerical solution of ordinary differential equations

The *analytic* solution of ordinary differential equations is an extensive topic, and no attempt is made to cover it here. However, many of the numerical procedures described in the preceding chapters can be applied to the *numerical* solution of such equations, and hence this introduction is included. Thus, in the best mathematical tradition, one can attack this frequently occurring problem with tools already discussed.

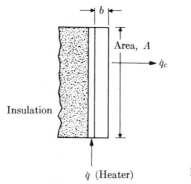

Fig. 9.1

To introduce the class of problems to be treated, we consider a simple heat-transfer problem. A solid rectangular plate of thickness b and surface area A (Fig. 9.1) is to be heated from the back. The heater supplies energy at a rate of $\dot{q}$ per unit time. Because the heater is insulated on one side, all of the energy is transferred to the plate, either raising the temperature of the plate, or being transferred from the outer surface to the surroundings by convection at a rate $\dot{q}_c$. The energy stored in the plate can be expressed in terms of its specific heat, (c_p), mass ($\rho A b$ where ρ is its density), and the rate of temperature rise T, in time, τ, as

$$\rho A b c_p \frac{dT}{d\tau} .$$

Applying Newton's law of cooling, the convective heat transfer can be expressed in terms of a heat-transfer coefficient h, the surface area A, and the difference between the surface temperature T and the temperature of the surroundings, T_0:

$$\dot{q}_c = hA(T - T_0).$$

An assumption has been made that the temperature of the plate is uniform, which would be true in the case of a thin plate having high thermal conductivity. Applying the first law of thermodynamics leads to the expression

$$\dot{q} = hA(T - T_0) + \rho A b c_p \frac{dT}{d\tau}.$$

This is an ordinary differential equation since the temperature, T, is a function of the single variable τ. It is helpful to simplify the equation by introducing the dimensionless temperature

$$\theta \equiv \frac{T - T_0}{(\dot{q}/hA)}.$$

Then

$$\frac{d\theta}{d\tau} = \frac{h}{\rho c_p b} (1 - \theta).$$

If we let

$$G = \frac{h}{\rho c_p b},$$

we are led to a general form of the problem we wish now to consider,

$$\frac{d\theta}{d\tau} = G(1 - \theta).$$

To complete the statement of the problem we require initial conditions; for example,

$$\text{when } \tau = \tau_0, \qquad \theta = \theta_0.$$

The problem can now be solved for the present case.

In this general class of problems, we are dealing with a lumped-parameter system in a transient condition. It is an initial-value problem, in that the initial state is prescribed and there is no prescribed end state. When values of the function (or its derivatives) are prescribed at the end state as well, it is a boundary-value problem. In this case, the equation is a first-order equation because only the first derivative is involved, and it is an ordinary differential equation because only a total derivative (i.e., no partial derivative) is present. Finally it is linear in the variables and the derivative. While all of the solutions in this chapter will be directed at ordinary differential equations, they are not confined to first-order or to linear differential equations.

In the example problem, the analytic solution is easily obtained by separating the variables and integrating:

$$\int \frac{d\theta}{1 - \theta} = G \int d\tau.$$

This produces the result

$$\ln (1 - \theta) = -G\tau + C,$$

where C is to be found by applying the initial conditions $\tau = \tau_0, \theta = \theta_0$. Pursuing this we find

$$C = \ln (1 - \theta_0) + G\tau_0.$$

When this is substituted into the solution of the differential equation, we obtain the result

$$\theta = 1 - (1 - \theta_0)e^{-G(\tau - \tau_0)}.$$

Further specializing the problem to require that $\tau_0 = 0$ and $\theta_0 = 0$, the solution is

$$\theta = 1 - e^{-G\tau}.$$

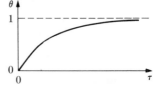

Fig. 9.2

Numerical values of θ can now be found for any value of τ, and presented in graphical or tabular form. A graph of the solution is shown in Fig. 9.2. Notice that for large τ the solution approaches $\theta = 1$ asymptotically. This is called the steady-state solution, and in this example is the point in time where the heat transferred from the surface, $\dot{q}_c$, is equal to the heating flux, $\dot{q}$. This would be the solution if $d\theta/d\tau \rightarrow 0$ in the differential equation.

Numerical solutions

In the preceding example, the analytic solution was easily obtained. It is important to note at the outset that the analytic form of solution is preferable if it can be obtained. Unfortunately, many, if not most, differential equations arising in practice cannot be so readily integrated. However, equations of the type

$$\frac{dy}{dx} = f(x), \qquad y = y_0 \quad \text{at} \quad x = x_0$$

(and this includes the example) can be integrated numerically by methods already described. Given x_0 and y_0, the problem is to compute

$$y_1 = y_0 + \int_{x_0}^{x_1} f(x)\, dx$$

$$y_2 = y_1 + \int_{x_1}^{x_2} f(x)\, dx$$

$$y_3 = y_2 + \int_{x_2}^{x_3} f(x)\, dx$$

$$\vdots$$

$$y_n = y_{n-1} + \int_{x_{n-1}}^{x_n} f(x)\, dx.$$

Since any value of the derivative which is a function of x alone can be readily

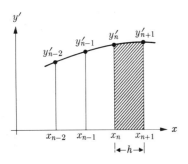

Fig. 9.3

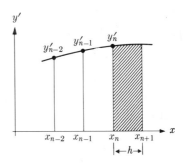

Fig. 9.4

computed, the interpolating polynomials which are integrated over the single interval could be composed from many values of $f(x)$. In the general form of the first-order equation, the derivative is a function of y as well as of x:

$$y' = \frac{dy}{dx} = f(x, y), \qquad y = y_0 \quad \text{at} \quad x = x_0,$$

and the derivative can be evaluated only where the values of y are known. In this instance, the interpolating polynomial to be used in the integration can be formed only from "past history," and another set of integration formulas is needed to deal with this case. With the assumption that the abscissas are equally spaced at intervals of width h, one step in the integration is

$$y_{n+1} = y_n + \int_{x_n}^{x_n+h} f(x, y)\, dx.$$

When the derivative does not depend on y, adjacent points may be used to form the polynomial (Fig. 9.3). Such integration formulas which include the endpoints of the integration interval are called *closed* formulas. In the more general case, the right endpoint cannot be used since the value of y is unknown there (Fig. 9.4). When the right endpoint is not used, the integration formulas are said to be *open*.

Open integration formulas

The process of deriving open integration formulas is the same one that produced the closed forms. One determines the indefinite integral of an equal-interval interpolating polynomial and then obtains different formulas by varying the number of terms retained in the integrated polynomial, as well as the number of intervals included in the integration. The backward-difference notation (∇) is useful since all known information precedes the point under consideration. After the transformation is made, $x = x_n + hs$, a single step is

$$y_{n+1} = y_n + h\int_0^1 f(x_n + hs, y)\, ds = y_n + h\int_0^1 y'_{n+s}\, ds.$$

When an integration over $p + 1$ intervals is considered (rather than over just one interval as shown), one step becomes

$$y_{n+1} = y_{n-p} + h \int_{-p}^{1} y'_{n+s} \, ds.$$

Representing the integrand with the backward-difference polynomial written with respect to the last known point y'_n gives

$$y_{n+1} = y_{n-p} + h \int_{-p}^{1} \left[y'_n + s\nabla y'_n + s(s+1) \frac{\nabla^2 y'_n}{2!} + s(s+1)(s+2) \frac{\nabla^3 y'_n}{3!} \right.$$
$$\left. + s(s+1)(s+2)(s+3) \frac{\nabla^4 y'_n}{4!} + \cdots \right] ds.$$

When the indefinite integral, a function of p, is evaluated for $p = 0$, the result is

$$y_{n+1} = y_n + h(y'_n + \tfrac{1}{2}\nabla y'_n + \tfrac{5}{12}\nabla^2 y'_n + \tfrac{3}{8}\nabla^3 y'_n + \tfrac{251}{720}\nabla^4 y'_n + \cdots).$$

For $p = 1$,

$$y_{n+1} = y_{n-1} + h(2y'_n + 0\nabla y'_n) + \frac{h^3}{3} y'''_n(\zeta)$$

and for $p = 3$,

$$y_{n+1} = y_{n-3} + h(4y'_n - 4\nabla y'_n + \tfrac{8}{3}\nabla^2 y'_n + 0\nabla^3 y'_n) + \tfrac{14}{45}h^5 y^{(5)}(\zeta).$$

Again the zero coefficients permit a jump in the order of the error term when an even number of intervals is used (that is, p is odd), and this fact accounts for the widespread use of these forms. Open formulas can be rewritten in terms of the ordinates, and the last, which is analogous to Simpson's rule in the closed form, becomes

$$y_{n+1} = y_{n-3} + \tfrac{4}{3}h(2y'_n - y'_{n-1} + 2y'_{n-2}) + \tfrac{14}{45}h^5 y^{(5)}(\zeta).$$

Assuming for the moment that a way has been provided to obtain the necessary preliminary values, one sees that the numerical solution by open formulas is a process of computing y_{n+1} from the preceding known values of the function and its derivatives. The computation may be based either upon the derivative values directly or upon backward differences which are formed from the derivative values. The latter approach has an advantage in that the first neglected difference can be used to produce an error estimate. As an illustration of one step of such a process, consider the differential equation

$$\frac{dy}{dx} = \frac{\pi}{180} \cos\left(\frac{\pi}{180} x \right), \qquad y = 0 \quad \text{at} \quad x = 0.$$

The solution is, of course, $y = \sin[(\pi/180)x]$. and for $h = 5$, the initial difference table (obtained before using the method) is:

x	y	y'	$\nabla y'$	$\nabla^2 y'$	$\nabla^3 y'$
0	0.00000	0.0174533			
5	0.08716	0.0173868	−0.0000665		
10	0.17365	0.0171882	−0.0001986	−0.0001321	
15	0.25882	0.0168587	−0.0003295	−0.0001309	+0.0000012

Using the single-interval formula ($p = 0$) with the first three differences retained, we have

$$y_{n+1} = y_n + h(y'_n + \tfrac{1}{2}\nabla y'_n + \tfrac{5}{12}\nabla^2 y'_n + \tfrac{3}{8}\nabla^3 y'_n),$$
$$y(20) = 0.25882 + 5\big(0.0168587 + \tfrac{1}{2}(-0.0003295) + \tfrac{5}{12}(-0.0001309)$$
$$+ \tfrac{3}{8}(0.0000012)\big)$$
$$= 0.34201.$$

At this point the next derivative value, $y'(20)$, is computed. If the derivative were a function of y as well as of x, which is not the case here, it could not be computed earlier since it would depend on $y(20)$:

$$y'(20) = 0.0174533 \cos\left(\tfrac{20}{180}\pi\right) = 0.0164007.$$

The next line of the difference table can now be formed:

0.0164007	−0.0004580	−0.0001285	0.0000024
—	—	—	—
0.0168587	−0.0003295	−0.0001309	0.0000012
‖	‖	‖	‖

20	0.34366	0.0164007	−0.0004580	−0.0001285	0.0000024 0.0000012

The rightmost difference is not used, but evaluating the neglected term, $\tfrac{251}{720}h\nabla^4 y'_n$, is useful to estimate the error:

$$\tfrac{251}{720}(5)(0.0000012) = 0.00002.$$

The actual error is $\sin 20° - 0.34201 = 0.34202 - 0.34201 = 0.000001$. In this case the integration appears to be quite precise and the evaluation of the first truncated term provides a good indication of the precision.

Euler's method

Before extending this numerical procedure to include both open and closed integration formulas, let us note that the simplest open form, neglecting all differences, is

$$y_{n+1} = y_n + hy'_n = y_n + hf(x_n, y_n).$$

Here $p = 0$. The truncation error is

$$\frac{h^2}{2!} y''(\zeta).$$

The application of this formula is called Euler's method. While it offers a simple method of evaluating differential equations, inherent difficulties with accumulated errors and lack of stability reduce its effectiveness. It simply predicts the value of y_{n+1}, assuming a straight-line segment having a slope given by the object function evaluated at the point (x_n, y_n). A solution is developed by marching forward in single steps, using previously computed values (with all their inherited error) to predict the next point.

Predictor-corrector methods

In this group of methods an open formula is used to predict a value of y_{n+1}, and then a closed formula is used, where the resultant y_{n+1} permits the computation of a corrected y_{n+1}. In fact, the closed formula may be repeatedly employed until two successive applications produce the same value. More often, however, the interval h is small enough so that after the prediction only one correction cycle is carried out for each point. A frequently used algorithm (Milne's method) uses a pair of three-point open and closed formulas. With $y' = f(x, y)$ one step in the procedure is shown below. Here p and c are used to indicate the predicted and corrected approximate y values.

$$p_{n+1} = y_{n-3} + \frac{4h}{3}[2y_n' - y_{n-1}' + 2y_{n-2}'], \qquad \text{Error: } \tfrac{28}{90}h^5 y^{(5)}(\zeta_1)$$

$$p_{n+1}' = f(x_{n+1}, p_{n+1}),$$

$$c_{n+1} = y_{n-1} + \frac{h}{3}[p_{n+1}' + 4y_n' + y_{n-1}'].$$

The corrector formula is based on Simpson's rule. The derivative value, p_{n+1}', used to produce the corrected value of the function, is approximate since it was computed using a predicted value for y. If it were the true value (i.e., $p_{n+1}' = y_{n+1}'$), we could write the corrected result as a y value.

$$y_{n+1} = y_{n-1} + \frac{h}{3}[y_{n+1}' + 4y_n' + y_{n-1}'] - \tfrac{1}{90}h^5 y^{(5)}(\zeta_2)$$

or

$$y_{n+1} = c_{n+1} - \tfrac{1}{90}h^5 y^{(5)}(\zeta_2).$$

Then since

$$y_{n+1} = p_{n+1} + \tfrac{28}{90}h^5 y^{(5)}(\zeta_2),$$
$$c_{n+1} - \tfrac{1}{90}h^5 y^{(5)}(\zeta_2) = p_{n+1} + \tfrac{28}{90}h^5 y^{(5)}(\zeta_1)$$

and

$$c_{n+1} - p_{n+1} = \tfrac{28}{90}h^5 y^{(5)}(\zeta_1) + \tfrac{1}{90}h^5 y^{(5)}(\zeta_2).$$

If $y^{(5)}$ is constant in the interval, then the right-hand side can be written

$$\tfrac{29}{90}h^5 y^{(5)}(\zeta_2)$$

where $y^{(5)}(\zeta_1) = y^{(5)}(\zeta_2)$. The value of c_{n+1} is larger than y_{n+1} by $\tfrac{1}{29}$ of the amount above, but this could be corrected if c_{n+1} were *modified*; that is,

$$c_{n+1(\text{modified})} = y_{n+1} = c_{n+1} - \tfrac{1}{29}(c_{n+1} - p_{n+1}).$$

This procedure is called a modified Milne's method. However, the premise for this development was that $p'_{n+1} = y_{n+1}$, which is not in general true. A slightly more extensive analysis would show that the modified value would be exact (i.e., $c_{n+1(\text{modified})} = y_{n+1}$) only if $y^{(5)}(\zeta) = 0$; that is,

$$f_y(x_{n+1}, \eta) = 0$$

where η is between p_{n+1} and y_{n+1}.

If the modification is actually used, it should be small relative to the value of y_{n+1}. As is the case with high-order differences, the modifying difference is a measure of the error of y_{n+1}. If it encroaches on the number of accurate figures that are desired in the solution, a smaller interval should be chosen. The type of error measured by such differences is that which is due to truncation in the computation of a single step. Since the terminus of one integration is the initial point of another, error produced in one step is propagated to the next. The analysis of propagation of error is a complicated task and will not be pursued here.

Milne's method is unstable for some types of object functions; that is, the solution may oscillate with increasing magnitude no matter how small the interval h is chosen. A modification due to Hamming is more stable than Milne's method. The penalty is a small increase in the truncation error. *Hamming's method* uses the same predictor as Milne's, but a modified corrector, resulting in the following expressions:

$$p_{n+1} = y_{n-3} + \tfrac{4}{3}h[2y'_n - y'_{n-1} + 2y'_{n-2}] + \tfrac{28}{90}h^5 y^{(5)}(\zeta_1),$$
$$c_{n+1} = \tfrac{1}{8}(9y_n - y_{n-2}) + \tfrac{3}{8}y'_{n+1} + 2y'_n - y'_{n-1} - \tfrac{1}{40}h^5 y^{(5)}(\zeta_2),$$

and

$$c_{n+1\ (\text{modified})} = c_{n+1} - \tfrac{9}{121}(c_{n+1} - p_{n+1}).$$

Notice that Milne's method is not *self-starting*. In order to compute y_1, ($n = 0$), knowledge of the function prior to y_0 is required, and it is not available. A first-degree predictor-corrector pair is useful in starting a solution, as the history of previous values is not needed for these simple formulas. Called the modified Euler method, the predictor formula is the Euler-method formula

$$p_{n+1} = y_n + hy'_n, \qquad \text{Error:} \ \frac{h^2 y''}{2}\,(\zeta_1).$$

The corrector formula corresponds to the trapezoidal rule for integration,

$$c_{n+1} = y_n + \frac{h}{2}(y_n' + p_{n+1}')$$

$$= y_n + \frac{h}{2}(f(x_n, y_n) + f(x_{n+1}, p_{n+1})).$$

The process of starting a solution from x_0, y_0 with this pair, using repeated application of the corrector, would be to compute

$$y_1^{(0)} = y_0 + hf(x_0, y_0) \quad \text{as} \quad y' = f(x, y),$$

$$y_1^{(1)} = y_0 + \frac{h}{2}(f(x_0, y_0) + f(x_0 + h, y_1^{(0)})),$$

$$y_1^{(2)} = y_0 + \frac{h}{2}(f(x_0, y_0) + f(x_0 + h, y_1^{(1)})).$$

The superscript indicates successive corrected values. When the change in two successive corrected values is sufficiently small, x and y are incremented and the process is repeated applying the predictor at (x_1, y_1). Although the modified Euler method could be employed to produce the complete solution it is more useful as a starter for another solution (e.g., Milne's method). In this case when y_3 has been found by the modified Euler method, Milne's method could be introduced for y_4 as $y_{n-3} = y_0$, which is available.

Milne's method, as well as a very useful method of starting a solution, is illustrated in detail by a program at the end of this chapter.

Taylor's series

Another method of starting the solution, or of solving it completely, is to write the object function as a Taylor series and simply evaluate the series at the desired points. The initial series expansion is made at the given point (x_0, y_0), and it is necessary to evaluate the derivatives at this point. The first derivative is given in a first-order equation, but the higher-order derivatives must be obtained by differentiation. The situation is complicated by the fact that the function $y' = f(x, y)$ depends, in general, on both x and y. Hence one must differentiate with respect to two variables, using the chain rule:

$$\frac{df}{dx} = \frac{\partial f}{\partial x} + \frac{\partial f}{\partial y}\frac{dy}{dx}.$$

In many practical instances, the differentiation quickly becomes very complicated, but for a simple example, such as the one below, it is straightforward. Given

$$y' = xy + 1 \quad \text{and at} \quad x = 0, \quad y = 1,$$

then at $x = 0$, $y' = 1$. The successive derivatives and their values at $x = 0$ can be determined:

$$\frac{dy'}{dx} = \frac{\partial y'}{\partial x} + \frac{\partial y'}{\partial y}\frac{dy}{dx}$$

or

$$y'' = y + xy' \qquad \text{at} \quad x = 0, \quad y'' = 1,$$

$$\frac{dy''}{dx} = \frac{\partial y''}{\partial x} + \frac{\partial y''}{\partial y}\frac{dy}{dx} + \frac{\partial y''}{\partial y'}\frac{dy'}{dx}$$

or

$$y''' = y' + y' + xy'' \qquad \text{at} \quad x = 0, \quad y''' = 2.$$
$$\vdots$$

Combining these function and derivative values to produce a Taylor series of the form

$$f(x_0 + h) = f(x_0) + hf'(x_0) + \frac{h^2}{2!}f''(x_0) + \frac{h^3}{3!}f'''(x_0) + \frac{h^4}{4!}f''''(x_0) + \cdots$$

we obtain

$$y(h) = 1 + h + \frac{h^2}{2} + \frac{h^3}{3} + \frac{h^4}{8} + \cdots$$

After tabulating the function for several points, we may find it desirable to expand about another point, say x_k, before continuing.

Iteration

It is also possible to construct a power-series solution by an iterative integration. While it produces the same result as the Taylor-series expansion, iteration sometimes can be employed when it is not possible to apply the Taylor expansion. The problem is rewritten in the form

$$\int_{y_0}^{y} dy = \int_{x_0}^{x} f(x, y)\, dx.$$

An initial trial solution is substituted in the right-hand integral and the integral evaluated to produce the improved approximation.

To demonstrate, return to the example

$$y' = xy + 1 \qquad \text{and at} \qquad x = 0, \quad y = 1.$$

Rearranging the equation and integrating the left side, we obtain the general form of the iteration:

$$y^{(i+1)} = 1 + \int_0^x (xy + 1)^{(i)}\, dx.$$

Making the substitution for the initial trial, we get

$$y^{(1)} = 1 + \int_0^x (0 + 1)\, dx = 1 + x,$$

and successive steps are

$$y^{(2)} = 1 + \int_0^x ((1+x)x + 1)\, dx = 1 + \frac{x^2}{2} + \frac{x^3}{3} + x$$

and

$$y^{(3)} = 1 + x + \frac{x^2}{2} + \frac{x^3}{3} + \frac{x^4}{8} + \frac{x^5}{15};$$

or, for the interval $x = 0$ to $x = h$,

$$y = 1 + h + \frac{h^2}{2} + \frac{h^3}{3} + \frac{h^4}{8} + \cdots,$$

which is the same result as the Taylor expansion. The procedure is sometimes called Picard's method of successive approximations.

The method of undetermined parameters

Application of Taylor series to the solution of initial-value problems produces approximate solutions in the form of analytical expressions. Accuracy is good in the neighborhood of the initial state, becoming poorer at the end of the interval. This characteristic can be seen in the curve of the error for a Taylor-series solution to the example problem (Fig. 9.5). The method of undetermined parameters produces an approximate solution in analytical form which sacrifices some accuracy near the initial condition, but offers more uniform accuracy throughout the interval. This method is similar to curve fitting, distributing errors in a similar way. A trial form of the solution is guessed, having several coefficients which are to be determined subject to some "best-fit" criterion (e.g., a least-squares criterion).

To illustrate, return to the previous example,

$$y' = xy + 1,$$

with initial conditions $x = 0$, $y = 1$. A trial solution is chosen. Knowledge of the behavior of the physical system can improve this choice, but for the present we shall employ the trial family (and indicate the trial y by y_t)

$$y_t = C_0 + C_1 x + C_2 x^2.$$

This must first be made to satisfy the initial condition. Hence, $C_0 = 1$. The parameters C_1 and C_2 can now be selected using a criterion for the best approximation to the solution. Define a residual as we have done in earlier chapters. In this case,

$$R = y_t' - f(x, y_t) = y_t' - xy_t - 1.$$

An exact solution will cause the residual to vanish. While it is not likely that an exact solution will be found, it is possible to choose coefficients which will make R

small in the interval. For the trial family assumed here,

$$R = C_1 + 2xC_2 - x(1 + C_1x + C_2x^2) - 1,$$

or

$$R = C_1(1 - x^2) + C_2(2x - x^3) - x - 1.$$

Any of the several criteria for curve fitting can be used to determine C_1 and C_2.

Two methods which employ points throughout the interval will be demonstrated here. The first is called Galerkin's method. For a "best fit," this method requires that the weighted average of the residual go to zero over the interval. A weighted average here is defined as

$$\frac{\int_a^b w(x)R(x)\, dx}{\int_a^b w(x)\, dx},$$

where $w(x)$ is a weighting function, and a and b are the limits of the interval. The weighted average is zero when the numerator is zero. Weighting functions are normally chosen as the same functions used in constructing the trial solution, in this case x and x^2. This provides as many equations as there are coefficients. The integrals are evaluated over the interval and set equal to zero. In this case for the interval $x = 0$ to $x = 1$ they are

$$\int_0^1 xR\, dx = C_1(.2500) + C_2(.4667) - .8333 = 0,$$

and

$$\int_0^1 x^2R\, dx = C_1(.1333) + C_2(0.3333) - .5833 = 0.$$

Solving these for the coefficients, and then substituting the latter into the trial solution for y, we get the approximate solution

$$y = 1. + 0.240x + 1.656x^2.$$

An alternate way of selecting coefficients is to employ a "least-squares" criterion. Parameters are chosen to minimize the square of the residual over the interval, just as was done in approximating functions. Forming the integrals and setting the derivatives with respect to the coefficients equal to zero, one obtains, for the example,

$$\frac{1}{2}\frac{\partial}{\partial C_1}\int_0^1 R^2\, dx = \int_0^1 R\frac{\partial R}{\partial C_1}\, dx$$
$$= C_1(.5333) + C_2(.4167) - .9167 = 0,$$

$$\frac{1}{2}\frac{\partial}{\partial C_2}\int_0^1 R^2\, dx = \int_0^1 R\frac{\partial R}{\partial C_2}\, dx$$
$$= C_1(.4167) + C_2(.6762) - 1.2167 = 0.$$

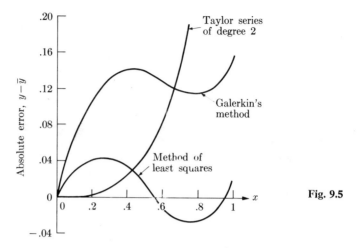

Fig. 9.5

Solving the set of equations for C_1 and C_2 we obtain the least-squares approximation

$$y = 1.0 + 0.6036x + 1.4273x^2.$$

A comparison of the absolute errors for these approximate solutions is shown for the interval $x = 0$ to $x = 1$ in Fig. 9.5. Absolute error is defined as $y - \bar{y}$, where $\bar{y}$ is the approximate ordinate at each x, and y is the ordinate given by the exact solution to the differential equation, which is

$$y = e^{x^2/2}\left(1 + \sqrt{\frac{\pi}{2}}\, \text{erf}\, \frac{x}{\sqrt{2}}\right).$$

Error curves are presented for Galerkin's method, the method of least squares, and the Taylor-series solution of degree two, for the example problem.

Runge-Kutta methods

The difficulty with Taylor's series as a solution-starting procedure is that the series is specific to the problem, and differentiation in more realistic cases may be extremely difficult. It would be very useful to be able to obtain values of the solution function with an accuracy equivalent to a specified number of terms in the Taylor series representation, and yet compute these values by using a weighted-sum-of-ordinates formula similar to those employed in numerical integration. The popular Runge-Kutta methods fulfill these requirements and also have the added advantage that the interval of integration, h, may be readily changed. Unlike the techniques which are based on the integration of equal-interval interpolating formulas, the Runge-Kutta procedures are *single-step*; they do not require any "past history" of values. When a past history is required, changing the value of h is complicated, and separate solution-starting procedures are required.

The derivation of the integration formulas is a little more complicated than it was for simple integration. Again this is due to the fact that the derivative is a function of both x and y:

$$y' = f(x, y).$$

It is worthwhile to go through the derivation of a Runge-Kutta formula, if for no other reason than that at first glance it seems difficult to make a sum of derivative evaluations correspond to terms of the Taylor series which contain higher-order derivatives.

Third-order Runge-Kutta derivation

The problem is: Given $y' = f(x, y)$, at $x = x_n$, $y = y_n$, produce a sum-of-ordinates integrating formula of the form

$$y_{n+1} \approx y_n + h(af_0 + bf_1 + cf_2),$$

where

$$
\begin{aligned}
f_0 &= f(x_n, y_n), \\
f_1 &= f(x_n + mh, y_n + mhf_0), \\
f_2 &= f(x_n + ph, y_n + phf_1) = f(x_n + ph, y_n + qhf_1 + (p - q)hf_0),
\end{aligned}
$$

such that the result is identical with the first four terms of the Taylor series,

$$y_{n+1} \approx y_n + hy'_n + \frac{h^2}{2} y''_n + \frac{h^3}{3!} y'''_n.$$

(The second form on the right in the third equation, introducing another parameter, q, simplifies the subsequent derivation.) One produces this equivalence by expanding f_0, f_1, f_2 and, from the immediately preceding formula, y'_n, y''_n, y'''_n, in terms of the two variables x and y and then equating the two expressions for y_{n+1}. The expansion of the last three is done by chain-rule differentiation as in the specific case of the previous section on Taylor series. The f's are expanded by using the two-variable Taylor series.

Taylor's series for two variables is most easily remembered from its operational form:

$$
\begin{aligned}
f(x + h, y + k) &= (e^{h(\partial/\partial x) + k(\partial/\partial y)}) f(x, y) \\
&= \left(1 + h\frac{\partial}{\partial x} + k\frac{\partial}{\partial y} + \frac{1}{2}\left(h^2 \frac{\partial^2}{\partial x^2} + 2hk \frac{\partial^2}{\partial x\, \partial y} + k^2 \frac{\partial^2}{\partial y^2} \right) + \cdots \right) f(x, y) \\
&= f + hf_x + kf_y + \left(\frac{h^2}{2} f_{xx} + hk f_{xy} + \frac{k^2}{2} f_{yy} \right) + \cdots
\end{aligned}
$$

Here the subscripted f's indicate partial derivatives of $f(x, y)$. Expanding f_0,

f_1, f_2 in this manner yields

$$f_0 = f,$$

$$f_1 = f + mhf_x + mhf \cdot f_y + \frac{m^2h^2}{2} f_{xx} + m^2h^2 f \cdot f_{xy} + \frac{m^2h^2f^2}{2} f_{yy} + \cdots$$

$$= f + mh(f_x + f \cdot f_y) + \frac{m^2h^2}{2} (f_{xx} + 2f_{xy} \cdot f + f_{yy} \cdot f^2) + \cdots,$$

$$f_2 = f + phf_x + (qhf_1 + (p - q)hf)f_y + \frac{p^2h^2}{2} f_{xx}$$

$$+ ph(qhf_1 - (p - q)hf)f_{xy} + (qhf_1 + (p - q)hf)^2 f_{yy} + \cdots$$

Substituting for f_1 permits, after a considerable manipulative effort, simplification to

$$f_2 = f + ph(f_x + f_y f) + \frac{p^2h^2}{2} (f_{xx} + 2f_{xy} + f_{yy}f^2)$$

$$+ mqh^2(f_x + f_y f)f_y + \cdots$$

The higher derivatives can be expressed in terms of partial derivatives by chain-rule differentiation:

$$y' = f = f(x, y),$$

$$y'' = f_x + f_y y' = f_x + f_y f,$$

$$y''' = f_{xx} + f_{xy}f + f_y f_x + (f_{xy} + f_{yy}f + f_y^2)f$$

$$= f_{xx} + 2f_{xy}f + f_{yy}f^2 + f_y(f_x + f_y f).$$

When the derivative expressions are replaced by single symbols,

$$f_x + f_y f = A,$$

$$f_{xx} + 2f_{xy}f + f_{yy}f^2 = B,$$

the two formulas that must correspond are: Taylor's series,

$$y_{n+1} \approx y_n + hf + \frac{h^2}{2} A + \frac{h^3}{6} B + \frac{h^3}{6} Af_y,$$

and the integration formula,

$$y_{n+1} \approx y_n + ahf + bh\left(f + mhA + \frac{m^2h^2}{2} B\right)$$

$$+ ch\left(f + phA + \frac{p^2h^2}{2} B + mqh^2 Af_y\right).$$

In order for this to be true, the coefficients of hf must correspond; that is,

$$a + b + c = 1.$$

Similarly the coefficients of h^2A, h^3B, and h^3Af_y must correspond:

$$bm + cp = \tfrac{1}{2},$$
$$bm^2 + cp^2 = \tfrac{1}{3},$$
$$cmq = \tfrac{1}{6}.$$

Since there are six variables in these four equations, the values of two variables may be selected arbitrarily. It is convenient to select $m = \tfrac{1}{2}$ and $p = 1$, since the evaluations of $f(x, y)$ will then be at x_n, $x_n + h/2$, and $x_n + h$. By using these values one can determine the remaining ones from the equations and write the integrating formula. For $m = \tfrac{1}{2}$ and $p = 1$,

$$a = \tfrac{1}{6}$$
$$b = \tfrac{2}{3}$$
$$c = \tfrac{1}{6} \quad \text{and} \quad y_{n+1} \approx y_n + \frac{h}{6}(f_0 + 4f_1 + f_2),$$
$$q = 2$$

where

$$f_0 = f(x_n, y_n),$$
$$f_1 = f\left(x_n + \frac{h}{2}, y_n + \frac{hf_0}{2}\right),$$
$$f_2 = f(x_n + h, y_n + 2hf_1 - hf_0).$$

Two fourth-order formulas

With correspondingly greater algebraic effort one can obtain fourth-order formulas. Again there is some free choice of parameters, and the simplest choice (Runge) is:

$$y_{n+1} \approx y_n + \frac{h}{6}(f_0 + 2f_1 + 2f_2 + f_3),$$
$$f_0 = f(x_n, y_n),$$
$$f_1 = f\left(x_n + \frac{h}{2}, y_n + \frac{hf_0}{2}\right),$$
$$f_2 = f\left(x_n + \frac{h}{2}, y_n + \frac{hf_1}{2}\right),$$
$$f_3 = f(x_n + h, y_n + hf_2).$$

Many other variants of these fourth-order formulas have been produced. One of them, the Gill version, was designed to save storage in digital computers using this algorithm. If m equations are being solved simultaneously, the method above requires the use of four linear arrays of length m, one array for y_n, one for the intermediate y-values, one for the values of the derivative f, and one in which the partial sums of the integrating formula are stored. The Gill version reduces this storage requirement to three arrays by eliminating the necessity for retaining y_n through the four evaluations. The formulas are reproduced here, although much of the motivation for their use has disappeared as high-speed storage in machines has increased.

$$\begin{cases} f_1 = f(x_n, y_n), \\[2mm] y_{n+1}^{(1)} = y_n + \dfrac{hf_1}{2}, \\[2mm] q_1 = f_1. \end{cases}$$

$$\begin{cases} f_2 = f\left(x_n + \dfrac{h}{2}, y_{n+1}^{(1)}\right), \\[2mm] y_{n+1}^{(2)} = y_{n+1}^{(1)} + h\left(1 - \dfrac{1}{\sqrt{2}}\right)(f_2 - q_1), \\[2mm] q_2 = (2 - \sqrt{2})f_2 + \left(-2 + \dfrac{3}{\sqrt{2}}\right)q_1. \end{cases}$$

$$\begin{cases} f_3 = f\left(x_n + \dfrac{h}{2}, y_{n+1}^{(2)}\right), \\[2mm] y_{n+1}^{(3)} = y_{n+1}^{(2)} + h\left(1 + \dfrac{1}{\sqrt{2}}\right)(f_3 - q_2), \\[2mm] q_3 = (2 + \sqrt{2})f_3 + \left(2 + \dfrac{3}{\sqrt{2}}\right)q_2. \end{cases}$$

$$\begin{cases} f_4 = f(x_n + h, y_{n+1}^{(3)}), \\[2mm] y_{n+1} = y_{n+1}^{(4)} = y_{n+1}^{(3)} + h\left(\dfrac{f_4}{6} - \dfrac{q_3}{3}\right). \end{cases}$$

The three arrays required are for y, f, and the "bridging q's." The latter is the temporary storage arranged so that the y_n-values need not be preserved.

Note that the original third-order formulation, as well as the fourth-order one, reduces to Simpson's rule when the derivative is a function of x alone. In such a case, f_1 and f_2 are equal, and, since the interval is effectively $h/2$, Simpson's rule is obtained.

Simultaneous ordinary differential equations

A higher-order differential equation, or a system of equations including some high-order members, may be reduced to a set of first-order equations by making a simple change of variable. An nth-order equation,

$$y^{(n)} = f(x, y, y', y'', \ldots, y^{(n-1)}),$$

may be transformed by letting

$$y = y_0,$$
$$y' = y_1,$$
$$y'' = y_1' = y_2,$$
$$y''' = y_1'' = y_2' = y_3,$$
$$\vdots$$
$$y^{(n)} = \cdots = y_{n-1}' = f(x, y_0, y_1, y_2, \ldots, y_{n-1}).$$

Such simultaneous systems can be handled by all the methods described. The computation proceeds in parallel; one makes one step for each equation before advancing to the next increment. When the derivative functions are evaluated, the current values of the array of the functional values (the y's) are used. In the fourth-order Runge-Kutta case, the increment might be described more accurately as a one-fourth step taken in parallel since on every intermediate step (there are four) all the equations are evaluated.

To illustrate the fourth-order Runge-Kutta procedure (Runge version), consider the two simultaneous equations:

$$\dot{y}_1 = y_2 + t \qquad \text{at} \quad t = 0, \quad y_1 = 0$$
$$\ddot{y}_2 = -y_2\dot{y}_2 - y_1 t^2 \qquad \qquad y_2 = 1$$
$$\dot{y}_2 = 1$$

The dot notation indicates differentiation with respect to t. For a second-order differential equation, the initial value of both the function and its first derivative are required.

Making the substitution $\dot{y}_2 = y_3$ results in a system of three first-order equations. The derivatives are also labeled f to conform with the earlier notation:

$$f_1 = \dot{y}_1 = y_2 + t \qquad \text{at} \quad t = 0, \quad y_1 = 0$$
$$f_2 = \dot{y}_2 = y_3 \qquad \qquad y_2 = 1$$
$$f_3 = \dot{y}_3 = -y_2 y_3 - y_1 t^2 \qquad \qquad y_3 = 1$$

Selecting $h = 0.2$, one evaluates the first of the Runge formulas. The parenthesized superscript indicates which Runge formula applies, while the subscript is the equation number. Initially, at $t = 0$,

$$\left.\begin{array}{l} y_1(0) = 0 \\ y_2(0) = 1 \\ y_3(0) = 1 \end{array}\right\} \quad \text{from which} \quad \left\{\begin{array}{l} f_1^{(1)} = 1 + 0 = 1 \\ f_2^{(1)} = 1 \\ f_3^{(1)} = -1 \times 1 = -1. \end{array}\right.$$

A trial value of y, indicated by the bar, is obtained and the second evaluation of the derivative functions is made by means of these trial values.   With $t = t(0) + h/2 = 0 + 0.2/2 = 0.1$,

$$\bar{y}_1 = 0 + \left(\frac{0.2}{2}\right) \times 1 = 0.1, \qquad f_1^{(2)} = 1.1 + 0.1 = 1.2,$$

$$\bar{y}_2 = 1 + \left(\frac{0.2}{2}\right) \times 1 = 1.1, \qquad f_2^{(2)} = 0.9,$$

$$\bar{y}_3 = 1 + \left(\frac{0.2}{2}\right) \times (-1) = 0.9, \qquad f_3^{(2)} = -1.1 \times 0.9 - 0.1 \times 0.01$$
$$= -0.991.$$

Continuing with $t = 0.1$, we have

$$\bar{y}_1 = 0 + \left(\frac{0.2}{2}\right) \times 1.2 = 1.2, \qquad f_1^{(3)} = 1.09 + 0.1 = 1.19,$$

$$\bar{y}_2 = 1 + \left(\frac{0.2}{2}\right) \times 0.9 = 1.09, \qquad f_2^{(3)} = 0.9009,$$

$$\bar{y}_3 = 1 + \left(\frac{0.2}{2}\right) \times (-0.991) \qquad f_3^{(3)} = -1.09 \times 0.9009 - 1.2 \times 0.01$$
$$= 0.9009, \qquad\qquad\qquad = -0.983181.$$

The last evaluation is based on $t = t(0) + h = 0.2$:

$$\bar{y}_1 = 0 + \left(\frac{0.2}{2}\right) \times 1.19 = 0.238, \qquad f_1^{(4)} = 1.18018 + 0.2 = 1.38018,$$

$$\bar{y}_2 = 1 + \left(\frac{0.2}{2}\right) \times 0.9009, \qquad f_2^{(4)} = 0.8033638,$$
$$= 1.18018$$

$$\bar{y}_3 = 1 + \left(\frac{0.2}{2}\right) \times (-0.983181) \qquad f_3^{(4)} = -1.18018 \times 0.8033638$$
$$= 0.8033638, \qquad\qquad\qquad -0.238 \times 0.04$$
$$= -0.9576339.$$

Substituting these results in the integration formula yields

$$y_1(0.2) = 0 + \frac{0.2}{6}\left(1 + 2(1.2) + 2(1.19) + 1.38018\right) = 1.009339,$$

$$y_2(0.2) = 1 + \frac{0.2}{6}\left(1 + 2(0.9) + 2(0.9009) + 0.8033638\right) = 1.793745,$$

$$y_3(0.2) = 1 + \frac{0.2}{6}\left(-1 + 2(-0.991) + 2(-0.983181) + (-0.9576339)\right)$$
$$= 0.143352.$$

This completes one step of the integration; the process is then repeated for the desired number of intervals.

Example programs

To illustrate these solution procedures, we present external functions which integrate a system of differential equations by the Milne and Runge-Kutta methods.

Example 1 Write external functions which integrate one step of a system of m first-order equations by the Milne method. The function should have two entries, PRED, which produces a predicted value of y_{n+1} for each equation, and CORR, which then computes the corrected value of y_{n+1}. The necessary parameters are m, two arrays, y and f, the current value of the independent variable, x, and the interval size, h. The two linear arrays should initially contain all the preceding values required by the method; these will be stored according to the following scheme. (The superscripts here indicate the equation number.)

$$y_1 \text{ to } y_m \quad \text{contain} \quad y_n^{(1)} \text{ to } y_n^{(m)}$$

$$y_{m+1} \text{ to } y_{2m} \quad \text{contain} \quad y_{n-1}^{(1)} \text{ to } y_{n-1}^{(m)}$$

$$y_{2m+1} \text{ to } y_{3m} \quad \text{contain} \quad y_{n-2}^{(1)} \text{ to } y_{n-2}^{(m)}$$

$$y_{3m+1} \text{ to } y_{4m} \quad \text{contain} \quad y_{n-3}^{(1)} \text{ to } y_{n-3}^{(m)}$$

$$f_1 \text{ to } f_m \quad \text{contain} \quad f_n^{(1)} \text{ to } f_n^{(m)}$$

$$f_{m+1} \text{ to } f_{2m} \quad \text{contain} \quad f_{n-1}^{(1)} \text{ to } f_{n-1}^{(m)}$$

$$f_{2m+1} \text{ to } f_{3m} \quad \text{contain} \quad f_{n-2}^{(1)} \text{ to } f_{n-2}^{(m)}$$

For uniformity, the calling program should reserve $4m$ locations for both the y- and f-arrays. This is computed in the calling program and transferred to the subroutine as NO. In summary, references to the function would be of the form

$$\text{CALL} \quad \text{PRED (M, NO, Y, F, X, H)}$$

and

$$\text{CALL} \quad \text{CORR (M, NO, Y, F, X, H)}.$$

The function should, in addition, increment the value of the independent variable x and return, as a direct result, the value of y_{n+1} for the last equation computed. This return will simplify the calling statements when a single first-order equation is being solved.

The flow chart and the program are presented in Fig. 9.6. The variable y_T is a temporary location used to retain a value which is to be replaced. In the program calling this routine, the statements computing the derivative functions, the f's, appear between calling PRED and calling CORR.

Example 2 Write an external function, to be called RKSUB., which has the same parameters as the preceding Milne routine, namely, m, y, f, x, and h, but uses the Runge formulas to integrate the differential equations. For m first-order simul-

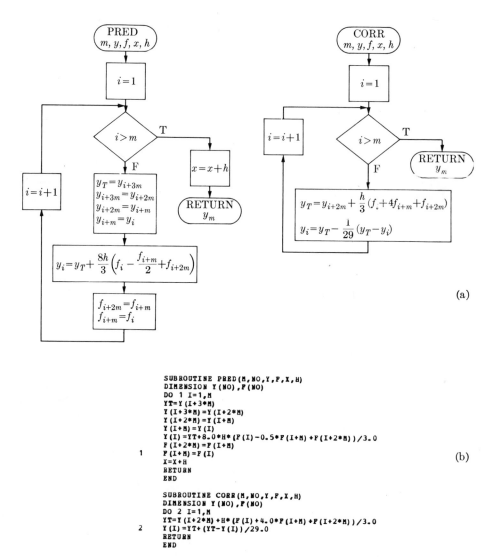

Fig. 9.6   Predictor-corrector subroutines. (a) Flow chart. (b) Program.

taneous equations, the following parameters are required: two arrays, y and f, for the current values of the functions and the derivatives, respectively, the current values of the independent variable x, and the interval size h. Here, however, the arrays need contain only the m current values since the Runge-Kutta procedure is a single-step method and does not depend on the preceding values other than the last.

The flow chart and the program are presented in Fig. 9.7. Note that two working arrays are needed by the external function: Y to store the original values

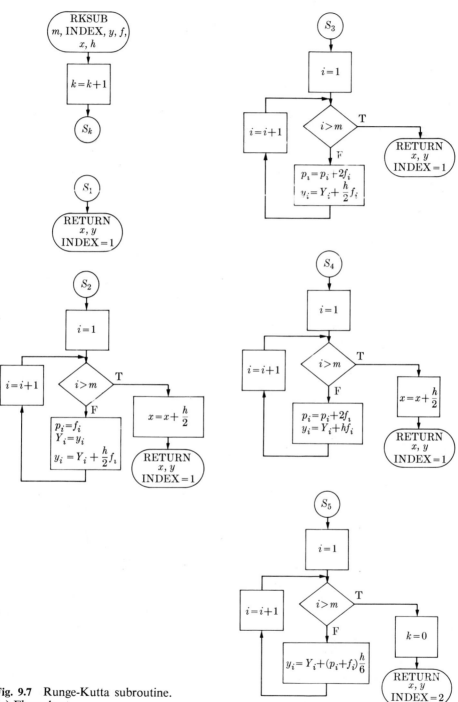

Fig. 9.7 Runge-Kutta subroutine.
(a) Flow chart.

```
SUBROUTINE RKSUB(M,K,INDEX,Y,F,X,H)
      DIMENSION CAPY(40),P(40),Y(M),F(M)
      K=K+1
      GO TO (1,2,3,4,5),K
2     DO 12 I=1,M
      P(I)=F(I)
      CAPY(I)=Y(I)
12    Y(I)=CAPY(I)+0.5*H*F(I)
11    X=X+0.5*H
1     INDEX=1
      RETURN
3     DO 13 I=1,M
      P(I)=P(I)+2.0*F(I)
13    Y(I)=CAPY(I)+0.5*H*F(I)
      INDEX=1
      RETURN
4     DO 14 I=1,M
      P(I)=P(I)+2.0*F(I)
14    Y(I)=CAPY(I)+H*F(I)
      GO TO 11
5     DO 15 I=1,M
15    Y(I)=CAPY(I)+(P(I)+F(I))*H/6.0
      INDEX=2
      K=0
      RETURN
      END
```

Fig. 9.7   (b) Program.

of y_n, and P to store the partial sums as terms are added in the process of producing y_{n+1}. When the integration is incomplete and additional evaluations of the derivative functions are required, the integer INDEX = 1 is directly returned; when the integration is complete, the integer INDEX = 2 is returned. These integer results may be used as statement label subscripts in the calling program to perform the necessary switching. The integer variable K, which is used to switch to the proper subsection of the program, is guaranteed to be initially zero by an initializing statement in the main program. The last substitution statement returns K to zero after an integration step has been completed.

The use of these two functions is illustrated by the following example.

Example 3 Write a main program which numerically solves the three simultaneous equations

$$f_1 = \dot{y}_1 = y_2 + t,$$
$$f_2 = \dot{y}_2 = y_3,$$
$$f_3 = \dot{y}_3 = -y_2 y_3 - y_1 t^2,$$

where at $t = 0$,

$$y_1 = 0, \qquad y_2 = 1, \qquad y_3 = 1.$$

Read the values of the interval size and the limits of the integration. Use the Runge-Kutta function to start the solution and, once enough values have been computed, continue the solution with the Milne function.

The first values of the functions, which are read as data, as well as the results of the first three applications of the Runge-Kutta method, must be transferred to the appropriate places in the y-array. Similarly, the derivative values must be stored for three points also. After this solution-starting phase, the remaining values are computed by the Milne procedure until the independent variable t reaches its final value.

Before the first use of the predictor routines the differential equations are evaluated again. Without these evaluations the most recently computed derivative values would be those obtained from the last stage of the Runge-Kutta evaluation which did not use the final function values for the interval.

The flow chart and the program are presented in Fig. 9.8. The derivative evaluations could be written as a function subprogram or, with a little more switching of statement labels, the same set of statements could be used for both procedures. Results for a sample set of initial conditions are also given in Fig. 9.8.

This program could be solved by means of the Runge-Kutta procedure alone, but the Milne process has several advantages. Only one derivative evaluation for

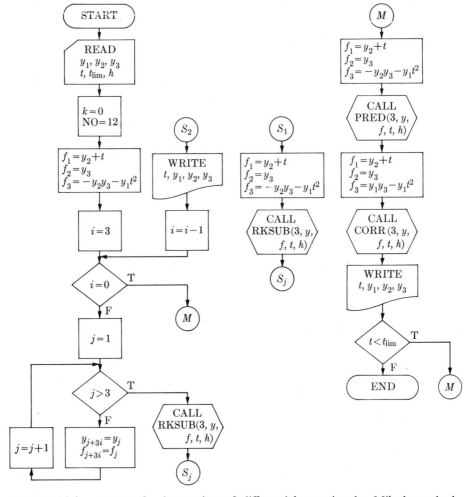

Fig. 9.8 Main program for integration of differential equation by Milne's method. (a) Flow chart.

```
C       MASTER MILNE
C       MAIN PROGRAM FOR THE SOLUTION ORDINARY DIFFERENTIAC
C       EQUATIONS USING MILNES METHOD
        DIMENSION Y(12),F(12)
        READ(5,20) Y(1),Y(2),Y(3),T,TLIM,H
20      FORMAT(6F9.0)
        WRITE(6,25)
25      FORMAT(5X,4HTIME,8X,4HY(1),8X,4HY(2),8X,4HY(3),//)
        WRITE(6,30) T,Y(1),Y(2),Y(3)
30      FORMAT(3X,F6.3,6X,F6.3,6X,F6.3,6X,F6.3,/)
        K=0
        NO=12
        F(1)=Y(2)+T
        F(2)=Y(3)
        F(3)=-Y(2)*Y(3)-Y(1)*T*T
        I=3
3       IF(I) 4,7,4
4       DO 5 J=1,3
        Y(J+I*3)=Y(J)
5       F(J+I*3)=F(J)
6       CALL RKSUB(3,K,INDEX,Y,F,T,H)
        GO TO (1,2),INDEX
1       F(1)=Y(2)+T
        F(2)=Y(3)
        F(3)=-Y(2)*Y(3)-Y(1)*T*T
        GO TO 6
2       WRITE(6,30) T,Y(1),Y(2),Y(3)
        I=I-1
        GO TO 3
7       F(1)=Y(2)+T
        F(2)=Y(3)
        F(3)=-Y(2)*Y(3)-Y(1)*T*T
        CALL PRED(3,NO,Y,F,T,H)
        F(1)=Y(2)+T
        F(2)=Y(3)
        F(3)=-Y(2)*Y(3)-Y(1)*T*T
        CALL CORR(3,NO,Y,F,T,H)
        WRITE(6,30) T,Y(1),Y(2),Y(3)
        IF(T.LT.TLIM) GO TO 7
        STOP
        END
                INPUT

   0.,1.,1.,0.,2.,.2,

             OUTPUT
```

TIME	Y(1)	Y(2)	Y(3)
0.0	0.0	1.000	1.000
0.200	0.294	1.132	0.852
0.400	0.643	1.238	0.710
0.600	1.042	1.312	0.553
0.800	1.419	1.447	0.279
1.000	1.944	1.410	0.026
1.200	2.391	1.432	-0.502
1.400	2.979	1.207	-1.142
1.600	3.436	0.930	-2.288
1.800	3.961	0.270	-4.033
2.000	4.243	-0.791	-7.339
2.200	4.376	-2.871	******

Fig. 9.8   (b) Program.

each integration step is required, and it is possible to monitor (by computing the difference between predicted and corrected values) the error introduced at each step. As mentioned before, the Runge-Kutta procedure is self-starting, and the interval size may be changed from step to step.

Error and stability

Truncation errors presented for several of the methods discussed are for a single step in the computation. When the next step in the computation is made, the error from the first step will be carried over. This will then add to the local step error. As the solution proceeds, the error propagates, and may become so large as to render the solution useless. Propagation of error in the solution depends on both the method employed and the form of the object function. To demonstrate this we shall examine the propagation of errors for Euler's method. We first define the local truncation error for a single step as

$$\epsilon_i = y_i - \bar{y}_i,$$

where y_i is the exact value and $\bar{y}_i$ is the approximate result at the ith step. For Euler's method the exact value of y_{i+1} would be

$$y_{i+1} = \bar{y}_{i+1} + \epsilon_{i+1} = y_i + hf(x_i, y_i) + \epsilon_{i+1}.$$

It was demonstrated earlier that

$$\epsilon_{i+1} = \frac{h^2}{2} y''(\zeta) + \text{Roundoff},$$

with

$$x_i \le \zeta \le x_{i+1}.$$

In terms of the error in y_i, the exact value of y_{i+1} becomes

$$y_{i+1} = (\bar{y}_i + \epsilon_i) + hf(x_i, \bar{y}_i + \epsilon_i) + \epsilon_{i+1}.$$

Expanding $f(x_i, \bar{y}_i + \epsilon_i)$ in a Taylor series in y_i about $\bar{y}_i$ and assuming ϵ_i is sufficiently small to drop terms of second degree and higher in ϵ_i, leads to the expression

$$f(x_i, \bar{y}_i + \epsilon_i) = f(x_i, \bar{y}_i) + \epsilon_i f_y(x_i, \bar{y}_i)$$

where $f_y = \partial f/\partial y$. Substituting this in the expression for y_{i+1} gives

$$y_{i+1} = \bar{y}_i + hf(x_i, \bar{y}_i) + \epsilon_i[1 + hf_y(x_i, \bar{y}_i)] + \epsilon_{i+1}.$$

It is useful to assume that the error in y_{i+1} results from the local truncation error, ϵ_{i+1}, plus the error propagated by the repeated computations, e_{i+1}. Then the exact value of y_{i+1} can be expressed as

$$y_{i+1} = \bar{y}_{i+1} + \epsilon_{i+1} + e_{i+1}.$$

Comparing the last two expressions gives a relation for the propagated error,

$$e_{i+1} = [1 + hf_y(x_i, \bar{y}_i)]\epsilon_i.$$

Repeating the process for the next step

$$e_{i+2} = [1 + hf_y(x_{i+1}, \bar{y}_{i+1})]\epsilon_{i+1} = [1 + hf_y(x_{i+1}, \bar{y}_{i+1})][1 + f_y(x_i, \bar{y}_i)]\epsilon_i,$$

and for the $(i + n)$th step

$$e_{i+n} = [1 + hf_y(x_i, \bar{y}_i)][1 + hf_y(x_{i+1}, \bar{y}_{i+1})] \cdots [1 + hf_y(x_{i+n-1}, \bar{y}_{i+n-1})]\epsilon_i.$$

It is evident that the local error, ϵ_i, is propagated by a factor of the form $[1 + hf_y]$. A sufficient condition for the error to diminish would be for f_y to be negative everywhere in the interval, and for h to be small. If f_y is everywhere positive or h is large, the error will grow, causing the solution to become unstable and computed magnitudes will increase without bound. Stability is ensured for $|1 + hf_y| < 1$. It is evident that the success of Euler's method is dependent upon the nature of the object function and the size of the interval used to step off the solution.

For the truncation error, the total accumulated error is the sum

$$\sum_i^{i+n} \epsilon_i = \frac{h^2}{2} \sum_i^{i+n} y''(x)_i.$$

This could lead to large errors if the sign of $y''(x_i)$ is constant over the range and the interval h is large.

Although the discussion above was confined to Euler's method, difficulties with propagation of error are common to numerical solutions of initial-value problems generally. If the solution error grows as the computation proceeds, the method is said to be *unstable*. Solution errors may initially oscillate with a damped oscillation and become stable, or they may oscillate and increase indefinitely (i.e., "blow up"). Stability of a solution depends on the method, the object function, and the interval size. Often the solution can be made stable by reducing the interval size, but this is not always the case. A method is said to be *convergent* if the numerical solution approaches the exact solution of the differential equation as h is reduced to zero. When a complex problem is being treated, it is not possible to know these things from the solution alone. A great deal can be learned concerning the behavior of a numerical solution by solving the same problem with several sizes of the interval h. A more sophisticated treatment of the stability of numerical solutions for initial-value problems can be developed by solving the finite-difference equations explicitly for the method employed, and the recursion relation for a specific problem. For a further discussion of this the reader is directed to other sources, for example, Crandall's *Engineering Analysis*.

In summary two broad classes of methods were discussed for the solution of ordinary differential equations of initial value problems:

1. Methods which produce as solutions formulas dependent upon a specific differential equation, which are applicable over the entire interval of integration; and

2. Methods which repetitively use generally applicable formulas to obtain numerical values of the solution function. The latter are suitable for machine computation.

In the first category are:

1. *Taylor series.* The required differentiation may be difficult to do; error is large at the end of the interval.

2. *Picard's method of successive integration.* It may be impossible to integrate the object function; this method produces the same result with the same error problems as does the Taylor series.

3. *Method of undetermined coefficients.* Galerkin's method or the method of least squares are as good as the approximation selected; errors are distributed over the interval; the algebra can be difficult.

In the second category are numerical integration procedures of several types.

1. *Predictor methods using Taylor series.* Only the first-degree case was treated (Euler's method) and this lacked accuracy; the method is often unstable and requires small steps for good accuracy.

2. *Predictor-corrector methods.* As these require the past history of more than one point they are not self-starting; the method may be unstable regardless of size of interval, but provides for error control since each step is checked; calculation is complicated.

3. *Runge-Kutta methods.* These algorithms are simple and single-step; for h sufficiently small, they are basically stable, and they are self-starting; computational time for each step, however, is longer than for predictor-corrector methods.

Exercises

1. Beginning with the expression

$$y_{n+1} = y_{n-p} + h \int_{-p}^{1} y'_{n+s} \, ds,$$

and representing the integrand with the backward-difference polynomial, carry out the integration and necessary mathematics to show, for $p = 3$, that

$$y_{n+1} = y_{n-3} + \tfrac{4}{3}h(2y'_n - y'_{n-1} + 2y'_{n-2}) + \tfrac{14}{45}h^5 y^{(5)}(\zeta).$$

2. Numerically solve $y' = y$ in the interval $(0, 1)$, where $y = 1$ at $x = 0$. Find the first values by using Taylor's series and then continue with the three-point open integration formula. Check the resulting values of the function y with the true solution.

3. Consider the initial-value problem

$$\frac{dx}{dt} = -x, \qquad t > 0,$$

$$x = 1, \qquad t = 0.$$

Using iteration, compare the zeroth, first, second, and third approximation to the true solution by plotting x versus t from 0 to 2.0.

4. Expand the function of Exercise 3 in a Taylor series about $t = 0$, and compare this to the solution given by Exercise 3.

5. Using a trial solution of the form

$$x = 1 + c_1 t + c_2 t^2,$$

solve Exercise 3 to determine the parameters c_1 and c_2 by:

a) Galerkin's method;

b) The method of least squares

6. Solve Exercise 3 by Euler's method, using time steps of 0.5 and 0.25.

7. By means of a graph of error versus t for $t = 0$ to $t = 2.0$, compare the error for the second-order Taylor series (Exercise 4); Galerkin's method; the method of least squares; and the Euler's method solution for the two time steps (Exercise 6). Error is the difference,

$$x_{\text{approx.}} - x_{\text{exact}}.$$

8. Alter the Milne predictor-corrector function to iterate on the closed integration; that is, repeat the computation of the new point y_{n+1}, using the corrector formula, until there is no change in successive values.

9. Use Euler's method as a predictor, with the modified Euler formula as a corrector, to solve numerically $y' = x^2 - y$ in the interval $x = 0$ to $x = 1$. The initial conditions are $x = 0$, $y = 1$. Let $h = 0.2$.

10. Derive the second-order Runge-Kutta formula starting with the expression

$$y_{n+1} = y_n + h(af_0 + bf_1),$$

where

$$f_0 = f(x_n, y_n),$$
$$f_1 = f(x_n + ph, y_n + qhf_0).$$

Obtain relations for a, b, p, and q. For the case that $b = \frac{1}{2}$, show that the result is the modified Euler formula

$$y_{n+1} = y_n + \tfrac{1}{2}h[f(x_n, y_n) + f(x_n + hf(x_n, y_n))];$$

this is sometimes called the *improved Euler method*.

11. Using the results of Exercise **10**, show that for $b = 1$, the second-order Runge-Kutta formula is

$$y_{n+1} = y_n + hf[x_n + \tfrac{1}{2}h, y_n + \tfrac{1}{2}hf(x_{n1}, y_n)].$$

This is sometimes called the *modified Euler* method or the *improved polygon method*.

12. Draw a flow chart of the Runge-Kutta-Gill algorithm.

13. By any appropriate numerical method with a step size $h = 0.1$, obtain a solution to the problem

$$\left.
\begin{aligned}
\frac{dx_1}{dt} &= x_2 \\[2mm]
\frac{dx_2}{dt} &= -x_1 - x_2^2
\end{aligned}
\right\} \quad t > 0,$$

$$\left.
\begin{aligned}
x_1 &= -1 \\
x_2 &= 1
\end{aligned}
\right\} \quad t = 0.$$

Carry out 5 steps.

14. Cast the problem

$$\frac{d^2x}{dt^2} = 1 - x - \epsilon x^3, \qquad t > 0,$$

with the initial conditions $t = 0$, $x = 0$, $(dx/dt) = 0$, into a form similar to that given in Exercise 10 with the appropriate initial conditions.

15. Numerically solve $y'' = -y$ in $(0, 4)$, given that $y = 0$ and $y' = 1$ at $x = 0$. Print the values of both y and y', and check these with the true solution.

16. If we apply Newton's law to a simplified rocket, the equation of motion of the rocket in vertical flight is

$$\frac{(M_0 - m_f t)}{g_0} \frac{du}{dt} = -cu^2 - \frac{(M_0 - m_f t)}{g_0} g + \frac{m_f V_j}{g_0},$$

where M_0 = (the initial mass of the rocket and fuel) = 20,000 lbm, and

m_f = rate of fuel burned = 200 lbm/sec,

t = time in seconds,

V_j = velocity of combustion products leaving the rocket nozzle = 12,000 ft/sec,

g_0 = conversion factor for mass = $32.2 \dfrac{\text{lbm-ft}}{\text{lbf-sec}^2}$,

u = rocket velocity, ft/sec,

c = drag coefficient for air resistance = $.0002 \dfrac{\text{lbf-sec}^2}{\text{ft}^2}$, and

g = acceleration of gravity = 32.2 ft/sec.

Write a third-order Runge-Kutta program and have it determine the trajectory (altitude) and velocity versus time. Assume that there is enough fuel for 10 seconds of thrust (2000 lbm). Mark this point in the trajectory; then continue to solve the problem with $m_f = 0$ for the remainder of the ascent until $u = 0$. At this point modify the equation for the vertical descent and continue the calculation until the flight is terminated.

17. Determine the limiting value of h which will ensure a stable solution of the problem

$$y' = x^2 - y$$

with the initial condition

$$x = 0, \quad y = 1,$$

using Euler's method. Try several steps with Euler's method for $h = 3$; $h = 2$; and $h = 1$.

partial differential equations

In previous chapters, systems were treated as if they were composed of a number of discrete elements. These were called *lumped-parameter* systems. One example consisted of several masses joined by springs. Now we shall consider systems that are best modelled by treating them as composed of an infinite number of elements. These are called continuous systems. A vibrating beam might best be treated as a continuous system. Both the discrete and continuous models are useful for the representation of real systems, and it is the job of the analyst to decide which approach is the more appropriate.

Analysis of continuous systems usually leads to sets of partial differential equations. Occasionally the partial differential equations can be simplified to ordinary differential equations. In most cases, the formulation results in a problem for which classical methods of solution are inadequate, but it is usually possible to solve the problem using numerical methods.

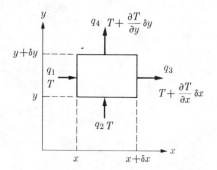

Fig. 10.1

To begin we shall formulate several example problems each representing a class of partial differential equations, which in turn will require a certain type of numerical solution. As a first example, the problem of transient heat conduction in a two-dimensional solid will be formulated. Fourier's law of heat conduction will be applied to a small element having constant thermal conductivity. A sketch of the element of thickness δz, width δx, and length δy, is shown in Fig. 10.1. The heat flux across each of the four boundaries is represented by the q's. These and the temperature at each boundary are shown on the figure. A zero gradient of temperature normal to the xy-plane allows us to treat the problem in two dimensions. The three-dimensional case is no more difficult. The net heat flux into the element is the sum of the individual heat fluxes across the four boundaries,

$$q_1 + q_2 - q_3 - q_4.$$

If the net heat flux is zero, a steady state exists. If it is not, the temperature of the element will change in time, t, at the rate $(\partial T/\partial t)$. Assuming constant specific heat c_p, and uniform and constant density ρ, the net accumulation of energy in the element, $\rho c_p(\partial T/\partial t) \times$ (volume) will equal the net heat flux; or, symbolically,

$$q_1 + q_2 - q_3 - q_4 = \rho c_p(\delta x\, \delta y\, \delta z)\frac{\partial T}{\partial t}.$$

This is a consequence of the First Law of Thermodynamics. Each of the heat-flux terms can be expressed in terms of the local temperature gradient by applying Fourier's law of heat conduction. Then for constant thermal conductivity k,

$$q_1 = -k(\delta y\, \delta z)\frac{\partial T}{\partial x},$$

$$q_2 = -k(\delta x\, \delta z)\frac{\partial T}{\partial y},$$

$$q_3 = -k(\delta y\, \delta z)\frac{\partial}{\partial x}\left(T + \frac{\partial T}{\partial x}\,\delta x\right),$$

$$q_4 = -k(\delta x\, \delta z)\frac{\partial}{\partial y}\left(T + \frac{\partial T}{\partial y}\,\delta y\right).$$

Substituting these expressions for the heat-flux terms, the first-law expression becomes a partial differential equation in $T(x, y, t)$:

$$\frac{\partial^2 T}{\partial x^2} + \frac{\partial^2 T}{\partial y^2} = \frac{\rho c_p}{k}\frac{\partial T}{\partial t}.$$

The group of properties on the right is called the thermal diffusivity, α, with the definition

$$\alpha = \frac{k}{\rho c_p},$$

having the dimensions (length)^2/time.

We have developed the *equation for unsteady heat conduction* in two dimensions. It is linear in T, and it is a second-order partial differential equation. Two sets of boundary conditions and an initial temperature distribution are required to state the problem completely.

For the same system in steady state, the time rate of change of the temperature is set to zero. The result is the *steady-state heat conduction equation*, more generally called Laplace's equation. In rectangular coordinates it is

$$\frac{\partial^2 T}{\partial x^2} + \frac{\partial^2 T}{\partial y^2} = 0.$$

As before, the equation is linear in T and of second order. If the right-hand side were not zero, but a function of x and y, the expression would become:

$$\frac{\partial^2 T}{\partial x^2} + \frac{\partial^2 T}{\partial y^2} = f(x, y);$$

this is called Poisson's equation. Both Laplace's equation and Poisson's equation must be accompanied by two pairs of boundary conditions. They will be used to demonstrate a second class of equations.

To introduce a third type of partial differential equation, consider the problem of a vibrating, taut string, Fig. 10.2(a). The string, of length L, is attached at both ends so that a tension T exists along its length. The mass of the string per unit of length is ρ, and the vertical displacement from its equilibrium position is $y(x, t)$, where x is the distance along the string to the element, and t is the time. Now look at the small element of string of length δx, as shown in Fig. 10.2(b). On each end of the element the tension T acting along the curved string is assumed constant. On the left side there is a component in the y-direction of $T(\partial y/\partial x)$. On the right side, due to curvature along the element, the vertical force is slightly different, being equal to

$$T\left(\frac{\partial}{\partial x}\right)\left(y + \frac{\partial y}{\partial x}\,\delta x\right).$$

Consequently, the net vertical force on the element of string is

$$T\frac{\partial^2 y}{\partial x^2}\,\delta x.$$

Applying Newton's second law, and neglecting transverse motion of the element, we see that this force will cause a vertical acceleration of the string, which, in terms of the y-coordinate and time t, is $(\partial^2 y/\partial t^2)$. As the element of string has a total mass $\rho\,\delta x$, Newton's law can be written for the element as

$$\frac{\partial^2 y}{\partial x^2} = \frac{\rho}{T}\frac{\partial^2 y}{\partial t^2}.$$

The second-order partial differential equation representing the behavior of this system is called the *wave equation*. The wave equation is linear and second-order, requiring for solution two boundary conditions and two initial conditions.

Classical solutions are known for all of these equations for a few physical systems. The difficulty is being able to introduce the boundary conditions for boundaries which do not correspond to the coordinates. In nearly all real problems

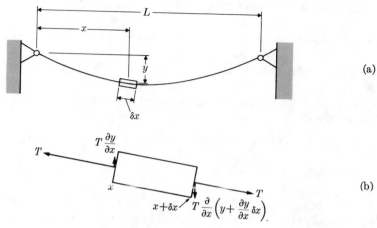

Fig. 10.2

it is necessary to resort to numerical procedures to obtain solutions. Again in numerical solutions, it is adopting the solution to fit the boundary conditions which presents the most difficulty. It is safe to say there is no limit to the problems in partial differential equations which can be solved numerically if enough computer storage and time are available.

In this chapter we shall employ the method of finite differences for the solution of partial differential equations. By treating the system as a large number of small but finite elements, it is possible to reduce the partial differential equations to algebraic equations. In fact the eventual algebraic problems will correspond directly to the formulation achieved in lumped-parameter systems governed by the same physical laws.

While this chapter will provide only an introduction to the solution of partial differential equations, it will be adequate for the student to be able to solve rather complex problems, perhaps more complex than he is able, at this stage of development, to formulate. It will also provide better physical insight into these problems than does a classical solution, as one soon observes that there is a close relationship between the behavior of the numerical solution and the physical system, and in the way that variables affect the behavior of each.

Because solutions are dependent upon the nature of boundary conditions, it is simpler to formulate them for specific physical and geometric properties. Consequently, problems will be related to specific physical systems (e.g., problems of heat conduction).

Classification of partial differential equations

Partial differential equations can be classified as linear if the dependent variable and all its derivatives enter the equations linearly. Classical treatments are generally confined to the solution of linear equations. When the equations are nonlinear, it is not possible to combine solutions in a linear way. With the advent of digital computers it has been possible to obtain solutions of nonlinear equations with reasonable ease.

Linear differential equations of order two, such as those presented earlier, fall into several categories which cover a great many situations encountered in physical systems. For the partial differential equation in the variable $\Phi(x, y)$, having the general form

$$a(x, y)\Phi_{xx} + b(x, y)\Phi_{xy} + c(x, y)\Phi_{yy} + f(x, y, \Phi_x, \Phi_y) = 0,$$

it is possible to classify according to the expression $b^2 - 4ac$.

1. The equation is called elliptic if $b^2 - 4ac < 0$ everywhere in the region, and $a > 0$. The boundary conditions are given as Φ or its normal derivative, or a linear combination of them, on the boundary of the region in which Φ is to be determined. The Laplace equation (steady-state heat conduction)

$$\frac{\partial^2 \Phi}{\partial x^2} + \frac{\partial^2 \Phi}{\partial y^2} = 0,$$

and Poisson's equations are examples of elliptic equations.

2. The equation is parabolic if

$$b^2 - 4ac = 0.$$

It is necessary for the solution that initial values of Φ be given everywhere and the value of Φ or its derivatives be known on the boundaries. An example of this is the equation of unsteady heat conduction

$$\alpha \frac{\partial^2 \Phi}{\partial x^2} - \frac{\partial \Phi}{\partial t} = 0,$$

where α is a constant.

3. The equation is hyperbolic if

$$b^2 - 4ac > 0.$$

Complete statement of these problems requires that Φ or its derivative be given at the boundaries, and Φ and its derivative be given everywhere at the initial state. The familiar wave equation is an example:

$$\frac{\partial^2 \Phi}{\partial x^2} - \frac{\rho}{T} \frac{\partial^2 \Phi}{\partial t^2} = 0.$$

As the coefficients a, b, and c may be functions of x and y, it is possible to have equations which are of one type in one region and of a different type in another. The classification of partial differential equations in this way is not a critical matter when we use numerical methods. It is more useful to identify the type of physical system which produces them. In this way, the form of the result can be anticipated, and the nature of the boundary conditions or initial conditions which will be required can be established.

Difference equations

Numerical solutions for partial differential equations can be developed by transforming the partial derivatives into finite-difference form. The ordinary derivative of a function is defined as

$$\frac{dy}{dx} = \lim_{\delta x \to 0} \frac{y(x + \delta x) - y(x)}{\delta x}.$$

While it is not possible to represent the derivative exactly for a finite element, it is possible to approximate it over a finite difference δx, in which δx is small but finite. We have already worked with divided differences and finite-difference representation of derivatives. We shall develop the finite-difference approximations for partial derivatives with Taylor series. This permits us to examine the nature of the truncation error. For a function of more than one variable (say $u(x, y)$) to be expanded in a Taylor series, it is necessary to employ partial derivatives of the function. Expansion of the function about a point x_0, y_0 in a Taylor series is written as

$$u(x, y) = u(x_0, y_0) + (x - x_0) \frac{\partial u}{\partial x} (x_0, y_0)$$

$$+ \frac{(x - x_0)^2}{2!} \frac{\partial^2 u}{\partial x^2} (x_0, y_0) + (y - y_0) \frac{\partial u}{\partial y} (x_0, y_0)$$

$$+ \frac{(y - y_0)^2}{2!} \frac{\partial^2 u}{\partial x^2} (x_0, y_0) + (x - x_0)(y - y_0) \frac{\partial^2 u}{\partial x \, \partial y} (x_0, y_0) + \cdots$$

$$+ \frac{1}{n!} \left[(x - x_0) \frac{\partial}{\partial x} + (y - y_0) \frac{\partial}{\partial y} \right]^n u(x_0, y_0).$$

The general nth term $(n > 2)$ is written as a bracketed operator raised to a power n and applied to $u(x_0, y_0)$.

Evaluating the series at $x = x_0 + \delta x$ and $y = y_0$ is the first step in obtaining the forward-difference approximation to the first partial derivative. Up to second-order terms the series is

$$u(x_0 + \delta x, y_0) = u(x_0, y_0) + \delta x \frac{\partial u}{\partial x} (x_0, y_0) + \frac{(\delta x)^2}{2} \frac{\partial^2 u}{\partial x^2} (x_0, y_0),$$

from which the finite-difference approximation to the first partial derivative can be found as

$$\frac{\partial u}{\partial x} (x_0, y_0) \approx \frac{u(x_0 + \delta x, y_0) - u(x_0, y_0)}{\delta x},$$

with the truncation error

$$\left(\frac{\delta x}{2}\right)^2 \frac{\partial^2 u}{\partial x^2}(\zeta, y_0), \qquad x_0 < \zeta < x_0 + \delta x.$$

A backward-difference approximation can be developed by evaluating the series at $x_0 - \delta x$ and y_0. Solving for the first derivative yields the approximation

$$\frac{\partial u}{\partial x}(x_0, y_0) = \frac{u(x_0, y_0) - u(x_0 - \delta x, y_0)}{\delta x},$$

with a truncation error of the same order as for the forward difference. By evaluating the series at $u(x_0 + \delta x, y_0)$ and $u(x_0 - \delta x, y_0)$, a central-difference approximation can be found by subtracting one series from the other. The resulting approximation is

$$\frac{\partial u}{\partial x}(x_0, y_0) = \frac{u(x_0 + \delta x, y_0) - u(x_0 - \delta x, y_0)}{2(\delta x)}.$$

Because of the difference in signs for the two expressions, the even terms drop out, and the truncation error is

$$\frac{(\delta x)^2}{3!} \frac{\partial^3 u}{\partial x^3}(\zeta, y_0).$$

Similar expressions can be found for $(\partial u/\partial y)(x_0, y_0)$ by evaluating the series at x_0 and $y_0 + \delta y$.

Adding the series evaluated at $x_0 + \delta x$, y_0 and $x_0 - \delta x$, y_0 provides the central-difference approximation for the second derivative

$$\frac{\partial^2 u}{\partial x^2}(x_0, y_0) = \frac{u(x_0 + \delta x, y_0) - 2u(x_0, y_0) + u(x_0 - \delta x, y_0)}{(\delta x)^2}$$

with truncation error

$$2\frac{(\delta x)^2}{4!} \frac{\partial^4 u}{\partial x^4}(\zeta, y_0).$$

It follows from a similar procedure that

$$\frac{\partial^2 u}{\partial y^2}(x_0, y_0) = \frac{u(x_0, y_0 + \delta y) - 2u(x_0, y_0) + u(x_0, y_0 + \delta y)}{(\delta y)^2}.$$

The cross derivative can be found by evaluating the series approximation at

$$x_0 + \delta x, \quad y_0 + \delta y;$$
$$x_0 + \delta x, \quad y_0 - \delta y;$$
$$x_0 - \delta x, \quad y_0 + \delta y;$$

and

$$x_0 - \delta x, \quad y_0 - \delta y.$$

Combining these to eliminate all but the cross-derivative term gives the approximation

$$\frac{\partial^2 u}{\partial y\, \partial x} = \frac{1}{4(\delta x\, \delta y)} [u(x_0 + \delta x, y_0 + \delta y) - u(x_0 - \delta x, y_0 + \delta y)$$

$$- u(x_0 + \delta x, y_0 - \delta y) + u(x_0 + \delta x, y_0 + \delta y)],$$

and the truncation error is of the order $(\delta x + \delta y)^2$. Finite-difference approximations for higher partial derivatives can be found by extending the procedures described above.

Computational molecules

Solution of partial differential equations by finite-difference procedures is done by dividing the domain into a grid with finite spacing in the coordinate directions. Spacing is given by δx, δy, or δt. The grid is chosen sufficiently small to control the size of the truncation errors, yet large enough so that the amount of calculation necessary does not exceed the capabilities of the computer. Finite-difference approximations can then be written at each intersection of these grid lines, which we shall call *grid points*, and the appropriate boundary conditions can be applied. This leads to a set of algebraic equations, hopefully linear, with the unknowns being the function at the grid point.

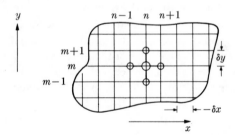

Fig. 10.3

As a demonstration of this, consider the Laplace equation

$$\frac{\partial^2 u}{\partial x^2} + \frac{\partial^2 u}{\partial y^2} = 0$$

applied to a domain. A domain, divided into a grid with spacing δx and δy, is shown in Fig. 10.3. Focus attention on the interior grid point (m, n). From the finite-difference expressions for the second derivatives it is evident that $(\partial^2 u/\partial x^2)$ depends on the values at points (m, n), $(m, n - 1)$, and $(m, n + 1)$. In the same way, the second derivative in y depends on the three points $(m - 1, n)$, (m, n), and $(m + 1, n)$. These five points have been circled in Fig. 10.3. To write the Laplace equation at the interior point (m, n) it is necessary to add the finite-difference expressions for the second derivatives in x and y, and then set the sum equal to zero, producing the equation

$$\frac{u_{m,n-1} - 2u_{m,n} + u_{m,n+1}}{(\delta x)^2} + \frac{u_{m+1,n} - 2u_{m,n} + u_{m-1,n}}{(\delta y)^2} = 0.$$

It is advisable to set δx equal to δy, so the equation can be simplified to

$$u_{m,n-1} + u_{m,n+1} + u_{m-1,n} + u_{m+1,n} - 4u_{m,n} = 0.$$

The relationship between the equation and the grid geometry can be presented diagrammatically. Because of its appearance the representation is called a *computational molecule*. It relates the location of a grid point to the weight it receives in the algebraic expression. For the finite-difference representation of the Laplace equation in rectangular coordinates, the computational molecule is shown in Fig. 10.4. It is easy to see how the expression depends on the central point and those four immediately adjacent. As the solution requires that this relation be satisfied for every point in the domain, an equation must be written with every point as a central point. This necessarily includes the boundaries. Expressions are written for boundary points so that they express the particular boundary condition. It often happens that the boundary is of irregular shape and does not correspond to natural grid points. Account of this must be taken, perhaps by distorting the grid size near the edge. This often complicated procedure will be discussed later on.

In order to apply some of these ideas, consider the extremely simple example of a domain having rectangular boundaries, with the value of the function prescribed on all of the boundaries. Throughout the domain the function must satisfy the Laplace equation. Problems of this sort might arise in two-dimensional conduction of heat with prescribed surface temperatures; plane sections in torsion; or two-dimensional inviscid flow, to mention three instances of application. A grid has been arranged as shown in Fig. 10.5. Points will be designated by double subscripts, with the first representing the location along the vertical, and the second along a horizontal. The problem is to find the values of the functions $u_{ij}(x, y)$ which satisfy the finite-difference form of the Laplace equation for interior points when values are prescribed along the boundaries. Let values given on the boundary be designated $g_{i,j}$. Specifically they are:

$$g_{0,1}; \quad g_{0,2}; \quad g_{0,3}; \quad g_{1,0}; \quad g_{1,4};$$
$$g_{2,0}; \quad g_{2,4}; \quad g_{3,1}; \quad g_{3,2}; \quad g_{3,3}.$$

Having set the grid spacing to make $\delta x = \delta y$, the governing equation to be satisfied for the particular point (m,n) is

$$u_{m+1,n} + u_{m-1,n} + u_{m,n-1} + u_{m,n+1} - 4u_{m,n} = 0.$$

As we know the numerical value of u on all boundary points, we need only write the expression for the remaining six interior points. With the constant values, $g_{i,j}$,

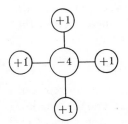

Fig. 10.4

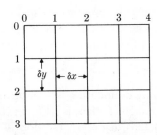

Fig. 10.5

moved to the right-hand side in each case, the six equations are:

Point	Equation
$(1, 1)$	$-u_{1,2} - u_{2,1} + 4u_{1,1} \qquad\quad = g_{0,1} + g_{1,0}$
$(1, 2)$	$-u_{1,1} - u_{1,3} - u_{2,2} + 4u_{1,2} = g_{0,2}$
$(1, 3)$	$-u_{1,2} - u_{2,3} + 4u_{1,3} \qquad\quad = g_{0,3} + g_{1,4}$
$(2, 1)$	$-u_{1,1} - u_{2,2} + 4u_{2,1} \qquad\quad = g_{2,0} + g_{3,1}$
$(2, 2)$	$-u_{1,2} - u_{2,1} - u_{2,3} + 4u_{2,2} = g_{3,2}$
$(2, 3)$	$-u_{1,3} - u_{2,2} + 4u_{2,3} \qquad\quad = g_{2,4} + g_{3,3}$

It is convenient to write these equations in matrix form as follows

$$
\begin{bmatrix}
4 & -1 & 0 & -1 & 0 & 0 \\
-1 & 4 & -1 & 0 & -1 & 0 \\
0 & -1 & 4 & 0 & 0 & -1 \\
-1 & 0 & 0 & 4 & -1 & 0 \\
0 & -1 & 0 & -1 & 4 & -1 \\
0 & 0 & -1 & 0 & -1 & 4
\end{bmatrix}
\begin{bmatrix}
u_{1,1} \\
u_{1,2} \\
u_{1,3} \\
u_{2,1} \\
u_{2,2} \\
u_{2,3}
\end{bmatrix}
=
\begin{bmatrix}
g_{0,1} + g_{1,0} \\
g_{0,2} \\
g_{0,3} + g_{1,4} \\
g_{2,0} + g_{3,1} \\
g_{3,2} \\
g_{2,4} + g_{3,3}
\end{bmatrix}
$$

Several characteristics of these equations are easily discerned and are worth noting. It is the nonzero boundary conditions which make the solution nontrivial, and it is their magnitude which makes one solution distinct from the next. Notice that the coefficient matrix is symmetric, is sparse, and has a strong diagonal. These characteristics are favorable for reducing the difficulties of solution, and they frequently occur in problems of this type. It is interesting to observe that the steady-state or equilibrium problem has, in the case of the continuous system, been reduced to a set of simultaneous algebraic equations as was the case for lumped-parameter systems. As we are treating the continuous problem as a large number of discrete elements, this is not surprising.

For any real problem, it is apparent that there will be a great many nodes in the grid, resulting in a large number of simultaneous equations. This favors computation schemes with minimal computer-storage requirements. Considering this and noting the presence of strong diagonal elements in the coefficient matrix, we find the Gauss-Seidel method appealing. To carry the example a bit further, let the value of the function on the top side of the rectangle be 100. That is,

$$g_{0,1} = g_{0,2} = g_{0,3} = 100.$$

The remaining boundaries will be zero. For a conduction problem this represents the case where one surface is set at one temperature (100 degrees) and the remaining sides set to zero degrees. Iteration of the equations proceeds from a set of trial values according to the formula

$$u_{i,j} = \tfrac{1}{4}(u_{i-1,j} + u_{i+1,j} + u_{i,j-1} + u_{i,j+1}).$$

Trial values for all interior points were assumed to be zero. Temperature distributions for the first three Gauss-Seidel iterations are given below

	100	100	100				100	100	100	
0	0	0	0	0		0	25	31.25	32.81	0
0	0	0	0	0		0	6.25	9.37	10.55	0
	0	0	0				0	0	0	

Trial distribution   After first iteration

	100	100	100				100	100	100	
0	34.38	44.14	38.67	0		0	38.77	48.46	40.56	0
0	10.94	16.41	13.77	0		0	12.70	18.73	14.92	0
	0	0	0				0	0	0	

After second iteration   After third iteration

For hand calculation, the method of relaxation is particularly useful for problems of this sort. One can comfortably manage dozens of nodal points, and the method is self-correcting. A few calculations have been carried out directly on a network of points in Fig. 10.6. Here the property to be evaluated (temperature) is written to the left of the vertical grid, and the residual written to the right of it. First guess (all zeros in this example) have been written just above the horizontal grid line, with the current residuals and changes in temperature listed below. Each step has been iterated three times with the values for each step listed. Current temperatures and residuals appear at the bottom.

An example computer program is developed at the end of this chapter which treats the specific problem of solving the Laplace equation in two dimensions with values of the function prescribed on the boundaries. The main program formulates the augmented matrix, and then calls the Gauss-Seidel subroutine to solve for values of the function.

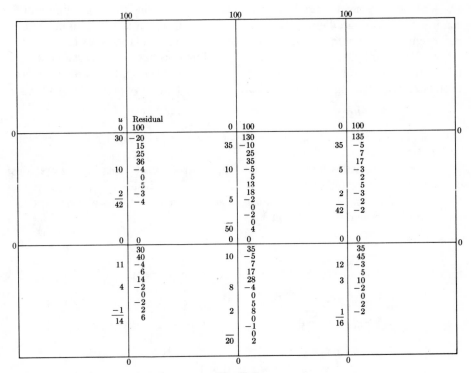

Fig. 10.6

Boundary conditions

In the example just treated, the function u had a prescribed value on all the boundaries. Furthermore, the boundaries correspond to lines in the grid with no difficulty. Plainly, this is not a typical situation. Let's first examine other types of boundary conditions, putting off the question of irregular boundaries until later.

It is helpful to attach physical significance to the boundary conditions. Consequently, heat flux and temperatures will be considered.

Insulated Boundary

An insulated boundary is one in which the heat flux across it is zero. From the law of heat conduction at the boundary,

$$\frac{q}{A} = -k\left(\frac{\partial T}{\partial x}\right)_{\text{Boundary}},$$

which implies, for the insulated boundary, $(\partial T/\partial x) = 0$ at the boundary; or, in general, the gradient of the function normal to the boundary is zero. The function on the boundary is now an unknown, and an equation must exist for each such point.

A plane of symmetry produces the same boundary condition. Looking at the condition of a zero gradient in this way is useful. One can think of an imaginary point reflecting the point one grid space in from the boundary. This is shown for a boundary point (M, N) in Fig. 10.7. The function has the same value at the imaginary point as at the point $(M, N - 1)$. Consequently, the finite difference expression representing Laplace's equation for the boundary point (M, N) is

$$4u_{M,N} - 2u_{M,N-1} - u_{M-1,N} - u_{M+1,N} = 0.$$

One could reach this same expression by writing an energy balance for an element at (M, N) with zero heat flux out of the element (see Fig. 10.1).

We can see how this boundary condition is introduced by treating the previous example problem for the case in which the lower edge is insulated. Now, there are three new unknowns, $u_{3,1}$, $u_{3,2}$, and $u_{3,3}$. Additional equations will be supplied by writing the boundary condition

$$\left(\frac{\partial u}{\partial y}\right) = 0$$

for these points. In matrix form the equations become

$$
\begin{bmatrix}
4 & -1 & 0 & -1 & 0 & 0 & 0 & 0 & 0 \\
-1 & 4 & -1 & 0 & -1 & 0 & 0 & 0 & 0 \\
0 & -1 & 4 & 0 & 0 & 0 & -1 & 0 & 0 \\
-1 & 0 & 0 & 4 & -1 & 0 & 0 & 0 & -1 \\
0 & -1 & 0 & -1 & 4 & -1 & 0 & -1 & 0 \\
0 & 0 & -1 & 0 & -1 & 4 & -1 & 0 & 0 \\
0 & 0 & 0 & 0 & 0 & -2 & 4 & -1 & 0 \\
0 & 0 & 0 & 0 & -2 & 0 & -1 & 4 & -1 \\
0 & 0 & 0 & -2 & 0 & 0 & 0 & -1 & 4
\end{bmatrix}
\begin{bmatrix}
u_{1,1} \\ u_{1,2} \\ u_{1,3} \\ u_{2,1} \\ u_{2,2} \\ u_{2,3} \\ u_{3,1} \\ u_{3,2} \\ u_{3,3}
\end{bmatrix}
=
\begin{bmatrix}
g_{0,1} + g_{1,0} \\ g_{0,2} \\ g_{0,3} + g_{1,4} \\ g_{2,0} \\ 0 \\ g_{2,4} \\ g_{3,0} \\ 0 \\ g_{3,4}
\end{bmatrix}
$$

It has been necessary to add values for $g_{3,0}$ and $g_{3,4}$ which will take the set value of their vertical boundaries. Notice that the symmetry of the mathematical problem is gone, as is that of the physical problem. Boundary conditions are not something to be applied to a solution, as in the classical mathematical treatment, but they become a part of the problem to be solved, being absorbed in the equations.

Prescribed Heat Flux

A prescribed heat flux across a boundary requires that the normal gradient of the function take a prescribed value, which is not zero. This is evident from the equation of conduction as

$$\frac{q}{A} = -k\left(\frac{\partial T}{\partial x}\right)_{\text{Boundary}}.$$

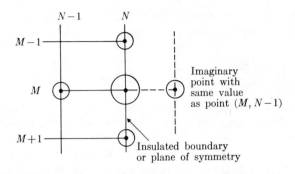

Imaginary
point with
same value
as point $(M, N-1)$

Insulated boundary
or plane of symmetry

Fig. 10.7

For a prescribed heat flux and known conductivity, a value of the temperature gradient can be found from the above. At each boundary point the finite-difference form of the first derivative can be formed and equated to the gradient. For the nodal point (M, N), with a gradient prescribed in the x-direction, the equation is

$$\frac{T_{M,N} - T_{M-1,N}}{\delta x} = \text{Prescribed gradient.}$$

Temperature at each such boundary point is an unknown, but there will be an additional equation for each point in the above form to provide a solvable set of equations.

Environment Prescribed

In some cases the boundary condition is determined by the state of the surroundings. This is equivalent to the case of constant heat flux, except that the heat flux is given in terms of the boundary temperatures. Heat transfer to the surroundings can occur by convection or radiation. In the case of convective heat-transfer the rate can be calculated by using the defined heat-transfer coefficient h, which is a function of fluid properties, geometry, and fluid motion. In this case, the convective heat-transfer from the surface of area A is

$$q_{\text{conv}} = hA \, \delta T,$$

where δT is the temperature difference between wall and surrounding gas or liquid. In the case of the nodal point (M, N) this is equivalent to writing, per unit of depth δz,

$$q_{\text{conv}} = h(\delta y)(T_{M,N} - T_0),$$

where T_0 is the temperature of the surrounding liquid. For simplicity we shall assume that T_0 is everywhere the same. As the heat-flux across the boundary must exactly equal the heat transferred to the surroundings, we can write

$$h(\delta y)(T_{M,N} - T_0) = -k(\delta y)\left(\frac{\partial T}{\partial x}\right)_{\text{Boundary}}.$$

As it is written here, it represents heat flow across a boundary in the yz-plane. When we express the right-hand side in finite difference form, the equation for the boundary is

$$\left(\frac{h(\delta x)}{k} + 1\right) T_{M,N} - T_{M,N-1} = \frac{h(\delta x)}{k} T_0.$$

Each boundary temperature is again an unknown. The additional equations necessary for the solution come from expressions like the above written at each boundary node.

In the case of heat transfer from the surface by radiation, a fourth power dependence on temperature enters, causing the algebraic equations, resulting from the finite difference approximations, to be nonlinear. Radiation heat-transfer is proportional to the surface area, the fourth power of temperature (absolute), and the geometry and condition of the radiating surface. If we represent the last dependency by a factor K, the heat transferred by radiation from the surface at point (M, N) per unit depth is

$$q_{\text{rad}} = K(\delta y)(T_{M,N}^4 - T_0^4).$$

Neglecting convection in this case, the radiation heat-transfer is set equal to the conduction into the boundary. When we introduce the finite-difference representation for the temperature gradient, the boundary condition equation becomes

$$T_{M,N}^4 \left(\frac{K(\delta x)}{k}\right) + T_{M,N} - T_{M,N-1} = \frac{K(\delta x)}{k} T_0^4.$$

These provide the additional equations to determine the unknown boundary temperature. In some cases both convection and radiation may play a part. In that case, the two heat flux terms would be added together.

While it is not our purpose in this section to present a discussion of heat transfer, it is necessary to consult the physical problem in order to form an appropriate model of the boundary conditions. The field of heat transfer provides versatile and simple examples. For other physical systems, governed by the Laplace equations, boundary conditions would be expressed in terms of the variable or its derivatives along the boundary in a very similar way. These would then be reduced to algebraic form by introduction of finite-difference approximations exactly as was done here.

Irregular boundaries

The second difficulty encountered in the finite-difference solution of partial differential equations is that in which the contour of the physical boundaries differs from the contour of the grid. In the example just treated the two coincided. It is frequently possible to select a grid system to make this happen (e.g., cylindrical coordinate for circular sections, and so on). Usually, however, the boundary is

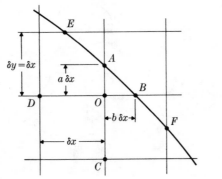

Fig. 10.8

irregular. One is shown in Fig. 10.8. In the case illustrated, the system boundary crosses the grid at points E, A, B, and F, which are not coincident with mesh points. When the finite-difference equation is written for the interior point O, account must be taken of the nonstandard spacing, $O - A$ and $O - B$, as well as the boundary conditions for A and B. We shall do this for the case where interior values of the function are governed by the Laplace equation and when the values of the function on the boundary are prescribed (i.e., values of u_A and u_B are given).

For the moment, assume an origin to be at O. Then the Taylor series expansion of the function $u(x, y)$ about O is

$$u = u_O + x \left(\frac{\partial u}{\partial x} \right)_O + y \left(\frac{\partial u}{\partial y} \right)_O + \frac{x^2}{2} \left(\frac{\partial^2 u}{\partial x^2} \right)_O$$
$$+ xy \left(\frac{\partial^2 u}{\partial x \, \partial y} \right)_O + \frac{y^2}{2} \left(\frac{\partial^2 u}{\partial y^2} \right)_O + \cdots$$

The subscript O indicates that evaluation of the function or its derivatives is to be made at point O. The function u can now be evaluated at the points A, B, C, and D by introducing the appropriate coordinate distances into the Taylor series. Referring to the figure, we may write:

$$u_A = u_O + a \, \delta x \left(\frac{\partial u}{\partial y} \right)_O + \frac{(a \, \delta x)^2}{2} \left(\frac{\partial^2 u}{\partial y^2} \right)_O + 0(\delta x)^3,$$

$$u_B = u_O + b \, \delta x \left(\frac{\partial u}{\partial x} \right)_O + \frac{(b \, \delta x)^2}{2} \left(\frac{\partial^2 u}{\partial x^2} \right)_O + 0(\delta x)^3,$$

$$u_C = u_O - \delta x \left(\frac{\partial u}{\partial y} \right)_O + \frac{(\delta x)^2}{2} \left(\frac{\partial^2 u}{\partial y^2} \right)_O + 0(\delta x)^3,$$

$$u_D = u_O - \delta x \left(\frac{\partial u}{\partial x} \right)_O + \frac{(\delta x)^2}{2} \left(\frac{\partial^2 u}{\partial x^2} \right)_O + 0(\delta x)^3.$$

The rightmost terms are meant to indicate terms of the order of $(\delta x)^3$ or higher.

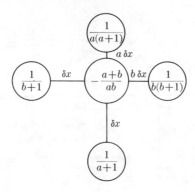

Fig. 10.9

By eliminating $(\partial u/\partial y)_0$ from the first and third equations, we may find a finite difference expression for approximating $(\partial^2 u/\partial y^2)_0$. Similarly the approximation for $(\partial^2 u/\partial x^2)_0$ can be obtained from a combination of the second and fourth equations. We may combine the two expressions and write the Laplace equation for the interior point O as

$$\left(\frac{\partial^2 u}{\partial x^2}\right)_0 + \left(\frac{\partial^2 u}{\partial y^2}\right)_0$$

$$= \frac{2}{(\delta x)^2}\left[\frac{u_A}{a(a+1)} + \frac{u_B}{b(b+1)} + \frac{u_C}{a+1} + \frac{u_D}{b+1} - \frac{a+b}{ab}u_0\right] + O(\delta x) = 0.$$

A computational molecule for point O is illustrated in Fig. 10.9.

Next consider the case where the boundary condition is given in terms of the normal derivative of the function at the boundary when the boundary is irregular. The procedure follows the same course as before. First the normal directions to the boundary must be established. These are shown as the angles α_A, α_O, and α_B in Fig. 10.10. Derivatives of u are known along these normals, n_A, n_O, and n_B. From the geometry it is simple to demonstrate that

$$\frac{\partial u}{\partial n} = \frac{\partial u}{\partial x}\cos\alpha + \frac{\partial u}{\partial y}\sin\alpha.$$

Expressions for the partial derivatives can be obtained by differentiating the Taylor-series expansion of u about the point O, with respect to x,

$$\frac{\partial u}{\partial x} = \left(\frac{\partial u}{\partial x}\right)_0 + x\left(\frac{\partial^2 u}{\partial x^2}\right)_0 + y\left(\frac{\partial^2 u}{\partial x\,\partial y}\right)_0 + \cdots,$$

and with respect to y,

$$\frac{\partial u}{\partial y} = \left(\frac{\partial u}{\partial y}\right)_0 + x\left(\frac{\partial^2 u}{\partial x\,\partial y}\right)_0 + y\left(\frac{\partial^2 u}{\partial y^2}\right)_0 + \cdots$$

With these expressions for the partial derivatives it is now possible to evaluate the

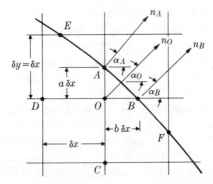

Fig. 10.10

derivative along the normal. At the point A, since the x above is zero, the normal derivative is

$$\left(\frac{\partial u}{\partial n}\right)_A = \left[\left(\frac{\partial u}{\partial x}\right)_O + a\,\delta x \left(\frac{\partial^2 u}{\partial x\,\partial y}\right)_O\right]\cos\alpha_A$$

$$+ \left[\left(\frac{\partial u}{\partial y}\right)_O + a\,\delta x \left(\frac{\partial^2 u}{\partial y^2}\right)_O\right]\sin\alpha_A + \cdots$$

Similarly at B:

$$\left(\frac{\partial u}{\partial n}\right)_B = \left[\left(\frac{\partial u}{\partial x}\right)_O + b\,\delta x \left(\frac{\partial^2 u}{\partial x^2}\right)_O\right]\cos\alpha_B$$

$$+ \left[\left(\frac{\partial u}{\partial y}\right)_O + b\,\delta x \left(\frac{\partial^2 u}{\partial x\,\partial y}\right)_O\right]\sin\alpha_B + \cdots;$$

and at O:

$$\left(\frac{\partial u}{\partial n}\right)_O = \left(\frac{\partial u}{\partial x}\right)_O \cos\alpha_O + \left(\frac{\partial u}{\partial y}\right)_O \sin\alpha_O.$$

Angles are known because the geometry is known. Through the boundary conditions, the left-hand sides of these expressions are known. There are then three expressions in the unknowns

$$\left(\frac{\partial u}{\partial x}\right)_O, \quad \left(\frac{\partial u}{\partial y}\right)_O, \quad \left(\frac{\partial^2 u}{\partial x^2}\right)_O, \quad \left(\frac{\partial^2 u}{\partial y^2}\right)_O \quad \text{and} \quad \left(\frac{\partial^2 u}{\partial x\,\partial y}\right)_O.$$

To provide the additional two equations, the Taylor-series expansion can be evaluated at points C:

$$u_C = u_O + \delta x \left(\frac{\partial u}{\partial x}\right)_O - \delta x \left(\frac{\partial u}{\partial y}\right)_O + \frac{(\delta x)^2}{2}\left(\frac{\partial^2 u}{\partial x^2}\right)_O$$

$$- (\delta x)^2 \left(\frac{\partial^2 u}{\partial x\,\partial y}\right)_O + \frac{(\delta x)^2}{2}\left(\frac{\partial^2 u}{\partial y^2}\right)_O + \cdots$$

A similar equation can be written at D. It is necessary to combine these to eliminate the first derivatives and the cross derivatives. Then the two second-derivative expressions can be combined to form the Laplace equation. This will yield an algebraic relation in terms of u_C, u_D, the grid spacing, boundary geometry, and values of the normal derivatives. Obviously this involves a great deal of arithmetic, and it must be repeated for every interior point near the boundary having this form of boundary condition.

A simpler but less accurate way of treating the irregular boundary is to write finite-difference equations which involve only interior points. Consequently, the computation molecule would, in this instance, be centered on D but not on point O. Boundary conditions at A and B would be transferred by linear interpolation to the point O. For the case of u prescribed along the boundary, interpolation would be made from D to B to produce a value u_{OB} as follows:

$$\frac{u_{OB} - u_D}{\delta x} = \frac{u_B - u_{OB}}{b\,\delta x},$$

from which

$$u_{OB} = \frac{b u_D + u_B}{1 + b}.$$

From C to A the interpolation produces u_{OA} as

$$u_{OA} = \frac{a u_C + u_A}{1 + b},$$

using an averaging process the value of u_O can be found as

$$u_O = \frac{u_{OA} + u_{OB}}{2}.$$

In this way the effect of the boundary condition has been transferred from an off-nodal point to a nodal point, and the solution can proceed as if the boundary coincided with the grid. Boundary conditions involving gradients of u could be transferred to point O using linear interpolation. Then the boundary can be treated as though it went through O with these modified boundary conditions existing at point O.

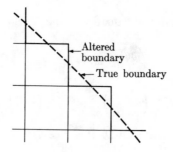

Altered boundary

True boundary

Fig. 10.11

A third and even simpler approach is to alter the boundary to fit the co-ordinates. A jagged boundary thus replaces the smooth natural boundary. An example of this is shown in Fig. 10.11, where the altered boundary (solid line), replaces the real boundary (dashed line). Once this is done the boundary conditions can be applied at points on the altered boundary, with some loss in accuracy.

Eigenvalue problems for continuous systems

The equation of motion of a vibrating string was developed in the first section of this chapter. Called the wave equation, it is a type of hyperbolic differential equation. It is

$$\frac{\partial^2 y}{\partial x^2} = \frac{\rho}{T} \frac{\partial^2 y}{\partial t^2}$$

where ρ is the density per unit length, T is the string tension, and t is time. For the problem to be fully stated, it is necessary to have a pair of boundary conditions and a pair of initial conditions. A string with fixed ends has the boundary conditions

$$x = 0, \quad y = 0,$$
$$x = L, \quad y = 0 \quad \text{for} \quad t > 0.$$

Initial conditions might describe the vertical displacement and velocity at zero time. For example,

$$t = 0, \quad y = f(x), \quad \frac{dy}{dt} = 0.$$

This second expression says that the velocity, (dy/dt), at every point along the string is initially zero. We will soon discover that this is an eigenvalue problem, just as were the corresponding problems of vibrating lumped-parameter systems. Eigenvalue problems in continuous systems also occur in the analysis of buckling of columns or plates, to cite one example.

As the string is composed of an infinite number of elements of length, it requires an infinite number of values of y to describe it. We have employed the continuous function $f(x)$ to do this for the initial condition. Consequently, the system is said to have an infinite number of degrees of freedom. From our experience with systems having a finite number of degrees of freedom, we would expect the vibrating string to have an infinite number of natural modes, and an infinite number of natural frequencies. While this is true, not all of these are of interest. Furthermore, a natural mode will execute simple harmonic motion at every point, with every point oscillating at the same frequency. This suggests that the problem can be simplified by describing the time dependence of the oscillation as a sinusoid, and writing the displacement as

$$y(x, t) = a(x) \sin (\omega t + \Phi),$$

where the amplitude a is a function of x only, ω is the natural frequency of the oscillation, and Φ is a phase angle. Substitution of the sinusoidal relation into the wave equation leads to the expression

$$-\frac{d^2a}{dx^2} = \frac{\rho\omega^2}{T}\,a.$$

Because the time dependence has been eliminated the equation has been reduced to an ordinary differential equation. The problem now is to find values for $a(x)$ and for the corresponding natural frequency ω which satisfy the equation and boundary conditions. There are an infinite number of each.

The governing equation can be made dimensionless by defining a dimensionless length

$$X = \frac{x}{L}\,,$$

and a dimensionless amplitude

$$A = \frac{a}{L}\,.$$

Introducing these definitions converts the equation of motion into the dimensionless form.

$$\frac{d^2A}{dX^2} + \lambda A = 0,$$

where $\lambda = \rho\omega^2 L^2/T$, a dimensionless frequency. The λ are the eigenvalues of the system. It is necessary to write the boundary conditions in dimensionless form:

$$X = 0, \quad A = 0;$$
$$X = 1, \quad A = 0.$$

To solve the problem numerically by finite-difference procedures, the string is divided into a finite number of elements, each of length δx (see Fig. 10.12). Finite difference equations are written for every interior point. It is not necessary to write an equation for the endpoints as these are fixed by the boundary conditions. As the second derivative enters, these will be included in the expressions at points 1 and 5 in this example. Then, for any interior point n, the finite-difference equation is

$$\frac{A_{n-1} - 2A_n + A_{n+1}}{(\delta x)^2} + \lambda A_n = 0.$$

At point 1 the boundary condition $A_0 = 0$ enters and the equation in finite-difference form is

$$\frac{-2A_1 + A_2}{\delta x^2} + \lambda A_1 = 0.$$

This will likewise occur at point 5. In an effort to be concise, we have employed

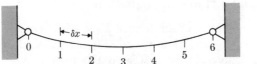

Fig. 10.12

only six elements in this presentation. When equations have been written there will be five algebraic equations in the unknowns A_1, A_2, A_3, A_4, A_5, and λ. In matrix form they are, for this example:

$$\begin{bmatrix} 2 - \lambda(\delta x)^2 & -1 & 0 & 0 & 0 \\ -1 & 2 - \lambda(\delta x)^2 & -1 & 0 & 0 \\ 0 & -1 & 2 - \lambda(\delta x)^2 & -1 & 0 \\ 0 & 0 & -1 & 2 - \lambda(\delta x)^2 & -1 \\ 0 & 0 & 0 & -1 & 2 - \lambda(\delta x)^2 \end{bmatrix} \begin{bmatrix} A_1 \\ A_2 \\ A_3 \\ A_4 \\ A_5 \end{bmatrix} = \begin{bmatrix} 0 \\ 0 \\ 0 \\ 0 \\ 0 \end{bmatrix}.$$

So the problem has now been cast into the same sort of eigenvalue problem we encountered with the lumped-parameter system. What has actually happened, of course, is that the continuous problem we set out to solve has been transformed, by the finite-difference approximation, to one having a discrete number of elements. Once again it is a problem of too few equations for the number of unknowns. Methods described in Chapter 8 are applicable for the solution of this eigenvalue problem. After finding the eigenvalues, it will sooner or later be necessary to incorporate the initial conditions to get a complete solution to the specific problem.

As a result of using a finite number of elements, only a finite number of frequencies and modes can be found; in this case, six. Choosing smaller elements will, in general, improve the accuracy of the approximation, and allow the determination of more modes and more frequencies (although this is often difficult, and not always necessary).

It is not difficult to extend these ideas to systems having more than one spatial dimension of interest. In these situations, equations of motion are not reduced to ordinary differential equations by the introduction of a sinusoidal time dependence. Rather they remain partial differential equations. There will be a greater number of unknowns for a given fineness of grid when the finite-differential equations are formulated.

Other systems lead to similar formulations with other types of boundary conditions. Vibrations in a beam with a built-in end (or buckling of a column with a built-in end) would require that the gradient of the deflection $(\partial y/\partial x)$ vanish at that boundary. For a more extensive discussion of eigenvalue problems in continuous systems, the reader is advised to consult other sources (e.g., S. H. Crandall's *Engineering Analysis*).

A parabolic partial differential equation

Next we turn to the numerical solution of a parabolic type of partial differential equation, by treating the equation of unsteady heat conduction developed at the beginning of this chapter. The system is continuous, with a time-dependent solution.

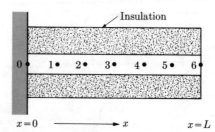

Fig. 10.13

As a specific example, the equation will be applied to the unsteady heat conduction in a long, slender rod, which is insulated from the environment along its entire length L. An illustration of the system is given in Fig. 10.13. Because it is slender, we can neglect temperature gradients normal to the x-direction. This, coupled with the insulated sides, reduces the problem to one in which temperature is a function of only the one spatial variable x. Consequently, the equation of unsteady heat conduction is reduced to the form

$$\frac{\partial^2 T}{\partial x^2} = \frac{\rho c_p}{k} \frac{\partial T}{\partial t}.$$

Two boundary conditions and an initial temperature distribution are needed to define the problem fully. Boundary conditions may be in the form of a pre-scribed temperature at one or both boundaries which might be time dependent, e.g.,

$$T(0, t) \qquad \text{or} \qquad T(L, t);$$

or they might be expressed in terms of the gradient of temperature, e.g.,

$$\frac{\partial T}{\partial x}(0, t) \qquad \text{or} \qquad \frac{\partial T}{\partial x}(L, t).$$

The initial condition must provide values of temperature for all x at time zero. In terms of x, t-coordinates there are known values of T along the line $t = 0$, and conditions are prescribed also along lines $x = 0$ and $x = L$. In this way the solution is bounded on three sides, and open on the fourth (see Fig. 10.14). A solution will be developed in the region labeled "Solution domain," proceeding in the direction of increasing time.

Before we examine the solution, the equation will be put into dimensionless form by the introduction of the following definitions:

$$u = \frac{T}{T_{\text{Ref}}}; \qquad \tau = \frac{t}{(\rho c_p / k)L^2}; \qquad \text{and} \qquad X = \frac{x}{L}.$$

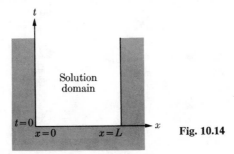

Fig. 10.14

The reference temperature T_{Ref} is taken as some constant temperature which relates to the system. It will be assigned a specific value later. With these inserted, the equation of unsteady heat conduction becomes

$$\frac{\partial^2 u}{\partial X^2} = \frac{\partial u}{\partial \tau}.$$

Boundary and initial conditions must also be made dimensionless.

Turning now to the finite-difference solution, the physical model is first divided into a number of discrete elements. In Fig. 10.13 this has been done for the rod to produce five interior points. The problem can be represented in X, τ-coordinates as a two-dimensional grid in the solution domain with lengthwise spacing δX and time spacing $\delta \tau$, and bounded on three sides as depicted in Fig. 10.15.

As a first approach to the finite-difference solution, the problem will be formulated in the most obvious way. That is, values of u will be computed one time step ahead for each point using presently known values of the function. This means the time derivative will be expressed as a forward difference. Referring to the partial differential equation, the finite-difference approximation for an interior point at spatial coordinate i and time coordinate j is

$$\frac{u_{i-1,j} - 2u_{i,j} + u_{i+1,j}}{(\partial X)^2} = \frac{u_{i,j+1} - u_{i,j}}{\delta \tau}.$$

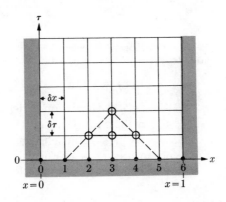

Fig. 10.15

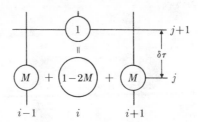

Fig. 10.16

Introducing the definition

$$M = \frac{\delta\tau}{(\delta X)^2},$$

we can find the value of u at the next time step, $j + 1$, by the expression

$$u_{i,j+1} = Mu_{i-1,j} + (1 - 2M)u_{i,j} + Mu_{i+1,j}.$$

A computational molecule for this calculation is shown in Fig. 10.16. This shows the location in X, τ-coordinates and the weighting for the temperatures which enter the calculation. Arithmetic operators have been drawn in the appropriate places. Beginning from the initial state the solution is marched forward in time, one step at a time. A temperature at a given space-time point is dependent upon three temperatures at the previous time step. This is demonstrated by the small circles in Fig. 10.15. The value of u at the third spatial point and the second time step was computed by $u_{2,1}$, $u_{3,1}$, and $u_{4,1}$. Tracing backwards to the initial state it is evident that only values lying within the dashed triangle have contributed to the value of $u_{3,2}$. The consequences of this will be discussed later.

Three aspects of this solution should now be considered. They are: the accuracy of the approximate solution; the treatment of the boundary conditions; and the behavior of the solution as it is "marched" forward.

An indication of the accuracy of the approximation (or discretization error), can be obtained by examining the Taylor-series approximation for the second derivatives. Retaining the truncation-error terms for the first and second derivatives, we can write

$$\left[\frac{\partial^2 u}{\partial X^2} - \frac{\partial u}{\partial \tau}\right]_{i,j} = \frac{u_{i-1,j} - 2u_{i,j} + u_{i+1,j}}{(\delta X)^2} - \frac{u_{i,j+1} - u_{ij}}{\delta\tau} + O(\delta\tau) + O(\delta X)^2.$$

Thus the local truncation error for a single step in the calculation is dependent upon the choice of grid size, or, in terms of orders of magnitudes, it is

$$O(\delta\tau) + O(\delta X)^2.$$

It must be remembered that any given calculation is dependent upon many preceding calculations as suggested by the dashed triangle (Fig. 10.15). Consequently, these local errors propogate, just as they did in solutions for transients in lumped-parameter systems, and the propogated error is larger than the local discretization error.

As a simple demonstration of the treatment of boundary conditions, consider the situation where one end of the rod, $x = 0$, is held at a given temperature T_{Ref}, which is constant in time, while the free end, $x = L$, is insulated,

$$\left(\frac{\partial T}{\partial x}\right)_{x=L} = 0.$$

Both conditions must be included in the finite-difference equations. In terms of the dimensionless variables, the boundary conditions can be written,

$$X = 0, \qquad u = 1;$$

$$X = 1, \qquad \left(\frac{\partial u}{\partial X}\right) = 0.$$

As the temperature at point O is constant, it is not necessary to center an equation there. At station 1 the recursion relation is

$$u_{1,j+1} = M + (1 - 2M)u_{1,j} + Mu_{2,j},$$

where the value $u_{0,j} = 1$ has been entered. Turning next to the free end condition, a zero gradient can be treated as a point of symmetry just as was done for the earlier heat-transfer problem. A point is imagined beyond point 6 which has the same value as u_5. Hence, for point 6, the recursion relation is

$$u_{6,j+1} = 2Mu_{5,j} + (1 - 2M)u_{6,j}.$$

Let us now proceed to some hand calculations for a few steps for several choices of M. Although the spatial increment has been chosen, the value of M can be varied by changing the size of the time increment. To carry out numerical calculations also requires an initial temperature distribution. Suppose the value of u everywhere is initially zero, and the value at O is suddenly (at $\tau = 0$) raised to 1. Calculations will now be made for the following cases: case 1, $M = 0.25$; case 2, $M = 0.5$; and case 3, $M = 1.0$.

Case 1. With $M = 0.25$ the central factor $(1 - 2M) = 0.5$, and the recursion relations become:

For point 1: $\qquad\qquad u_{1,j+1} = 0.25 + 0.5u_{1,j} + 0.25u_{2,j};$
for points $i = 2, 3, 4, 5$: $\quad u_{i,j+1} = 0.25u_{i-1,j} + 0.5u_{i,j} + 0.25u_{i+1,j};$
and for point 6 $\qquad\qquad u_{6,j+1} = 0.5u_{5,j} + 0.5u_{6,j}.$

Calculations are given for ten successive time steps in Table 10.1. It is evident that the temperature pulse requires several steps to affect the entire length of the rod; in this case, six. This is not the way one would expect the physical system to behave. However, the solution is stable, and the temperature at any given spatial location increases gradually with each passing time step. If enough steps were taken it is expected that the numerical solution would converge to the steady-state solution (in this case $u = 1$ everywhere).

Table 10.1

Finite-Difference Solution with $M = 0.25$

Point / Time step	0	1	2	3	4	5	6
0	1.	0	0	0	0	0	0
1	1.	0.2500	0	0	0	0	0
2	1.	0.3750	0.0625	0	0	0	0
3	1.	0.4531	0.1250	0.0156	0	0	0
4	1.	0.5078	0.1797	0.0391	0.0039	0	0
5	1.	0.5488	0.2266	0.0655	0.0117	0.0001	0
6	1.	0.5811	0.2669	0.0923	0.0222	0.0030	0.00005
7	1.	0.6072	0.3018	0.1184	0.0349	0.0070	0.0015
8	1.	0.6290	0.3323	0.1434	0.0488	0.0126	0.0042
9	1.	0.6478	0.3592	0.1670	0.0634	0.0196	0.0084
10	1.	0.6637	0.3833	0.1892	0.0758	0.0278	0.0140

Case 2. Here the value of M has been purposely chosen to cause the middle term in the recursion relation to vanish. It represents a time step twice as large as in case 1. Each projected value at a point is the average value of the two points adjacent. The procedure, sometimes called Schmidt's method, gives the relations for this case as:

For point 1: $\quad u_{1,j+1} = 0.5 + 0.5u_{2,j};$

for points $i = 2, 3, 4, 5:$ $\quad u_{i,j+1} = 0.5u_{i-1,j} + 0.5u_{i+1,j};$

and for point 6: $\quad u_{6,j+1} = u_{5,j}.$

Numerical results for ten time steps are given in Table 10.2. Although the solution is stable, there is an oscillation of the result at each point. This is contrary to the physical behavior of the system. Remembering that the elapsed time for ten steps in the second case is twice as large as in the first, values at step 5 in the second can be compared to values at the tenth step in the first. There is reasonable agreement between the two sets. However, the pulse of temperature has not yet reached the end of the rod at this time in case 2. Case 2 is less effective than case 1 as the solution oscillates, and takes more elapsed time to affect the entire rod. With a digital computer it is not likely that a time step this large would be chosen.

Case 3. Even more difficulty occurs when M is set to 1. Now the central term has a negative sign with the recursion relations being:

For point 1: $\quad u_{1,j+1} = 1 - u_{i,j} + u_{2,j};$

for points $i = 2, 3, 4, 5:$ $\quad u_{i,j+1} = u_{i-1,j} - u_{i,j} + u_{i+1,j};$

and for point 6: $\quad u_{6,j+1} = 2u_{5,j} - u_{6,j}.$

Table 10.2

Finite-Difference Solution with $M = 0.5$

Point Time step	0	1	2	3	4	5	6
0	1.	0	0	0	0	0	0
1	1.	0.5	0	0	0	0	0
2	1.	0.5	0.25	0	0	0	0
3	1.	0.625	0.25	0.125	0	0	0
4	1.	0.625	0.375	0.125	0.0625	0	0
5	1.	0.6875	0.375	0.2188	0.0625	0.0312	0
6	1.	0.6875	0.4531	0.2188	0.1250	0.0312	0.0312
7	1.	0.7266	0.4531	0.2890	0.1250	0.0781	0.0312
8	1.	0.7266	0.5078	0.2890	0.1835	0.0781	0.0781
9	1.	0.7539	0.5078	0.3456	0.1835	0.1308	0.0781
10	1.	0.7539	0.5498	0.3456	0.2382	0.1308	0.1308

Table 10.3

Finite-Difference Solution with $M = 1.0$

Point Time step	0	1	2	3	4	5	6
0	1.	0.	0.	0.	0.	0.	0.
1	1.	1.	0.	0.	0.	0.	0.
2	1.	0.	1.	0.	0.	0.	0.
3	1.	2.	−1.	1.	0.	0.	0.
4	1.	−2.	4.	−2.	1.	0.	0.
5	1.	7.	−8.	7.	−3.	1.	0.
6	1.	−14.	22.	−18.	11.	−4.	2.
7	1.	37.	−54.	51.	−33.	16.	−10.

In Table 10.3 results for seven time steps are given. It is immediately obvious that the solution is oscillating and unstable.

Stability

We have observed that solutions of the finite-difference representation of the parabolic partial differential equation may be oscillatory, and in some cases become unbounded as the number of time steps becomes large. A solution which becomes unbounded is said to be unstable. It is clear that for this problem, stability depends on the size of the space and time steps selected, or specifically upon the size of M. It is desirable to know beforehand whether a calculation will be unstable, but this is usually too difficult to be practical. It is possible to develop, for this case, a fairly simplified stability criterion. Stability will depend upon the form of the equation, coupled with the nature of the boundary conditions and certain critical parameters.

We shall examine, then, the stability of the finite-difference solution for the partial differential equation

$$\frac{\partial^2 u}{\partial x^2} = \frac{\partial u}{\partial t}.$$

Lower case x and t have been used for the space and time coordinates. It is necessary to specify the initial condition. An attempt has been made to make it a very general one by choosing one of the form

$$u(x, 0) = e^{ipx},$$

where i is $\sqrt{-1}$. There are two reasons for choosing this form of initial condition. First, a general function $f(x)$ can be expressed as a Fourier series with components of e^{ipx}. Thus the form can be used as a general form of initial condition to represent virtually any $f(x)$. Secondly, it is easy to introduce this form of initial condition into the exact solution of the differential equation. It is limited to cases where x extends to infinity.

An exact solution of the differential equation can be found by assuming a product form of solution $u(x, t) = X(x)T(t)$. Substituting this into the partial differential equation, and separating the variables, we find as a solution:

$$u(x, t) = Ce^{\kappa x + \kappa^2 t},$$

where κ is a separation constant.

Introducing the initial condition

$$u(x, 0) = Ce^{\kappa x} = e^{ipx},$$

we conclude:

$$C = 1, \quad \kappa = ip.$$

Hence the solution to the partial differential equation which satisfied the assumed initial condition is

$$u(x, t) = e^{ipx - p^2 t} = e^{ipx}g(t).$$

The finite-difference equation can be treated in much the same way by introducing a trial solution of the same form, namely, at the length position x,

$$u(x, t) = e^{ipx}g(t).$$

Now the trial solution can be substituted into the finite-difference form of the partial differential equation, evaluating the solution at the appropriate intervals. The result of this is

$$\frac{g(t + \delta t)e^{ipx} - g(t)e^{ipx}}{\delta t} = g(t)\frac{[e^{ip(x + \delta x)} - 2e^{ipx} + e^{ip(x - \delta x)}]}{(\delta x)^2}.$$

From this,

$$g(t + \delta t) = g(t) + g(t)M[e^{ip\delta x} - 2 + e^{-ip\delta x}],$$

with M being defined as before. The term in brackets can be simplified as

$$e^{ip\delta x} - 2 + e^{-ip\delta x} = [e^{(1/2)ip\delta x} - e^{-(1/2)ip\delta x}]^2 = -4\sin^2 \tfrac{1}{2}p \; \delta x.$$

Then

$$g(t + \delta t) = g(t)[1 - 4M\sin^2 \tfrac{1}{2}p \; \delta x].$$

In order for the assumed form of the solution to satisfy the initial condition it is necessary that

$$g(0) = 1.$$

Beginning at that point, the solution can be developed for several successive steps from the finite-difference relation:

$$g(\delta t) = (1)[1 - 4M\sin^2 \tfrac{1}{2}p \; \delta x]$$
$$g(2 \; \delta t) = \quad [1 - 4M\sin^2 \tfrac{1}{2}p \; \delta x]^2$$
$$\vdots$$
$$g(n \; \delta t) = \quad [1 - 4M\sin^2 \tfrac{1}{2}p \; \delta x]^n$$

This can now be substituted in the trial solution to express the value of the function u at length position x after n time steps. Mathematically this is

$$u(x, n \; \delta t) = e^{ipx}[1 - 4M\sin^2 \tfrac{1}{2}p \; \delta x]^n.$$

Recalling the relation

$$e^{-\epsilon} \approx 1 - \epsilon$$

for small ϵ, we can simplify the above expression for small δx as

$$1 - 4M\sin^2 \tfrac{1}{2}p \; \delta x \approx 1 - Mp^2(\delta x)^2$$
$$\approx e^{-Mp^2(\delta x)^2}.$$

With this approximation, the solution for u when δx is made small becomes

$$u(x, t) = e^{ipx}e^{-Mp^2(\delta x)^2 n}$$
$$= e^{ipx - p^2 t}$$

since, with the definition of M,

$$Mn(\delta x)^2 = n(\delta t) = t.$$

Thus we have demonstrated that the finite-difference solution tends to the solution of the partial differential equation when δx tends to zero.

Next we examine the question of stability of the finite-difference solution for a specific choice of δx. For the solution to remain bounded, and hence to be stable, it is necessary that

$$\left| \frac{g(n \; \delta t)}{g((n - 1) \; \delta t)} \right| \leq 1,$$

which, for all p and a prescribed δx, is equivalent to the requirement that

$$|1 - 4M \sin^2 \tfrac{1}{2}p\, \delta x| \leq 1.$$

It is possible for $\sin^2 \tfrac{1}{2}p\, \delta x$ to equal unity but not to exceed it. With this limitation, then, the procedure will be stable for $M \leq \tfrac{1}{2}$.

Strictly speaking, this proof is only applicable for the case wherein x extends to infinity, due to the choice of initial conditions. Results for a finite range of x are generally independent of boundary conditions, except near the critical value of M (i.e., $M = \tfrac{1}{2}$). Examining the results of the numerical example, we see that the solution was stable for $M = 0.25$ and $M = 0.5$, but was unstable for $M = 1$, as the analysis predicts. In general, analyses to predict limits of stability in finite-difference solutions are not this simple, and can often be done best by comparing computer results when a critical parameter is varied.

The implicit form

In the method of solution of the parabolic partial differential equation described in the last two sections, new values of u were dependent upon past history in the manner illustrated diagrammatically in Fig. 10.15. Furthermore, it was demonstrated that for some choices of interval size, the method could be unstable. It was necessary in this specific instance to choose $M < \tfrac{1}{2}$, which can be done for a fixed increment δx only by reducing the time step, and this in turn necessitates more computations for a given time interval. This method is sometimes called the *explicit method*.

One should not infer that the explicit method is not useful, but it is possible to alleviate the above difficulties with an alternate calculation scheme, called the *implicit method*. In this scheme, finite-difference relations are written to include values of u at the advanced point in time in the representation of $(\partial^2 u/\partial X^2)$ as well as for present values. Each expression is included in the finite-difference equation by using a weighting factor θ. Evaluating the time derivative, as was done in the explicit method, the finite-difference equation is

$$\frac{u_{i,j+1} - u_{i,j}}{\delta \tau} = \left(\frac{1}{(\delta X)^2}\right)[\theta(u_{i-1,j+1} - 2u_{i,j+1} + u_{i+1,j+1})$$
$$+ (1 - \theta)(u_{i-1,j} - 2u_{i,j} + u_{i+1,j})],$$

where the weighting factor is $0 \leq \theta \leq 1$. The computational molecule for this equation is shown in Fig. 10.17.

It is evident that if $\theta = 0$, the equation reverts to the explicit form. A choice of $\theta = 1$ leads to the formula of O'Brien, Hyman, and Kaplan, which is written

$$-Mu_{i-1,j+1} + (1 + 2M)u_{i,j+1} - Mu_{i+1,j+1} = u_{i,j}.$$

A diagram of the computation molecule for this form is shown in Fig. 10.18.

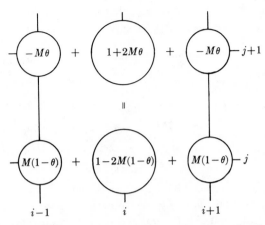

Fig. 10.17

An expression of the above form must be written out for the first time step at every point in the spatial grid. Boundary conditions will be incorporated into the equations. There will then be a set of simultaneous equations, in terms of the unknown values of u at the next time step. With as many equations as there are unknowns, the set can then be solved by a reduction procedure, and the process repeated for the next time step.

Let's return now to the numerical example incorporating boundary conditions as before. For the first time step, $j = 1$, the equations will be as follows:

For point 1: $\qquad -M + (1 + 2M)u_{1,1} - Mu_{2,1} = 0$

(here the boundary condition $u_0 = 1$ has been included);

for points $i = 2, 3, 4, 5$: $\quad -Mu_{i-1,1} + (1 + 2M)u_{i,1} - Mu_{i+1,1} = 0$;

and for point 6: $\qquad -2Mu_{5,1} + (1 + 2M)u_{6,1} = 0$

(here the condition of the insulated boundary has been included as in the explicit solution).

Zeros appear on the right-hand side because, with the exception of u_0, the initial value of the u's is everywhere zero. Because u_0 is nonzero, the right-hand

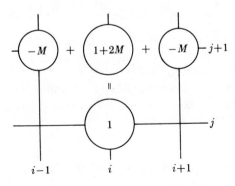

Fig. 10.18

sides will not all be zero, so the set is solvable for the $u_{i,1}$. More generally, the set of equations to be solved is, in matrix form,

$$
\begin{bmatrix}
1+2M & -M & 0 & 0 & 0 & 0 \\
-M & 1+2M & -M & 0 & 0 & 0 \\
0 & -M & 1+2M & -M & 0 & 0 \\
0 & 0 & -M & 1+2M & -M & 0 \\
0 & 0 & 0 & -M & 1+2M & -M \\
0 & 0 & 0 & 0 & -2M & 1+2M
\end{bmatrix}
\begin{bmatrix}
u_{1,j+1} \\
u_{2,j+1} \\
u_{3,j+1} \\
u_{4,j+1} \\
u_{5,j+1} \\
u_{6,j+1}
\end{bmatrix}
=
\begin{bmatrix}
u_{1,j}+M \\
u_{2,j} \\
u_{3,j} \\
u_{4,j} \\
u_{5,j} \\
u_{6,j}
\end{bmatrix}
$$

The set can first be solved for $u_{i,1}$ ($j = 0$). With j increased by 1, these become the right-hand sides for the next calculation, and the process is repeated. Because the finite-difference forms of the derivatives are the same as those used in the explicit method, the local truncation error is again $O(\delta t) + O(\delta x)^2$.

Returning next to the question of stability of the implicit method, with $\theta = 1$, the procedure parallels the one used for the explicit method. Substituting a solution having the form

$$ u(x, t) = e^{ipx} g(t) $$

into the finite-difference equation, we obtain the relation

$$ g(t) = g(t + \delta t)(1 + 4M \sin^2 \tfrac{1}{2}p\, \delta x). $$

It is possible, following the same procedure as before, to conclude that the method converges to the solution of the differential equation for small δx. If we require, for stability of the stepwise solution at any x, that

$$ \left| \frac{g(t + \delta t)}{g(t)} \right| \leq 1, $$

we meet the restriction

$$ \left| \frac{1}{1 + 4M \sin^2 \tfrac{1}{2}p\, \delta x} \right| \leq 1 $$

for a solution to be stable. It is evident that this will be true for any choice of M. Hence the implicit method is stable for all M when it is used to solve the transient heat-conduction equation.

It is possible to examine the stability of the finite-difference equation for all θ for heat-conduction equations. Following the same procedure as before, it can be shown that the formula is stable if

$$ 2M \leq (1 - 2\theta)^{-1} $$

in the range $0 \leq \theta < \tfrac{1}{2}$, and that it is stable for all M in the range $\tfrac{1}{2} \leq \theta \leq 1$.

The Crank-Nicolson method

Another useful method can be developed from the general recursion formula with $\theta = \frac{1}{2}$. The relation is

$$\frac{u_{i,j+1} - u_{i,j}}{\delta\tau} = \left(\frac{1}{2(\delta X)^2}\right) [(u_{i-1,j+1} - 2u_{i,j+1} + u_{i+1,j+1})$$

$$+ (u_{i-1,j} - 2u_{i,j} + u_{i+1,j})].$$

This is the formula of Crank and Nicolson, and, as $\theta = \frac{1}{2}$, is stable for all M when applied to the transient heat-conduction equation.

It is helpful in examining this method to treat it as a two-step calculation. Using the explicit method, values of $u_{i,j+1/2}$ are computed from the equation

$$\frac{u_{i,j+1/2} - u_{i,j}}{\frac{1}{2}\delta\tau} = \frac{u_{i-1,j} - 2u_{i,j} + u_{i+1,j}}{(\delta X)^2}.$$

Then, with the $u_{i,j+1/2}$ known, the values of $u_{i,j+1}$ are computed using the implicit relation

$$\frac{u_{i,j+1} - u_{i,j+1/2}}{\frac{1}{2}\delta\tau} = \frac{u_{i-1,j+1} - 2u_{i,j+1} + u_{i+1,j+1}}{(\delta X)^2}.$$

Because of their linearity, these two equations can be added together to produce the Crank-Nicolson formula. Local truncation error has been improved because the actual time step is halved, and the calculation is only slightly more complicated.

Actual solution with the Crank-Nicolson method is implicit. It is accomplished by formulating a set of simultaneous equations, and solving for the $u_{i,j+1}$, just as for the previous method. The equations can be formed by referring to Fig. 10.17 with $\theta = \frac{1}{2}$. Putting unknowns on the left, the equation at an interior point i is written

$$-\left(\frac{M}{2}\right)u_{i-1,j+1} + (1 + M)u_{i,j+1} - \left(\frac{M}{2}\right)u_{i+1,j+1}$$

$$= \left(\frac{M}{2}\right)u_{i-1,j} + (1 - M)u_{i,j} + \left(\frac{M}{2}\right)u_{i+1,j}.$$

While the right-hand side appears more complicated than for the simple implicit method, it should be noted that none of the terms on the right is unknown.

Parabolic partial differential equations in two spatial coordinates

When the system is two-dimensional, the dimensionless heat-conduction equation becomes

$$\frac{\partial^2 u}{\partial x^2} + \frac{\partial^2 u}{\partial y^2} = \frac{\partial u}{\partial t}.$$

A second-order partial-differential equation, it is three-dimensional in $u(x, y, t)$. Consequently two additional boundary conditions are required to describe the system.

An explicit method of solution can be formulated by replacing the spatial derivatives by their finite-difference approximations at each point on the spatial grid. Triple subscripts are required; i referring to the x-direction, j to the y-, and k to the time coordinate. In finite-difference form, the partial-differential equation is

$$\frac{u_{i,j,k+1} - u_{i,j,k}}{\delta t} = \frac{u_{i-1,j,k} - 2u_{i,j,k} + u_{i+1,j,k}}{(\delta x)^2} + \frac{u_{i,j-1,k} - 2u_{i,j,k} + u_{i,j+1,k}}{(\delta y)^2}.$$

Setting $\delta x = \delta y$ and solving for $u_{i,j,k+1}$, we obtain

$$u_{i,j,k+1} = Mu_{i-1,j,k} + Mu_{i+1,j,k} + Mu_{i,j-1,k} + Mu_{i,j+1,k} + (1 - 4M)u_{i,j,k},$$

where the definition of M is retained as

$$M = \frac{\delta t}{(\delta x)^2} = \frac{\delta t}{(\delta y)^2}.$$

Beginning with the initial distribution of $u(x, y, 0)$, the calculation can be marched forward in time as was the case for the equation in one spatial coordinate. Of course the equations must be modified at boundaries to incorporate the boundary conditions properly. For this method to be stable when applied to the heat-conduction equation, it is necessary that

$$\delta t \leq \frac{1}{2[(1/(\delta x)^2) + (1/(\delta y)^2)]};$$

or, for the usual case of $\delta x = \delta y$, this simplifies to $M \leq \frac{1}{4}$. Proof of this is omitted.

An implicit method can be developed for the two-dimensional system which is stable for all M. For the case $\delta x = \delta y$, the implicit form can be written

$$-Mu_{i-1,j,k+1} - Mu_{i+1,j,k+1} - Mu_{i,j-1,k+1} - Mu_{i,j+1,k+1} + (1 + 4M)u_{i,j,k+1}$$
$$= u_{i,j,k}.$$

A complete set of equations must be formed and reduced to find values of u at the $(k + 1)$th time step. Now there are as many as five unknowns in the equation for a single location, whereas, in the one-dimensional case, there were never more than three. As a result, the solution is made a little more difficult.

A method called the *alternating-direction method* makes it possible to work with coefficient matrices which have a maximum of three nonzero terms per row. In this method a set of values is determined at a time $[t + (\delta t/2)]$ in the x-direction only. Then values are found for the full-time step by means of a second equation which solves in the y-direction only. These equations are similar to the two cited previously for the Crank-Nicolson method, except that only the x-derivative is treated at the half-time increment. Forming the equation for the first half of the time step yields the expression

$$\frac{u_{i,j,k+1/2} - u_{i,j,k}}{\delta t/2} = \frac{u_{i-1,j,k+1/2} - 2u_{i,j,k+1/2} + u_{i+1,j,k+1/2}}{(\delta x)^2}$$

$$+ \frac{u_{i,j-1,k} - 2u_{i,j,k} + u_{i,j+1,k}}{(\delta y)^2},$$

and for the second half of the time step, the expression

$$\frac{u_{i,j,k+1} - u_{i,j,k+1/2}}{\delta t/2} = \frac{u_{i-1,j,k+1/2} - 2u_{i,j,k+1/2} + u_{i+1,j,k+1/2}}{(\delta x)^2}$$

$$+ \frac{u_{i,j-1,k+1} - 2u_{i,j,k+1} + u_{i,j+1,k+1}}{(\delta y)^2}.$$

Rewriting these for the case that $\delta x = \delta y$ and $M = \delta t/(\delta x)^2$, with unknowns appearing on the left side of each equation, we get

$$-\left(\frac{M}{2}\right)u_{i-1,j,k+1/2} + (1 + M)u_{i,j,k+1/2} - \left(\frac{M}{2}\right)u_{i+1,j,k+1/2}$$

$$= \left(\frac{M}{2}\right)u_{i,j-1,k} + (1 - M)u_{i,j,k} + \left(\frac{M}{2}\right)u_{i,j+1,k};$$

$$-\left(\frac{M}{2}\right)u_{i,j-1,k+1} + (1 + M)u_{i,j,k+1} - \left(\frac{M}{2}\right)u_{i,j+1,k+1}$$

$$= \left(\frac{M}{2}\right)u_{i-1,j,k+1/2} + (1 - M)u_{i,j,k+1/2} + \left(\frac{M}{2}\right)u_{i+1,j,k+1/2}.$$

Notice that both of these equations have only three unknowns. A solution is accomplished by reducing the first set of equations to determine values at the half-time interval. These are then inserted into the second set of equations, which are then reduced to give values of u at the end of the full time increment. It is obviously more complicated than the previous method as two sets of equations must be formed and solved to produce results at the end of a full time step.

Boundary conditions for parabolic equations

Boundary conditions for this type of problem are similar to those discussed in connection with systems in steady state. They are formulated in just the same way for the parabolic partial-differential equation. Irregular boundaries were discussed in that same connection. As the discussion does not need altering for the present situation, it will not be repeated here.

An added note on eigenvalue problems

Consider again the wave equation

$$\frac{\partial^2 u}{\partial x^2} - \frac{\partial^2 u}{\partial t^2} = 0.$$

Based on the treatment of parabolic differential equations, one might wish to formulate the solution to the wave equation in finite-difference form and treat it as an initial-value problem. Using an explicit procedure, the expression would be

$$u_{i,j+1} = \beta^2 u_{i-1,j} + 2(1 - \beta^2)u_{i,j} + \beta^2 u_{i+1,j} - u_{i,j-1},$$

where $\beta^2 = \delta t^2/\delta x^2$.

The computational molecule is illustrated in Fig. 10.19. It is evident that two initial conditions must be provided to begin the solution, as each step requires not only present values, but one from the previous time. In addition to the initial displacements, $u(x, 0)$, a condition such as $(\partial u/\partial t)(x, 0) = v(x)$ would be needed.

A computation using this procedure would provide a history of the displacements for this specific set of initial conditions. No knowledge of the eigenvalues (natural frequencies in this problem) would be forthcoming. As the resulting behavior of the displacement would actually be a combination of many modes (or frequencies), it would be difficult to separate them, or to ascertain when a series of oscillations had gone through a complete cycle. Knowledge of the eigenvalues is important if one is to gain an understanding of the behavior of the system. They provide a set of functions in which the behavior of the system can be expressed most conveniently. Each eigenvalue provides a component of the behavior. These components can then be combined to describe the behavior of the system for any initial state.

In summarizing this all-too-brief discussion, it should be noted that knowledge of the eigenvalues is important in understanding the behavior of these systems.

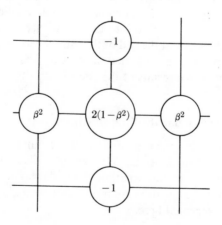

Fig. 10.19

Furthermore, eigenvalues are not present in the general formulation of the problem, but are introduced by assigning some special behavior to the mathematical model. In the case of the vibrating string, this happened when the displacement was assigned a sinusoidal dependence on time. From this it was evident that the natural frequency provided the basis for the eigenvalues.

A computer program

To illustrate how finite-difference solutions are programmed, a relatively simple example is presented. It treats the Laplace equation in a two-dimensional system in which the function, in this case temperature, is prescribed on the boundaries. A main program is written which identifies each point and writes the appropriate finite-difference equation for it. Eventually the augmented matrix is formed, and a subroutine is called to solve the set of equations for the steady-state values of temperature. The program then prints out the results in the form of a temperature map. The looping procedures used are neither unique nor the most compact. They have been chosen to illustrate the procedure in a relatively simple form.

Example Write a computer program which calculates the steady-state temperature distribution in a *square* pipe, as shown in Fig. 10.20a, using a finite-difference approximation to the Laplace equation. Temperatures are prescribed along both the outer and inner boundaries. Temperatures normal to the plane shown are assumed constant, permitting a two-dimensional model to be used. A geometry is assumed for the ratio of the lengths of the sides b/a which is always suitable for using a square mesh.

Because the pipe is square, it is only necessary to treat one-eighth of the cross section, shown shaded in Fig. 10.20a. This introduces the two lines of symmetry shown as dash-dot lines on the figure.

Mesh points are identified in Fig. 10.20b. There are m points in the first row, and n rows (excluding the outer boundary). In the illustration, $m = 6$ and $n = 3$. There are a total of n_0 interior points to be treated. It is necessary to compute n_0 from the entries m and n. Temperatures at each interior point are desig-

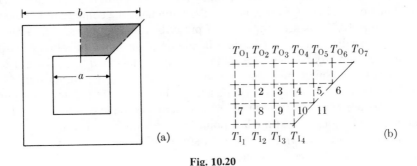

(a)

(b)

Fig. 10.20

nated T_i where the subscript indicates the location of the point using the numbering system illustrated in Fig. 10.20b. Outer wall temperatures are given as T_{Oi}; inner wall temperatures are T_{Ii}. It is necessary for the main program to identify each point and store the appropriate coefficients to form the finite-difference equation which approximates the Laplace equation

$$\frac{\partial^2 T}{\partial x^2} + \frac{\partial^2 T}{\partial y^2} = 0.$$

Account must be taken for points on the boundaries, and the equations modified accordingly. As the number of unknowns is n_0, the augmented matrix contains $[(n_0) \times (n_0 + 1)]$ coefficients. When all are entered, the program calls the subroutine GSMETH (Gauss-Siedel routine developed in Chapter 6) to obtain values of the steady-state temperatures. These are then returned to the main program and written. In order to employ GSMETH, a test for convergence, ϵ, must be supplied, along with a set of trial values of temperature. A single value called TRIAL is supplied for every point.

A flow chart, along with the main program and a sample of printed results are shown in Figs. 10.21a, 10.21b, and 10.21c.

It is the function of the main program to read in the appropriate information, to form and store the augmented matrix, to call the subroutine, and to print out the results. These sections are shown separately in the flow chart. As the first row is treated separately from the other rows, two sections are shown. This is not essential, and it is quite apparent from the flow chart that the two sections are very similar. The first row was treated separately so that the program could be modified easily for a new condition on the outer boundary, as suggested in an exercise at the end of the chapter.

Each point is treated in turn, being identified by the appropriate number i, and the appropriate coefficients a_{ij}, for the ith equation are read in. Hence the coefficient a_{ii} is a diagonal of the coefficient matrix. Each coefficient is first set to zero; then by a series of tests it is identified, and then altered to -1 or $+4$ or left equal to zero for interior points, in accordance with the computational molecule, Fig. 10.22a. It is also necessary to determine whether the point is next to a boundary, or on a line of symmetry. If it is next to the boundary, the computation molecule shown in Fig. 10.22b is used for cases of prescribed temperatures. The program must set the coefficient a_{i,n_0+1} (the right-hand side of the finite-difference equation) equal to the boundary temperature. Otherwise the coefficient a_{i,n_0+1} is set to zero.

In the case of a mesh point on a line of symmetry, the computational molecule shown in Fig. 10.22c controls the coefficients to be stored. In the first row, points of symmetry are at points l and m. For the other rows the first point in the row is identified by an index k, and the last point by l. Each new value of l is found by incrementing a dummy variable l_0 when the row is changed. When l is updated, it is necessary to store the old value as l_1.

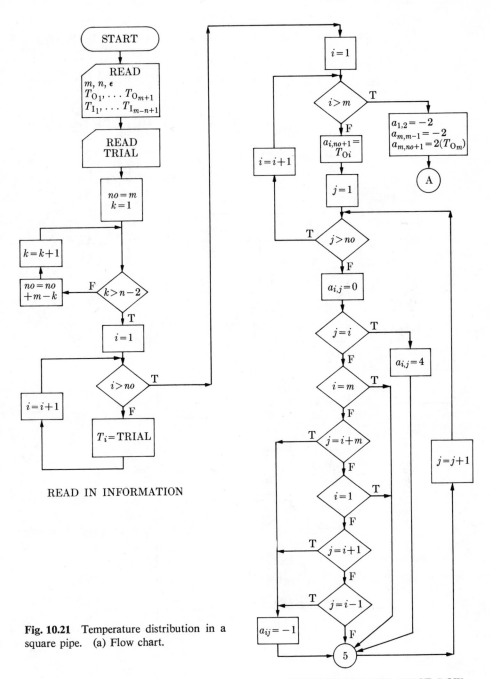

READ IN INFORMATION

Fig. 10.21 Temperature distribution in a square pipe. (a) Flow chart.

SET COEFFICIENTS FOR FIRST ROW

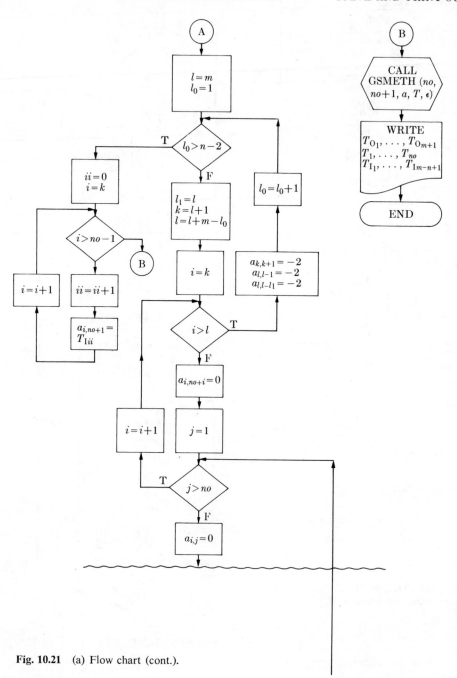

Fig. 10.21   (a) Flow chart (cont.).

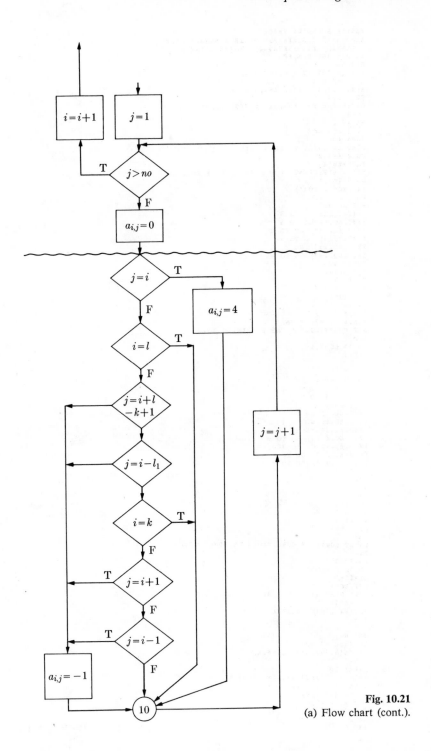

Fig. 10.21
(a) Flow chart (cont.).

```
C       MASTER BOUNDARY VALUE
C       TEMPERATURE DISTRIBUTION IN A SQUARE PIPE
        DIMENSION T(25),A(25,26),TO(26),TI(26)
        READ(5,100) M,N,EPS
        MP1=M+1
        READ(5,101) (TO(I),I=1,MP1)
        MMNP1=M-N+1
        READ(5,101) (TI(I),I=1,MMNP1)
        READ(5,102) TRIAL
C       DETERMINE THE NUMBER OF MESH POINTS
        NO=M
        NM2=N-2
        DO 1 K=1,NM2
 1      NO=NO+M-K
C       READ IN TRIAL TEMPERATURES
        DO 2 I=1,NO
 2      T(I)=TRIAL
C       SET COEFFICIENTS FOR FIRST ROW OF POINTS
        DO 5 I=1,M
        A(I,NO+1)=TO(I)
        DO 5 J=1,NO
        A(I,J)=0.
        IF(J.EQ.I) GO TO 3
        IF(I.EQ.M) GO TO 5
        IF(J.EQ.I+M) GO TO 4
        IF(I.EQ.1) GO TO 5
        IF(J.EQ.I+1) GO TO 4
        IF(J.EQ.I-1) GO TO 4
        GO TO 5
 3      A(I,J)=4.
        GO TO 5
 4      A(I,J)=-1.
 5      CONTINUE
        A(1,2)=-2.
        A(M,M-1)=-2.
        A(M,NO+1)=2.*TO(M)
C       SET COEFFICIENTS FOR REMAINING ROWS
        L=M
        DO 11 LO=1,NM2
        K=L+1
        L1=L
        L=L+M-LO
        DO 10 I=K,L
        A(I,NO+1)=0.
        DO 10 J=1,NO
        A(I,J)=0.
        IF(J.EQ.I) GO TO 8
        IF(I.EQ.L) GO TO 10
        IF(J.EQ.I+L-K+1) GO TO 9
        IF(J.EQ.I-L1) GO TO 9
        IF(I.EQ.K) GO TO 10
        IF(J.EQ.I+1) GO TO 9
        IF(J.EQ.I-1) GO TO 9
        GO TO 10
 8      A(I,J)=4.
        GO TO 10
 9      A(I,J)=-1.
 10     CONTINUE
        A(K,K+1)=-2.
        A(L,L-1)=-2.
        A(L,L-L1)=-2.
 11     CONTINUE
C       RESET BOUNDARY CONDITIONS ON INNER WALL
        II=0
        NOM1=NO-1
        DO 12 I=K,NOM1
        II=II+1
 12     A(I,NO+1)=TI(II)
        CALL GSMETH(NO,NO+1,A,T,EPS,INDEX)
        MP1=M+1
        WRITE(6,103) (TO(I),I=1,MP1)
        K=1
        L=M
        NM1=N-1
        DO 15 LO=1,NM1
        WRITE(6,103) (T(I),I=K,L)
        K=L+1
 15     L=L+M-LO
        WRITE(6,103) (TI(I),I=1,MMNP1)
 100    FORMAT(2I9,F9.0)
 101    FORMAT(10F9.0)
 102    FORMAT(F9.0)
 103    FORMAT(//,2X,10F10.4)
        STOP
        END
```

Fig. 10.21 (b) Program.

INPUT

```
8,4,.001,
  100.,100.,100.,100.,100.,100.,100.,100.,
  200.,200.,200.,200.,200.,
  100.,
```

OUTPUT

100.0000	100.0000	100.0000	100.0000	100.0000	100.0000	100.0000	100.0000	100.0000
119.4193	119.2370	118.5972	117.2150	114.6783	110.8951	107.2417	103.6208	
139.2030	138.9316	137.9365	135.5847	130.6029	121.6602	114.4509		
159.5297	159.3498	158.6327	156.5840	150.4885	130.6917			
200.0000	200.0000	200.0000	200.0000	200.0000				

Fig. 10.21   (c) Input and results.

Boundary temperatures are supplied with T_{0i} for the outer surface, and T_{Ii}, on the inner surface. These can vary along the surfaces, but they do not in the example computed here.

As an illustration of how the coefficients are selected at specific mesh points, let us examine some examples. In these examples the outer surface temperature, T_0, is everywhere 100, and T_I is everywhere 200. With $m = 6$ and $n = 3$, the number of points to be treated is $n_0 = 11$. These are numbered in Fig. 10.20b. Point $i = 3$ is adjacent to the outer boundary. Referring to the coefficients in the computational molecule of Fig. 10.22b, the third row of coefficients stored in the augmented matrix are:

$$0. \quad -1. \quad 4. \quad -1. \quad 0. \quad 0. \quad 0. \quad 0. \quad -1. \quad 0. \quad 0. \quad 100.$$

There are 12 ($n_0 + 1$) coefficients in each row. For the point $i = 11$, which is on a line of symmetry, the computation is based on a molecule similar to the one shown in Fig. 10.22c. The coefficients for the eleventh row are:

$$0. \quad 0. \quad 0. \quad 0. \quad -2. \quad 0. \quad 0. \quad 0. \quad 0. \quad -2. \quad 4. \quad 0.$$

As the first point, $i = 1$, is both near the outer boundary and on a line of symmetry, its coefficients are:

$$4. \quad -2. \quad 0. \quad 0. \quad 0. \quad 0. \quad -1. \quad 0. \quad 0. \quad 0. \quad 0. \quad 100.$$

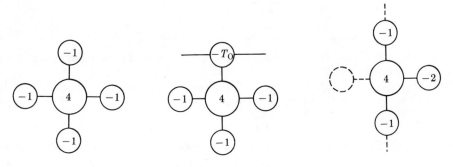

Fig. 10.22

It is suggested that these be verified, and that the reader examine the program to determine the tests used to locate each coefficient and store its proper value.

An example was run for this geometry with $m = 8$, $n = 4$, $T_I = 200$, $T_O = 100$, $\epsilon = 0.001$, and TRIAL $= 100$. There was a total of 21 mesh points, which required that GSMETH solve 21 simultaneous equations. Results are printed in Fig. 10.21c. A format was used to cause the program to write the temperatures in positions which corresponded to their location in the square pipe.

Exercises

1. Write in finite-difference form the Poisson equation

$$\frac{\partial^2 u}{\partial x^2} + \frac{\partial^2 u}{\partial y^2} = xy + y^2.$$

2. A problem in fluid mechanics has been formulated mathematically in the expression

$$\frac{\partial^2 V}{\partial Z^2} + W \frac{\partial V}{\partial Z} + S^2 \left[\frac{\partial^2 V}{\partial r^2} + \frac{1}{r} \frac{\partial V}{\partial r} - \frac{V}{r^2} \right] = 0.$$

a) What type of partial-differential equation is this?
b) Describe the additional information you require for its solution.
[*Note:* W and S are constants.]

3. An additional modification has been made to the equation in Exercise 2. It is noted that the term $(\partial^2 V / \partial Z^2)$ can be neglected as it is small compared to the other terms.
a) What type of partial-differential equation is this new expression?
b) Describe the additional information you require for its solution.

4. Consider the problem of heat-transfer from a fin of uniform cross section (see Fig. 10.23).

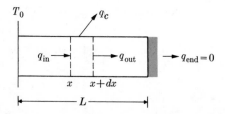

Fig. 10.23

The fin is of length L, and is circular in cross section, of radius R. For the element shown, heat is conducted in: $q_{in} = -k\pi R^2 (\partial T / \partial X)$. Heat is conducted out: $q_{out} = -k\pi R^2 (\partial / \partial x)(T + (\partial T / \partial x) dx)$. There is heat transfer by convection to the surroundings, which are at T_a. We can express this as

$$q_c = hA(T - T_a),$$

where the surface area $A = 2\pi R\, dx$, and h is the coefficient of heat transfer, and is taken as constant.

Thus for the element dx, steady-state conditions require

$$q_{in} - q_{out} - q_c = 0$$

or

$$k\pi R^2 \frac{\partial^2 T}{\partial x^2} - 2\pi Rh(T - T_a) = 0.$$

However, as $T = T(x)$,

$$\frac{d^2 T}{dx^2} = \frac{2h}{kR}(T - T_a).$$

Defining a temperature difference, $\theta = T - T_a$, the expression becomes

$$\frac{d^2\theta}{dx^2} - \frac{2h}{kr}\theta = 0.$$

Formulate the finite-difference equations for this problem, including the boundary conditions

$$x = 0, \qquad T = T_0 \qquad \text{(a constant)},$$
$$x = L, \qquad q_{end} = 0 \qquad \text{(insulated end)}.$$

5. Write a program which formulates the system of equations for the solution of Exercise 4, using the Gauss-Jordan subroutine, SLEQ1, to produce the solution. The main program will also write the results. Apply the program to the case where $L = 1$ ft, $k = 10$ Btu/hr-ft-°F, $h = 3$ Btu/hr-ft^2-°F, $R = 0.5$ in, $T_0 = 200$°F, and $T_a = 80$°F. Use more than 10 nodal points in the calculation. *Note:* As the matrix is tridiagonal a more efficient reduction method could be used. See Chapter 6, Ex. 17.

6. Modify the example program for steady-state conduction in a square pipe (page 283) so it solves for the case where the outer boundary condition is given in terms of the gradient. This is the case of heat convection from the outer surface, and leads to the condition on the outer boundary expressed as

$$\frac{\partial T}{\partial y} = -\frac{h}{k}(T - T_a).$$

Set $T_a = 70$°F, $h = 2$, and $k = 1$, and use the program to compute the temperature distribution. The inner boundary will be at a constant temperature of 200°F as before. It is simpler to introduce the temperature difference $\theta = T - T_a$ as the variable.

7. Return now to Exercise 4. Treat the transient case where initially the fin has some prescribed temperature distribution, then a change is made, and we examine the temperature history in the fin. The governing equation now is

$$\frac{\partial^2 T}{\partial x^2} - \frac{2h}{kR}(T - T_a) = \frac{\rho c_p}{k}\frac{\partial T}{\partial \tau},$$

where $\rho =$ density (lbm/ft^3), $c_p =$ specific heat (Btu/lbm · °F), and $\tau =$ time (hrs). Introducing the definition

$$\theta = T - T_a,$$

the equation becomes

$$\frac{\partial^2 \theta}{\partial x^2} - \frac{2h}{kR} \theta = \frac{\rho c_p}{k} \frac{\partial \theta}{\partial \tau} .$$

This is a parabolic partial-differential equation. Formulate the finite-difference equations.

8. Now consider Exercise 7 with the following boundary and initial conditions:

At $\tau = 0$,

$$x \geq 0, \qquad \theta = 0.$$

At $\tau > 0$,

$$x = 0, \qquad \theta = 120;$$

$$x = L, \qquad \frac{\partial \theta}{\partial x} = 0.$$

Solve the finite-difference equations numerically for the following relation between time steps and spatial steps.

a) $M > (2N + 2)$
b) $M < \sqrt{N^2 + 1}$
c) $\sqrt{N^2 + 1} < M < (2N + 2)$ where

$$M = \frac{k \, \delta \tau}{\rho c_p} (\delta x)^2 \qquad \text{and} \qquad N = \frac{h \, \delta x}{k} .$$

9. The transient heat-conduction equation in cylindrical coordinates with axial symmetry is

$$\frac{\partial^2 u}{\partial r^2} + \frac{1}{r} \frac{\partial u}{\partial r} = \frac{1}{\alpha} \frac{\partial u}{\partial t} ,$$

where α is the thermal diffusivity.

a)   Write the boundary conditions and initial conditions for a pipe of inner radius r_1 and outer radius r_2. Initially the temperature is everywhere 0. Suddenly the internal surface is raised to a temperature of 150 and held there. The outer surface is insulated.

b)   Formulate the finite-difference equations in explicit form, writing an equation for each boundary and for a single interior point.

10. Write the finite-difference equations for Exercise 9 in implicit form at the boundaries and for an interior point. Indicate how the computation would be started.

11. Write a computer program to formulate the finite-difference equations for Exercise 9 in implicit form, using a Gauss-Jordan or Gauss-Seidel subroutine for the solution. Carry out a series of calculations from the initial and boundary conditions of Exercise 9. Use $\alpha = 0.01$ (ft)^2/hr, $r_1 = 0.5$ ft, and $r_2 = 1.0$ ft.

bibliography

bibliography

Aitken, A. C., "The evaluation of the latent roots and latent vectors of a matrix," *Proc. Royal Soc. Edinburgh*, **62,** pp. 269–304 (1937).

Arden, B., *An Introduction to Digital Computing.* Reading, Mass.: Addison-Wesley, 1963.

Calingaert, P., *Principles of Computation.* Reading, Mass.: Addison-Wesley, 1965.

Carnahan, B., H. A. Luther, and J. O. Wilkes, *Applied Numerical Methods.* New York: John Wiley and Sons, 1969.

Conti, S. D., *Elementary Numerical Analysis.* New York: McGraw-Hill Book Co., 1965.

Crandall, S. H., *Engineering Analysis.* New York: McGraw-Hill Book Co., 1956.

Crank, J., and P. Nicolson, "A practical method for numerical evaluation of solutions of partial differential equations of heat-conduction type," *Proc. Cambridge Phil. Soc.,* **32,** pp. 50–67 (1947).

Dwyer, P. S., *Linear Computations.* New York: John Wiley and Sons, 1951.

Forsythe, G. E., and C. B. Moler, *Computer Solution of Linear Algebraic Systems.* Englewood Cliffs, N.J.: Prentice-Hall, 1967.

Grove, W. E., *Brief Numerical Methods.* Englewood Cliffs, N.J.: Prentice-Hall, 1966.

Hamming, R. W., *Numerical Methods for Scientists and Engineers.* New York: McGraw-Hill Book Co., 1962.

Hamming, R. W., "Stable predictor–corrector methods for ordinary differential equations," *Journal of the ACM,* **6,** pp. 37–47 (1959).

Hastings, C., Jr., *Approximations for Digital Computers.* Princeton, N.J.: Princeton Univ. Press, 1955.

Henrici, P., *Elements of Numerical Analysis.* New York: John Wiley and Sons, 1964.

Hildebrand, F. B., *Methods of Applied Mathematics.* Englewood Cliffs, N.J.: Prentice-Hall, 1952.

Hildebrand, F. B., *Introduction to Numerical Analysis.* New York: McGraw-Hill Book Co., 1956.

Householder, A. S., *The Theory of Matrices in Numerical Analysis.* New York: Blaisdell Publishing Co., 1964.

Householder, A. S., *Principles of Numerical Analysis.* New York: McGraw-Hill Book Co., 1953.

Ingersoll, L. R., O. J. Zobel, and A. C. Ingersoll, *Heat Conduction.* Madison, Wis.: Univ. of Wisconsin Press, 1954, pp. 58–77.

Isaacson, E., and H. B. Keller, *The Analysis of Numerical Methods.* New York: John Wiley and Sons, 1966.

James, M. L., G. M. Smith, and J. C. Wolford, *Applied Numerical Methods.* Scranton, Pa.: International Textbook Co., 1967.

Jennings, W., *First Course in Numerical Methods.* New York: The Macmillan Co., 1964.

Kunz, K. S., *Numerical Analysis.* New York: McGraw-Hill Book Co., 1957.

Kuo, S. S., *Numerical Methods and Computers.* Reading, Mass.: Addison-Wesley, 1965.

Lee, J. A. N., *Numerical Analysis for Computers.* New York: Reinhold Publishing Co., 1966.

McCormick, J. M., and M. G. Salvadori, *Numerical Methods in FORTRAN.* Englewood Cliffs, N.J.: Prentice-Hall, 1964.

McCracken, D., and W. S. Dorn, *Numerical Methods and FORTRAN Programming.* New York: John Wiley and Sons, 1964.

Milne, W. E., *Numerical Solutions of Differential Equations.* New York: John Wiley and Sons, 1953.

Noble, B., *Numerical Methods*, Vols. 1 and 2. London: Oliver and Boyd, 1964.

O'Brien, G., M. Hyman, and S. Kaplan, "A study of the numerical solution of partial-differential equations," *J. Math. Phys.* **29,** pp. 233–251 (1951).

Organick, E. I., *A FORTRAN IV Primer.* Reading, Mass.: Addison-Wesley, 1966.

Ralston, A., *A First Course in Numerical Analysis.* New York: McGraw-Hill Book Co., 1965.

Salvadori, M. G., and M. L. Baron, *Numerical Methods in Engineering.* Englewood Cliffs, N.J.: Prentice-Hall, 1961.

Scarborough, J. B., *Numerical Mathematical Analysis.* Baltimore: John Hopkins Press, 1962.

Scheid, F., *Theory and Problems of Numerical Analysis* (Schaum's Outline Series). New York: McGraw-Hill Book Co., 1968.

Smith, G. D., *Numerical Solutions of Partial Differential Equations.* London: Oxford Univ. Press, 1965.

Southwell, R. V., *Relaxation Methods in Theoretical Physics.* London: Oxford Univ. Press, 1946.

Stanton, R. G., *Numerical Methods for Science and Engineering.* Englewood Cliffs, N.J.: Prentice-Hall, 1961.

Weeg, G. P., and G. B. Reed, *Introduction to Numerical Analysis.* Waltham, Mass.: Blaisdell Publishing Co., 1966.

Willers, F. A., *Practical Analysis.* New York: Dover Publications, Inc., 1948.

appendices

appendices

a table of flow chart conventions used in the text

READ
a, e, x Input statement listing items to be read into program.

WRITE
a, e, x Output statement listing items to be written.

$y = \tan x + ax^3$ A major processing statement showing algebraic operations.

$I = 3$ An auxiliary processing statement or substitution.

$x > y$ T   F   A decision showing flow path for true and false conditions.

START   A terminal which indicates the beginning or end of a program or a point of interruption.

2   A connector showing entry or exit to another part of the flow chart where there is a connector with the same number.

CALL
PRED $(3, y, f, t, h)$ A call by the main program for a subroutine to be executed and return its results.

roots and coefficients for Gauss-Legendre quadrature according to the formula

$$\int_{-1}^{1} F(t)\, dt \cong \sum_{i=0}^{n} a_i F(t_i)$$

Roots (t_i)	Coefficients (a_i)

(Two-point formula)

$n = 1$

±0.57735 02691 89626 — 1.00000 00000 00000

(Three-point formula)

$n = 2$

0.00000 00000 00000 — 0.88888 88888 88889
±0.77459 66692 41483 — 0.55555 55555 55556

(Four-point formula)

$n = 3$

±0.33998 10435 84856 — 0.65214 51548 62546
±0.86113 63115 94053 — 0.34785 48451 37454

(Five-point formula)

$n = 4$

0.00000 00000 00000 — 0.56888 88888 88889
±0.53846 93101 05683 — 0.47862 86704 99366
±0.90617 98459 38664 — 0.23692 68850 56189

(Six-point formula)

$n = 5$

±0.23861 91860 83197 — 0.46791 39345 72691
±0.66120 93864 66265 — 0.36076 15730 48139
±0.93246 95142 03152 — 0.17132 44923 79170

Roots (t_i)	Coefficients (a_i)

(Ten-point formula)

$n = 9$

Roots (t_i)	Coefficients (a_i)
±0.14887 43389 81631	0.29552 42247 14753
±0.43339 53941 29247	0.26926 67193 09996
±0.67940 95682 99024	0.21908 63625 15982
±0.86506 33666 88985	0.14945 13491 50581
±0.97390 65285 17172	0.06667 13443 08688

(Fifteen-point formula)

$n = 14$

Roots (t_i)	Coefficients (a_i)
0.00000 00000 00000	0.20257 82419 25561
±0.20119 40939 97435	0.19843 14853 27111
±0.39415 13470 77563	0.18616 10001 15562
±0.57097 21726 08539	0.16626 92058 16994
±0.72441 77313 60170	0.13957 06779 26154
±0.84820 65834 10427	0.10715 92204 67172
±0.93727 33924 00706	0.07036 60474 88108
±0.98799 25180 20485	0.03075 32419 96117

index

ABCDE79876543210